Withdrawn
University of Waterloo

The Physics of Phase Transitions

Springer
Berlin
Heidelberg
New York
Barcelona
Hong Kong
London
Milan
Paris
Tokyo

Physics and Astronomy ONLINE LIBRARY

http://www.springer.de/phys/

Advanced Texts in Physics

This program of advanced texts covers a broad spectrum of topics which are of current and emerging interest in physics. Each book provides a comprehensive and yet accessible introduction to a field at the forefront of modern research. As such, these texts are intended for senior undergraduate and graduate students at the MS and PhD level; however, research scientists seeking an introduction to particular areas of physics will also benefit from the titles in this collection.

P. Papon
J. Leblond
P.H.E. Meijer

The Physics of Phase Transitions

Concepts and Applications

Tranlated from the French by S. L. Schnur

With 175 Figures

Springer

Pierre Papon
Jacques Leblond
École Supérieure
de Physique et de Chimie Industrielles
de Paris (ESPCI)
Laboratoire de Physique Thermique
10 rue Vauquelin
75005 Paris, France
papon@pmmh.espci.fr

Paul H.E. Meijer
Catholic University of America
Department of Physics
Washington, DC 20064, USA

Translator
S. L. Schnur
Concepts Unlimited
6009 Lincolnwood Court
Burke, VA 22015-3012, USA

Translation from the French language edition of
Physique des transitions de phases, concepts et applications
by Pierre Papon, Jacques Leblond and Paul H.E. Meijer
© 1999 Editions Dunod, Paris, France

This work has been published with the help of the
French Ministère de la Culture - Centre national du livre

The cover picture shows dentrites from a Co–Sm–Cu alloy, by courtesy of Prof. W. Kurz, Laboratoire de métallurgie Physique, Ecole Polytechnique fédérale de Lausanne, Lausanne, Switzerland.

Library of Congress Cataloging-in-Publication Data

Papon, Pierre, 1939-
[Physique des transitions de phases. English]
The physics of phase transitions : concepts and applications / P. Papon, J. Leblond, P.H.E. Meijer.
p. ; cm. -- (Advanced texts in physics, ISSN 1439-2674)
Includes bibliographical references and index.
ISBN 3540432361 (alk. paper)
1. Phase transformations (Statistical physics) I. Leblond, Jacques. II. Meijer, Paul Herman Ernst, 1921- III. Title. IV. Series.

QC175.16.P5 P36 2002
530.4'74--dc21

2002276534

ISSN 1439-2674
ISBN 3-540-43236-1 Springer-Verlag Berlin Heidelberg New York

This work is subject to copyright. All rights are reserved, whether the whole or part of the material is concerned, specifically the rights of translation, reprinting, reuse of illustrations, recitation, broadcasting, reproduction on microfilm or in any other way, and storage in data banks. Duplication of this publication or parts thereof is permitted only under the provisions of the German Copyright Law of September 9, 1965, in its current version, and permission for use must always be obtained from Springer-Verlag. Violations are liable for prosecution under the German Copyright Law.

Springer-Verlag Berlin Heidelberg New York
a member of BertelsmannSpringer Science+Business Media GmbH

http://www.springer.de

© Springer-Verlag Berlin Heidelberg 2002
Printed in Germany

The use of general descriptive names, registered names, trademarks, etc. in this publication does not imply, even in the absence of a specific statement, that such names are exempt from the relevant protective laws and regulations and therefore free for general use.

Typesetting by the authors using a Springer TeX macro package
Cover design: *design & production* GmbH, Heidelberg

Printed on acid-free paper SPIN 10782785 56/3141/mf 5 4 3 2 1 0

Foreword

We learned in school that matter exists in three forms: solid, liquid and gas, as well as other more subtle things such as the fact that "evaporation produces cold." The science of the states of matter was born in the 19th century. It has now grown enormously in two directions:

1) The transitions have multiplied: first between a solid and a solid, particularly for metallurgists. Then for magnetism, illustrated in France by Louis Néel, and ferroelectricity. In addition, the extraordinary phenomenon of superconductivity in certain metals appeared at the beginning of the 20th century. And other superfluids were recognized later: helium 4, helium 3, the matter constituting atomic nuclei and neutron stars... There is now a real zoology of transitions, but we know how to classify them based on Landau's superb idea.

2) Our profound view of the mechanisms has evolved: in particular, the very universal properties of fluctuations near a critical point – described by Kadanoff's qualitative analysis and specified by an extraordinary theoretical tool: the renormalization group.

Without exaggerating, we can say that our view of condensed matter has undergone two revolutions in the 20th century: first, the introduction of quantum physics in 1930, then the recognition of "self-similar" structures and the resulting scaling laws around 1970.

It would be naïve to make too much of these advances: despite all of this sophistication, we are still very unsure about certain points – for example, the mechanism governing superconducting oxides or the laws of the glass transition. However, a body of doctrines has been formed, and it is an important element of scientific culture in the year 2000.

This knowledge is generally expressed solely in works dedicated to only one sector. The great merit of the book by Drs. Papon, Leblond and Meijer is to offer a global introduction, accessible to students of physics entering graduate school. Of course, given the enormous expansion of the subject, many older readers will regret not finding their pet subject here (for example, in my case demixing transitions involving linear polymers). However, the panorama is broad enough for the young public targeted here: I hope the book will guide them soundly.

I wish it great success.

Paris, France
December 2001

P. G. de Gennes
Professor at the Collège de France
Director of the Ecole Supérieure de
Physique et Chimie Industrielles de Paris

Preface

This book takes up and expands upon our teachings on thermodynamics and the physics of condensed matter at the School of Industrial Physics and Chemistry and Diplôme d'Etudes Approfondies in Paris and at the Catholic University of America in Washington D.C. It is intended for graduate students, students in engineering schools, and doctoral students. Researchers and industrial engineers will also find syntheses in an important and constantly evolving field of materials science.

The book treats the major classes of phase transitions in fluids and solids: vaporization, solidification, magnetic transitions, critical phenomena, etc. In the first two chapters, we give a general description of the phenomena, and we dedicate the next six chapters to the study of a specific transition by explaining its characteristics, experimental methods for investigating it, and the principal theoretical models that allow its prediction. The major classes of application of phase transitions used in industry are also reported. The last three chapters are specifically dedicated to the role of microstructures and nanostructures, transitions in thin films, and finally, phase transitions in large natural and technical systems. Our approach is essentially thermodynamic and assumes familiarity with the basic concepts and methods of thermodynamics and statistical physics. Exercises and their solutions are given, as well as a bibliography.

Finally, we would we like to thank J. F. Leoni who assisted in the preparation of the manuscript and the drawings and diagrams and Dr. S. L. Schnur who put much effort into translating the book as well as J. Lenz and F. Meyer from Springer-Verlag who provided helpful advice in publishing the book. We are also grateful to our colleague Prof. K. Nishinari, from Osaka City University, for his valuable comments on our manuscript.

Paris, France, *Pierre Papon*
Paris, France, *Jacques Leblond*
Washington, D.C., U.S.A., *Paul H. E. Meijer*
December 2001

Preface

Contents

Principal Notation

A	Area
$\boldsymbol{B}$	Magnetic induction
C_p	Specific heat at constant pressure
c_p	Specific heat at constant pressure per unit of mass
C_v	Specific heat at constant volume
c_v	Specific heat at constant volume per unit of mass
d	Intermolecular distance
$D(\varepsilon)$	Density of states
e	Elementary charge
E	Energy
$\boldsymbol{E}$	Electric field
f	Free energy per unit of mass, radial or pair distribution function
F	Free energy (Helmholtz function)
$\boldsymbol{F}$	Force
g	Free enthalpy per unit of mass or volume
G	Free enthalpy (Gibbs function)
$g(E)$	Degeneracy factor
H	Enthalpy
h	Enthalpy per unit of mass or volume, Planck's constant
$\boldsymbol{H}$	Magnetic field, Hamiltonian
$\mathcal{H}$	Hamiltonian
$\boldsymbol{j}$	Current density per unit of surface
J	Flux, grand potential
$\boldsymbol{k}$	Wave vector
k	Boltzmann constant
L	Latent heat
l	Latent heat per unit of mass or volume, length
m	Mass
M	Molecular weight
$\boldsymbol{M}$	Magnetization
n	Particle density (N/V)
N	Number of particles
N_0	Avogadro's number
p	Pressure

$\boldsymbol{p}$ Momentum
P Order parameter, probability
$\boldsymbol{P}$ Electric polarization
q Position variable
Q, q Quantity of heat
$\boldsymbol{r}$ Distance
R Ideal gas constant
s Entropy per unit of mass or volume
S Entropy
t Time
T Absolute temperature (Kelvin)
T_C Critical temperature
U Internal energy
u Internal energy per unit of mass or volume, pair-potential
V Volume
v Velocity, variance
W Number of states, work
w Probability distribution
x Concentration
X Extensive variable
Y Intensive variable (field)
z Coordination number
Z Partition function, compressibility factor
α Volume expansion coefficient
β Reciprocal temperature parameter, $1/kT$
χ Magnetic susceptibility, helical pitch
Δ, δ Increase in a variable
ε Elementary particle energy, $|T - T_C|/T_C$
γ Surface tension
η Viscosity
Ξ Grand partition function
Θ Debye temperature
κ Compressibility
λ Wavelength, thermal conductivity
Λ de Broglie thermal wavelength
μ Chemical potential
ν Frequency
ρ Density
τ Relaxation time
ω Acentric factor, frequency
Ψ Thermodynamic potential, wave function
ξ Correlation length
Ω Grand potential
$\Omega(E)$ Number of accessible states

Table of Principal Constants.

Avogadro's number	N_0	6.02205×10^{23}
Boltzmann's constant	k	1.38066×10^{-23} J K^{-1}
Gas constant	R	8.31141J K^{-1} mole^{-1}
Planck's constant	h	6.62618×10^{-34} J s
Standard atmosphere	p_0	1.01325×10^5N m^{-2}
Triple point of water	T_0	273.16 K
Electron charge	e	1.60219×10^{-19} C
Electron mass	m_e	9.10953×10^{-31} kg
Bohr magneton ($eh/4\pi m_e$)	μ_B	0.927408×10^{-23}A m^2
kT at 300 K	–	4×10^{-21}J $= 1/40$ eV

Energy: 1 Joule $= 10^7$ ergs $= 0.2389$ cal $= 9.48 \times 10^{-4}$ btu

Pressure: 1 Pascal $=$ 1 Newton m^{-2} $= 10^{-5}$ bar $=$ 10 dynes cm^{-2}

1 Thermodynamics and Statistical Mechanics of Phase Transitions

1.1 What is a Phase Transition?

Any substance of fixed chemical composition, water H_2O, for example, can exist in homogeneous forms whose properties can be distinguished, called **states**. Water exists as a gas, a liquid, or a solid, ice. These three states of matter (solid, liquid, and gas) differ in density, heat capacity, etc. The optical and mechanical properties of a liquid and a solid are also very different. By applying high pressures to a sample of ice (several kilobars), several varieties of ice corresponding to distinct crystalline forms can be obtained (Fig. 1.1). In general, for the same solid or liquid substance, several distinct arrangements of the atoms, molecules, or particles associated with them can be observed and will correspond to different properties of the solid or liquid, constituting **phases**. There are thus several phases of ice corresponding to distinct crystalline and amorphous varieties of solid water. Either an isotropic phase or a **liquid crystal** phase can be obtained for some liquids, they can be distinguished by their optical properties and differ in the orientation of their molecules (Fig. 1.2). Experiments thus demonstrate **phase transitions** or **changes of state**. For example: a substance passes from the liquid state to the solid state (solidification); the molecular arrangements in a crystal are modified by application of pressure and it passes from one crystalline phase to another. Phase transitions are physical events that have been known for a very long time. They are encountered in nature (for example, condensation of drops of water in clouds) or daily life; they are also used in numerous technical systems or industrial processes; evaporation of water in the steam generator of a nuclear power plant is the physical process for activating the turbines in electric generators, and melting and then solidification of metals are important stages of metallurgical operations, etc.

Phase transitions manifested by the appearance of new properties of matter, for example, ferromagnetism and superconductivity, have also been observed; new phases or new states whose properties have important applications, appear below a **critical temperature**. These phase transitions are not always induced by modification of atomic or molecular arrangements but in the case of ferromagnetism and superconductivity, by modification of electronic properties. In general, a transition is manifested by a series of associated physical events. For most of them, the transition is accompanied

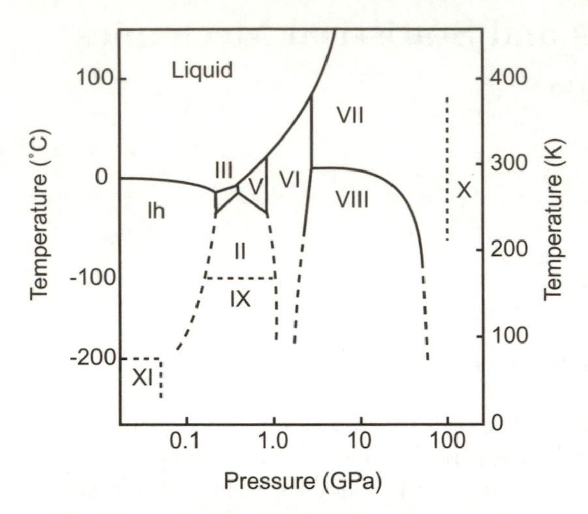

Fig. 1.1. Phase diagram of ice. Eleven crystalline varieties of ice are observed. A twelfth form XII was found in the 0.2–0.6 GPa region. "Ordinary" ice corresponds to form Ih. Ices IV and XII are metastable with respect to ice V (C. Lobban, J. L. Finney, and W. F. Kuhs, Nature, 391, 268 (1998), copyright 1998 Macmillan Magazines Limited).

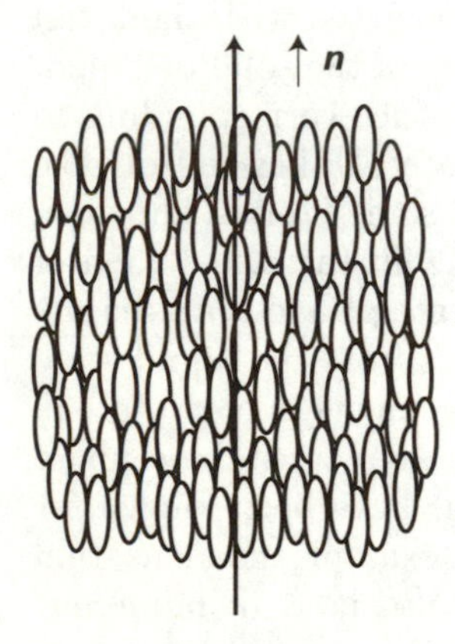

Fig. 1.2. Nematic liquid crystal. The arrangements of molecules in a nematic liquid crystal are shown in this diagram; they are aligned in direction $\boldsymbol{n}$.

by latent heat and discontinuity of a state variable characterizing each phase (density in the case of the liquid/solid transition, for example). It has also been observed that an entire series of phase transitions takes place with no latent heat or discontinuity of state variables such as the density, for example. This is the situation encountered at the critical point of the liquid/gas transition and at the Curie point of the ferromagnetic/paramagnetic transition. The thermodynamic characteristics of phase transitions can be very different. Very schematically, there are two broad categories of transitions:

those associated with latent heat on one hand, and those not involving latent heat on the other hand.

It is also necessary to note that a phase transition is induced by acting from the outside to modify an intensive thermodynamic variable characterizing the system: temperature, pressure, magnetic or electric field, etc. This variable is coupled with an extensive variable (for example, pressure and volume are coupled) in the sense of classic thermodynamics.

We also know from experience that a phase transition begins to appear on the microscopic scale: small drops of liquid whose radius can be smaller than one micron appear in the vapor phase before it is totally condensed in liquid form. This is **nucleation**. In the same way, solidification of a liquid, a molten metal, for example, begins above the solidification temperature from microcrystallites, crystal nuclei of the solid phase. For a polycrystalline solid such as a ceramic, the mechanical properties are very strongly dependent on the size of the microcrystallites.

In going to the atomic or molecular scale, repulsive and attractive forces between atoms or molecules intervene to account for the properties of the substance; the intermolecular forces determine them and explain cohesion of a solid or liquid involved in melting and evaporation phenomena in particular. In the case of a liquid like water, the intervention of hydrogen bonds between the molecules explains the abnormal properties of this liquid (for example, its density maximum at 4°C and the fact that the density of the solid phase is lower than the density of the liquid phase). In general, phase transitions are a central problem of materials science: the relationship between the macroscopic properties and the microscopic structure of a material.

Finally, returning to the thermodynamic approach to the phenomena, we know from experience that there are situations in which, beginning with a liquid phase, this state can be maintained below the solidification point of the substance considered (water, for example); we then have a supercooled liquid, corresponding to a metastable thermodynamic state. If the supercooled liquid is silica, we will then observe solidification of the liquid in the form of glass: this is the **glass transition**. An unorganized, that is, noncrystalline, solid state has been obtained with specific thermodynamic, mechanical, and optical properties which do not correspond to a thermodynamic state in equilibrium. The phase transition is produced without latent heat or change in density.

The world of phase transitions is still filled with unknowns. A new form of carbon was identified for the first time in 1985, **fullerene** (abbreviation for buckmunsterfullerene, in fact). Fullerene, corresponding to the stoichiometric composition C_{60}, is a spherical species of carbon molecules that can be obtained in solid form (for example, by irradiation of graphite with a powerful laser), with a crystal structure of face–centered cubic symmetry. Although a phase diagram has been calculated for C_{60} that predicts the existence of a liquid phase, this has not been demonstrated experimentally.

We thus see the very wide variety of phase transitions that can be encountered with different types of substances and materials involving a large number of properties and phenomena. The study of phase transition phenomena and their applications is the subject of this book. We will first consider the applications of phase transitions to technical and natural systems in each chapter of the book as a function of their specificity.

We will leave aside a fourth state of matter, the **plasma state**, which has very specific properties; plasma is a gas composed of charged particles (electrons or ions). It is obtained by electric discharges in gases at temperatures between several thousand and several million Kelvin. Plasmas are thus produced in extreme conditions not encountered in current conditions on Earth. Plasmas can be kept confined in a container by a magnetic field, this is the principle of **tokamaks**, and they can also be produced by bombarding a target (deuterium, for example) with a very powerful laser beam. This is the method of **inertial confinement**. Plasmas are also found in the stars.

1.2 Thermodynamic Description of Phase Transitions

If we consider the two condensed states of matter (solid and liquid), the forces between atoms or molecules (or the potentials from which they derive) determine the structure of the matter and its evolution in time, in a word, its dynamics.

Intermolecular forces contribute to cohesion of a liquid and a solid, for example. Within a solid, the interactions between the magnetic moments of the atoms, when they exist, or between electric dipoles, contribute to the appearance of phenomena such as ferromagnetism or ferroelectricity.

We can thus study phase transition phenomena by utilizing intermolecular potentials or interactions between particles; this is particularly the approach of quantum statistics, which is the most complex. We can also hold to a description using classic thermodynamics to attempt to determine phase transitions. In principle, we will first explain the simplest approach.

1.2.1 Stability and Transition – Gibbs–Duhem Criterion

A phase transition occurs when a phase becomes unstable in the given thermodynamic conditions, described with intensive variables (p, T, H, E etc.). At atmospheric pressure ($p = 1$ bar), ice is no longer a stable solid phase when the temperature is above 0°C; it melts, and there is a solid/liquid phase transition. It is thus necessary to describe the thermodynamic conditions of the phase transition if we wish to predict it.

We can describe the thermodynamic state of a system or material with the thermodynamic potentials classically obtained with a Legendre transformation. These thermodynamic potentials can also be calculated with quantum statistics if the partition function of the system is known. These potentials are expressed by extensive and intensive state variables, which characterize the system. The choice of variables for studying and acting on it determines the potential. We note that in working with variables (T, V), it is necessary to use the **free energy** F; the system will be investigated with the **free enthalpy** G (also called the Gibbs function) if the system is described with variables (p, T).

In thermodynamics, it is possible to show that a stable phase corresponds to the minimum of potentials F and G. More generally, by imagining virtual transformations Δ of thermodynamic quantities X from equilibrium, we have the stability criterion for this equilibrium situation, written as:

$$\Delta U + p\Delta V - T\Delta S \geq 0 \tag{1.1}$$

where $\Delta U, \Delta V$, and ΔS are virtual variations of internal energy U, volume V and entropy S from equilibrium. This is the **Gibbs–Duhem stability criterion**.

We can easily deduce from (1.1) that a stable phase is characterized by a minimum of potentials F (with constant T and V), G (with constant T and p), H (with constant S and p), U (with constant S and V), and by a maximum of the entropy (with constant U and V).

Condition (1.1), which can be used to find the equilibrium stability criterion, should be rigorously examined. This criterion and its variants can be used to specify the equilibrium conditions. Important physical states of matter, the glassy state, for example, suggest that the equilibrium state of a presumably stable system can be modified by application of a perturbation (thermal or mechanical shock). Moreover, it has been found that water kept in the liquid phase at a temperature below 0°C instantaneously solidifies if an impurity is added to the liquid phase or if a shock is induced in its container (a capillary tube, for example).

The conditions prevailing for the system when applying Gibbs–Duhem criterion (1.1) must thus be specified.

Equilibrium, in the broadest sense of the term, corresponds to the entropy maximum, and for all virtual infinitesimal variations in the variables, $\delta S = 0$.

However, we can distinguish between the following situations:

- The conditions ($\delta S = 0, \delta^2 S < 0$) are satisfied regardless of the virtual perturbations of the variables. If ΔS is written in the form of an expansion in the vicinity of equilibrium:

$$\Delta S = \delta S + 1/2\ \delta^2 S + 1/3!\ \delta^3 S + 1/4!\ \delta^4 S + \dots \tag{1.2}$$

  where terms $\delta^2 S, \delta^3 S, \delta^4 S$... are second-, third-, and fourth-order differentials with respect to the state variables. We thus have $\delta^2 S, \delta^3 S, \delta^4 S... < 0$; **the equilibrium is stable**.
- The conditions ($\delta S = 0, \delta S^2 < 0$) are verified for all virtual perturbations, but the condition $\Delta S < 0$ is violated for certain perturbations (in other words, we can have $\delta^3 S, \delta^4 S > 0$); **the equilibrium is metastable**.

- Certain perturbations satisfy the condition $\delta^2 S > 0$; **the equilibrium is unstable**.

We have introduced the notion of **metastability** of equilibrium which is in a way a thermodynamic state intermediate between stability and instability. Liquid water at a temperature below 0°C typically corresponds to a metastable thermodynamic state: it is called supercooled. Similary a substance maintained in a liquid state above its boiling point is also in a metastable state: it is **superheated**.

The condition $\boldsymbol{\delta^2 S = 0}$ gives the metastability limit of the equilibrium. When a material undergoes a transformation from an initial stable equilibrium state that satisfies this condition, it passes from metastability to instability and a phase transition is then observed.

The curve corresponding to this limiting condition of metastability is called the **spinodal**. The analytical shape of this curve can also be determined by writing the limiting condition of metastability with other thermodynamic potentials: $\delta^2 G = 0, \delta^2 F = 0$.

In the case of a material with only one chemical constituent and isotropic molecules, the free enthalpy G must be used to describe its properties if the equilibrium is modified by acting on variables (p, T). Function $G(p, T)$ can be represented by a surface in three–dimensional space; one state of the system (fixed p, T) corresponds to a point on this surface of coordinates (G, p, T). Assume that the material can exist in the form of two solid phases (solid 1–solid 2), one liquid phase, and one gas phase. We will then have four surface parts corresponding to these four phases with potentials G_{S1}, G_{S2}, G_1, G_V which are intersected along the lines; the potentials along these lines are by definition equal and thus the corresponding phases coexist (Fig. 1.3).

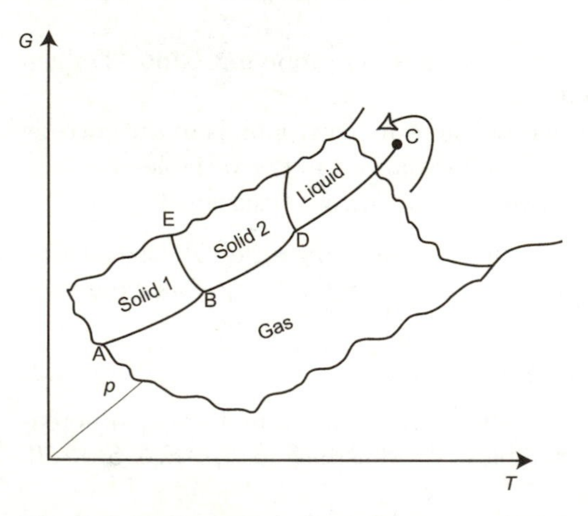

Fig. 1.3. Surface representing the free enthalpy $G(T, p)$. The liquid–gas coexistence curve has a terminal point which is critical point C.

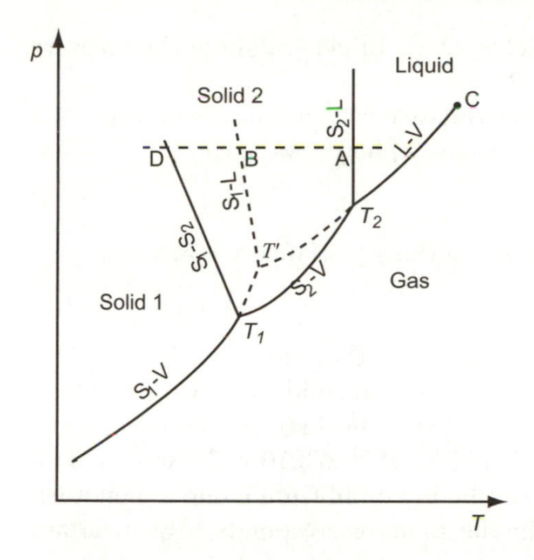

Fig. 1.4. Phase diagram with metastable phases.The dashed lines delimit the zones of existence of metastable phases; they intersect at point T′, which is a triple point where metastable solid 1, solid 2, and the gas coexist. Line T′B is the supercooling limit of the liquid; it then crystallizes in form S_1.

Direct application of the Gibbs–Duhem criterion indicates that the state of stable equilibrium corresponds to the phase which has the smallest potential (minimum of G). When these lines are crossed, **the material undergoes a phase transition**. At point C, the liquid and gas phases are totally identical. This is a singular point called **critical point**. Three phases can coexist at points B and D because the lines of coexistence have a common point at the intersection of three surfaces: these are the **triple points**.

It is useful to project the lines of coexistence AB, BE, BD, DC on plane (p, T), as the phase diagrams at equilibrium representing the different phases of the material in this plane are obtained in this way (Fig. 1.4).

1.2.2 Phase Diagrams

The preceding examples can be generalized and the different phases in which a material can exist will be represented by diagrams in a system of coordinates X_1 and X_2 (p and T, for example); these are the **phase diagrams**.

It is first necessary to note that the **variance** v of a system or a material is defined as the number of independent thermodynamic variables that can be acted upon to modify the equilibrium; v is naturally equal to the total number of variables characterizing the system minus the number of relations between these variables.

The situation of a physical system in thermodynamic equilibrium is defined by N thermal variables other than the chemical potentials (pressure, temperature, magnetic and electric fields, etc.). In general, only the pressure and temperature intervene and $N = 2$.

If the system is heterogeneous with c constituents (a mixture of water and alcohol, for example) that can be present in φ phases, we have:

$$v = c + N - \varphi \tag{1.3}$$

If the constituents are involved in r reactions, variance v is written

$$v = c - r + N - \varphi \tag{1.4}$$

For example, in the case of a pure substance that can exist in three different phases or states (water as vapor, liquid, and solid), $c = 1$, $N = 2$, and (1.4) shows that there is only one point where the three phases can coexist in equilibrium ($v = 0$); this is the triple point ($T = 273.16$ K for water). Two phases (liquid and gas, for example) can be in equilibrium along a monovariant line ($\varphi = 2$, $v = 1$). A region in the plane corresponds to a divariant monophasic system ($v = 2$).

As for the liquid/gas critical point, it corresponds to a situation where the liquid phase and the vapor phase become identical (it is no longer possible to distinguish liquid from vapor). We will have a critical point of order p when p phases are identical. We must write $r = p - 1$ criticality conditions conveying the identity of the chemical potentials. Then v is written:

$$v = c + N - p - (p - 1) = c - 2p + N + 1 \tag{1.5}$$

Since v must be positive, we should have: $c \geq 2p - N - 1$. If $N = 2$, the system must have a minimum of $2p - 3$ components in order to observe a critical point of order p.

For a transition corresponding to an "ordinary" critical point, only one constituent is sufficient for observing such a point; if $p = 2$, $c = 1$ and $v = 0$. For a tricritical point ($p = 3$), three constituents are necessary; the critical point in a ternary mixture is invariant ($v = 0$). The existence of such points in ternary mixtures such as n $C_4H_{10} - CH_3COOH - H_2O$ has been demonstrated experimentally.

In the case of a pure substance, the diagrams are relatively simple in planes (p, V), (p, T), or (V, T) corresponding to classic phase changes (melting or sublimation of a solid, solidification or vaporization of a liquid, condensation of a vapor, change in crystal structure). These are diagrams of the type shown in Fig. 1.4 representing four possible phases for the same substance in plane (p, T).

The situation is obviously much more complex for systems with several constituents. We will only mention the different types of diagrams encountered in general.

For simplification, consider a system with two constituents (binary system) A and B which can form a solid or liquid mixture (alloy or solid solution). Several phases can be present in equilibrium. To characterize this system, we introduce a concentration variable: x_A and x_B are the mole fractions of constituents A and B in the mixture ($x_A + x_B = 1$). Calculation of the free enthalpy G of the mixture as a function of x_A and x_B at any pressure p and T allows determining the stable thermodynamic phases by applying the Gibbs–Duhem criterion. If G_A^0 and G_B^0 designate the molar free enthalpies of elementary substances A and B and G_A and G_B are the molar free enthalpies of A and B in the mixture, the free enthalpy G_m of the mixture is written:

$$G_m = x_A\, G_A + x_B G_B \tag{1.6}$$

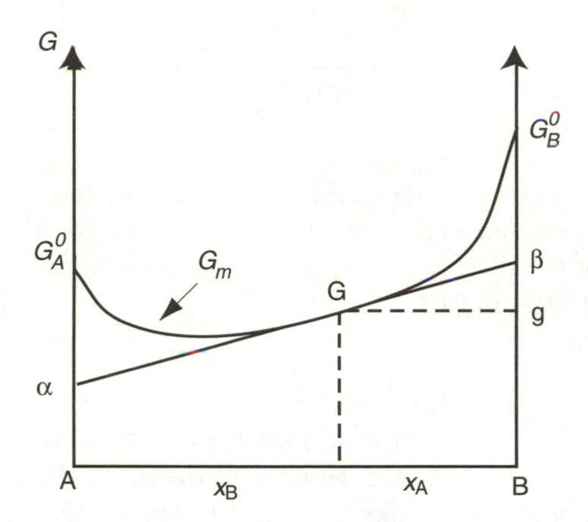

Fig. 1.5. Phase diagram of binary mixture A–B. The alloy is composed of two solids A and B of respective mole fractions x_A and x_B. The free enthalpy corresponding to composition (x_A, x_B) is represented by point G. A diagram of this type is obtained for each temperature ($x_A + x_B = 1$).

The corresponding diagram for temperature T and pressure p is shown in Fig. 1.5.

Using classic thermodynamics, we can show that the intersections of the tangent to curve $G_m(x_B)$ with vertical axes A and B (corresponding to pure solids A and B) are points with coordinates G_A and G_B, and the slope of the tangent is equal to the difference in chemical potentials $\mu_A - \mu_B$ of B and A in the mixture. This is a general property of phase diagrams.

In fact, phase diagrams especially have the advantage of allowing us to discuss the conditions of the existence and thus the stability of multiphasic systems as a function of thermodynamic variables such as temperature, pressure, and composition. This situation is illustrated with the diagrams corresponding to a mixture of two constituents A and B which are completely or partially soluble or miscible in each other. We have diagrams of the type shown in Figs. 1.6 and 1.7 in planes (T, x) at fixed pressure. Here x is the mole fraction of a constituent, B, for example.

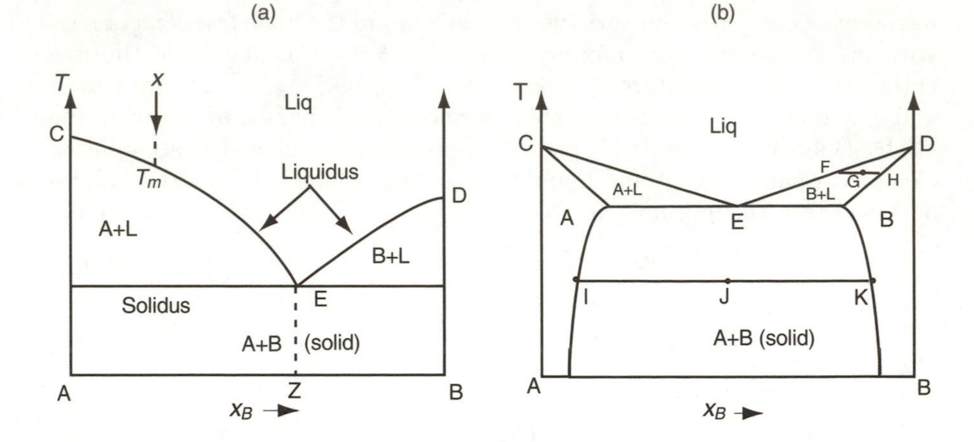

Fig. 1.6a,b. Phase diagram for a binary mixture of two solids. At point E, we have an invariant system whose composition corresponds to a eutectic mixture. Line CED is called the liquidus. The lower part of the diagram corresponds to a solid phase A + B. Points A and B respectively correspond to $x_B = 0$ and $x_B = 1$.

In the case of a solid solution (Fig. 1.6), if the liquid is cooled from composition point x, constituent A begins to solidify at temperature T_m on curve CE. This coexistence curve is called the liquidus. Continuing to decrease the temperature, the fraction of A that solidifies increases. At point E, the system is invariant ($v = 0$). At E, constituent B solidifies in turn; E is the eutectic point, and pure solids A and B can coexist there with the liquid of composition z. This point also corresponds to the lowest temperature at which the solid can exist. The diagram shown in Fig. 1.6b is a variant of the diagram in Fig. 1.6a. Point G corresponds to a two-phase system: a solid phase at point H and a liquid phase at point F. Moreover, point J corresponds to the two solid phases A and B represented by points I and K. The regions denoted A and B are bivariant; in fact, solid phases A and B are in equilibrium with their vapor, and these regions are deformed when the pressure is varied.

Figure 1.7 is another illustration of this type of situation in the particular case of the classic system consisting of the mixture H_2O and NaCl (called cooling mixture, also used for salting snow-covered roads). Several phases can be present: a vapor phase, a solution of NaCl in H_2O, ice and a solid phase NaCl, and an intermediate solid phase NaCl – $2H_2O$, which is a dihydrate.

Another slightly more complex situation is where a liquid phase and a solid phase can combine to form another solid phase in a two-component system. This transformation can be represented symbolically as: liquid + solid 1 $\Rightarrow$ solid 2.

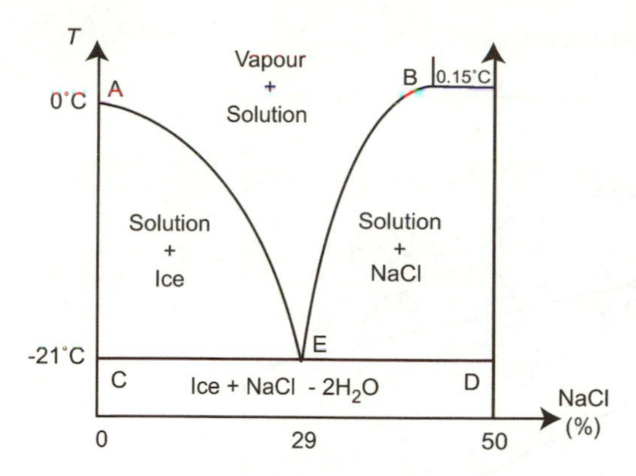

Fig. 1.7. Water–salt cooling mixture. The concentration of salt, NaCl, is shown on the abscissa. The regions of the plane such as ABE are bivariant ($v = 2$). The system is monovariant ($v = 1$) along curves AE and EB; three phases (solution, ice, and vapor or dihydrate, solution and its vapor) can be in equilibrium. At point E, the system is invariant ($v = 0$); it corresponds to the eutectic mixture at $T_0 = -21°C$ with a 29% concentration of NaCl. If $T >= -21°C$, the mixture is not in equilibrium: the ice melts.

The carbon–iron system, with different kinds of alloys which are steels is a classic example of the situation shown in Fig. 1.8. These are **peritectic phases** or systems. In the diagram in Fig. 1.8, it would seem that when a liquid solution with a carbon content of less than 2.43% is cooled, it separates into a solid solution δ and a liquid. When the temperature of 1493°C is reached, the liquid and solid phase δ form a new solid phase called γ iron, which is stable below this temperature and within certain concentration limits.

If the phase rule permits determining the regions of stability for the different phases of a system (for example, a binary alloy), it nevertheless does not allow calculating the fractions of liquid and solid phases in equilibrium. This can be done with the so-called **lever** rule, illustrated for the binary copper–nickel alloy in Fig. 1.9. If x designates the concentration of copper in the binary alloy at temperature T_0, we can show that the mole fraction f_L in liquid form is simply given by the ratio $f_L = a/(a + b)$.

The diagrams are similar for isothermally or isobarically vaporized solutions (distillation). Their use will be explained in Chapter 4 (Fig. 4.9). To conclude these general comments on phase diagrams, it is necessary to emphasize once more that a complete study of the conditions of stability of heterogeneous systems (binary, ternary, etc. mixtures of liquids or solids in which several phases can exist) and thus phase transitions requires having a free enthalpy–composition diagram as well. The Gibbs–Duhem criterion in fact indicates that the phase with the lowest free enthalpy is stable.

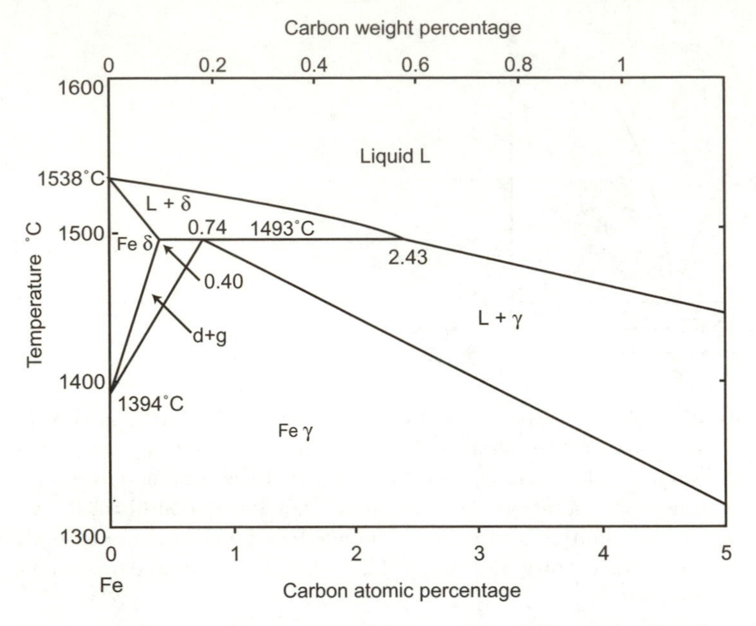

Fig. 1.8. Iron–carbon phase diagram. The different kinds of steel obtained are indicated. These are called **peritectic phases**. For example, if the carbon content is less than 2.43%, we have a solid solution γ in equilibrium with the liquid.

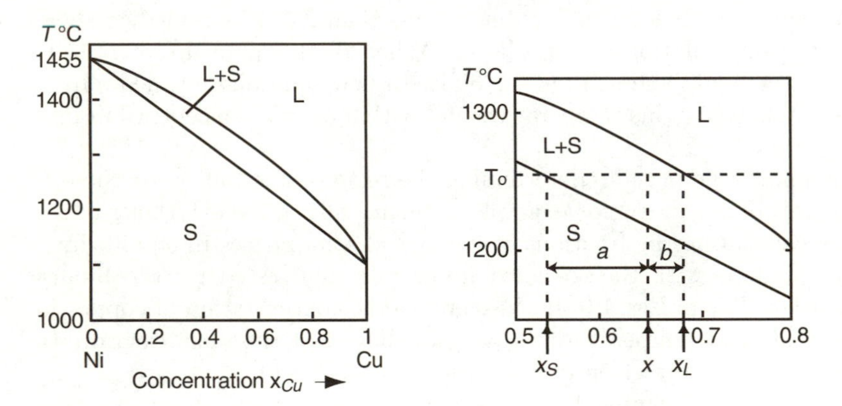

Fig. 1.9. Application of the lever rule. The fractions of liquid and solid phases in equilibrium in a Cu–Ni alloy are determined (enlarged phase diagram). The mole fraction in liquid form at T_0, f_L is given by the ratio $a/(a+b)$.

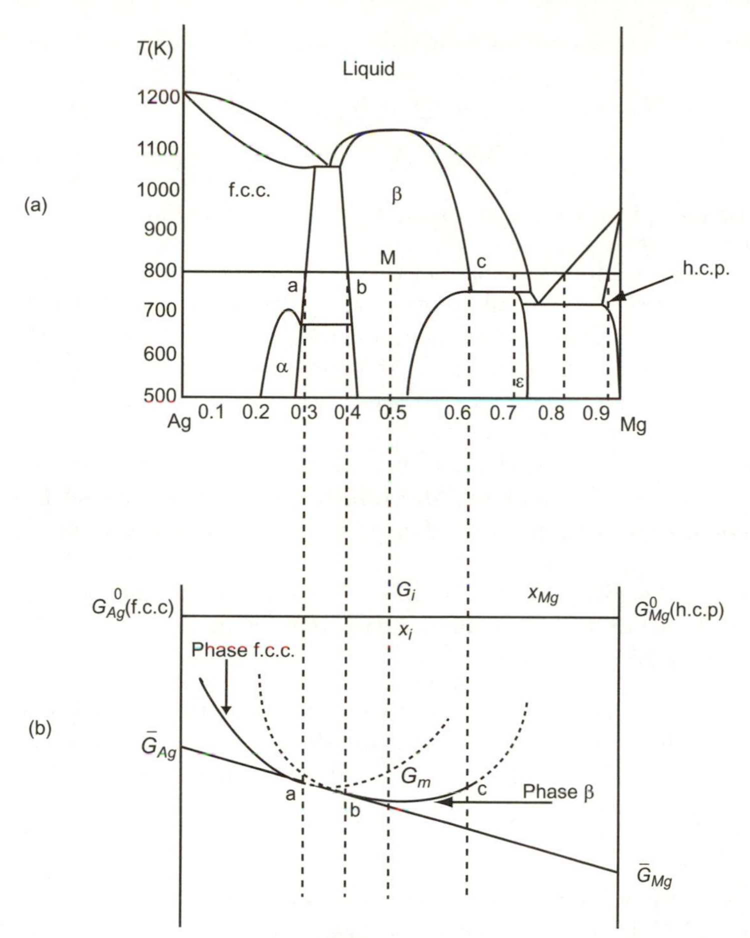

Fig. 1.10. Phase diagram of Mg–Ag alloy. (**a**) represents the different crystalline forms of the solid solution: f.c.c., β (of the CsCl type), h.c.p. (hexagonal close-packed), ϵ. (**b**) corresponds to the $G(x)$ diagrams at 800 K for phases β and f.c.c.; they have a common tangent at a and b. The chemical potentials μ_{Ag} and μ_{Mg} are identical for these two phases, which are thus in equilibrium. G_iG_m is the free enthalpy of formation of phase β of composition x_i from pure solids. The segments of the curves in dashed lines correspond to metastable phases β and f.c.c.. The concentration of Mg is designated by x.

Figure 1.10 corresponding to the binary silver–magnesium alloy describes an example of the reasoning that can result from finding several solid phases of different symmetry. At a given temperature, 800 K, for example, the diagram representing G_m for the system as a function of the molar concentration of magnesium x can be plotted. Point M in (T, x)-plane corresponds to phase β (simple cubic system characteristic of crystals of the CsCl type). The $G(x)$ diagrams corresponding to the β and face-centered cubic phases (f.c.c.) are also shown. Point G_i corresponds to unmixed systems Ag and Mg; for composition x_i, point G_m in the diagram corresponding to phase β represents the solid solution, and G_iG_m is thus the change in free enthalpy for formation of this solid solution.

1.2.3 Thermodynamic Classification of Phase Transitions

Evaporation of a liquid exhibits two distinct thermodynamic behaviors on the liquid–vapor coexistence curve: at all points of the curve except for terminal point C (called the critical point), latent heat and discontinuity of density are simultaneously observed in the transition. At point C, on the contrary, we continuously pass from the liquid phase to the vapor phase: there is neither latent heat nor density discontinuity.

We can thus schematically say that two types of phase transitions can exist: **transitions with latent heat on one hand, and transitions without latent heat on the other hand**. This is a thermodynamic classification.

More generally, the physicist P. Ehrenfest proposed a classification of phase transitions based on the thermodynamic potentials in 1933.

Ehrenfest proposed distinguishing:

- **First order transitions** which are accompanied by discontinuities of thermodynamic quantities such as the entropy and density, themselves associated with the first derivatives of thermodynamic potentials. For example,

$$S = -\left(\frac{\partial G}{\partial T}\right)_p \qquad V = \left(\frac{\partial G}{\partial p}\right)_T \qquad H = \frac{\partial(G/T)}{\partial(1/T)} \tag{1.7}$$

  In the phase transition, these quantities corresponding to first order derivatives of potential G are discontinuous (the latent heat is associated with discontinuity of the entropy).
- **Second order transitions** for which the thermodynamic potentials and their first order derivatives are continuous, while some second derivatives with respect to state variables are reduced to zero or approach infinite asymptotically at the transition point. In this way, we can write:

$$\begin{aligned} \frac{C_p}{T} &= -\left(\frac{\partial^2 G}{\partial T^2}\right)_p = \left(\frac{\partial S}{\partial T}\right)_p \\ \kappa_T V &= -\left(\frac{\partial^2 G}{\partial p^2}\right)_T = -\left(\frac{\partial V}{\partial p}\right)_T \end{aligned} \tag{1.8}$$

  C_p and κ_T are respectively the specific heat at constant pressure and the compressibility at constant temperature. These thermodynamic quantities actually approach infinity at the liquid/gas critical point or at the superfluid transition point in liquid helium (at 2.17 K in helium 4).
  For these transitions, we continuously pass from one phase to another without being able to really speak of the coexistence of the two phases: at the

liquid/gas critical point, the liquid phase can no longer be distinguished from the gas phase (their densities are strictly equal).

These very different thermodynamic behaviors can be demonstrated experimentally by directly or indirectly studying the thermodynamic behavior of characteristic physical quantities in the vicinity of the transition (density, latent heat, etc.). The absence of latent heat in the transition is a good discriminatory criterion that allows classifying it as a second order (or even higher order) transition.

The situations corresponding to these two categories of transitions are simply represented in the diagrams (Fig. 1.11).

This classification can be extended to phase transitions higher than second order, generally called **multicritical transitions**, with the definitions initially proposed by Ehrenfest. These transitions will be characterized as follows:

- the thermodynamic potentials are continuous at the transition point;
- certain second- and higher-order derivatives of the thermodynamic potentials with respect to the state variables are reduced to zero at the transition point.

The corresponding transition points are called multicritical points. More generally, we will call a point where p **phases become identical** (that is, they cannot be distinguished) the critical point of order p. Using the phase rule, we showed that the system must have a minimum of $2p-3$ components for a multicritical point of order p to be observed (1.5).

In fact, although the Ehrenfest classification of phase transitions has the great merit of revealing the similarities between phenomena as different as magnetism, ferroelectricity, superconductivity, and the liquid/gas transition at the critical point, it is nevertheless limited to a thermodynamic view of the phenomena. Although this is undoubtedly important, it is nevertheless not sufficient. A physicist such as L. D. Landau noted in 1937 that a phase transition without latent heat was accompanied by a change in symmetry (with the exception of the liquid/gas transition at the critical point, which is special). In the case of a magnetic material, this has no permanent magnetic moment above its Curie point (magnetic state); below this temperature, on the contrary, it has permanent magnetization oriented in a certain direction (ferromagnetic state). We say that the **symmetry was broken** in the transition: below the Curie point, the material is only invariant due to rotation around an axis oriented in the direction of the magnetization.

Questions of symmetry are thus very important in studying phase transition phenomena. Landau associated the notion of **order parameter** with these considerations on the changes or breaks in symmetry accompanying a phase transition phenomenon. The order parameter is in general a physical quantity of extensive character which is zero in the most symmetric (or most disordered) phase and non-zero in the least symmetric (or ordered) phase.

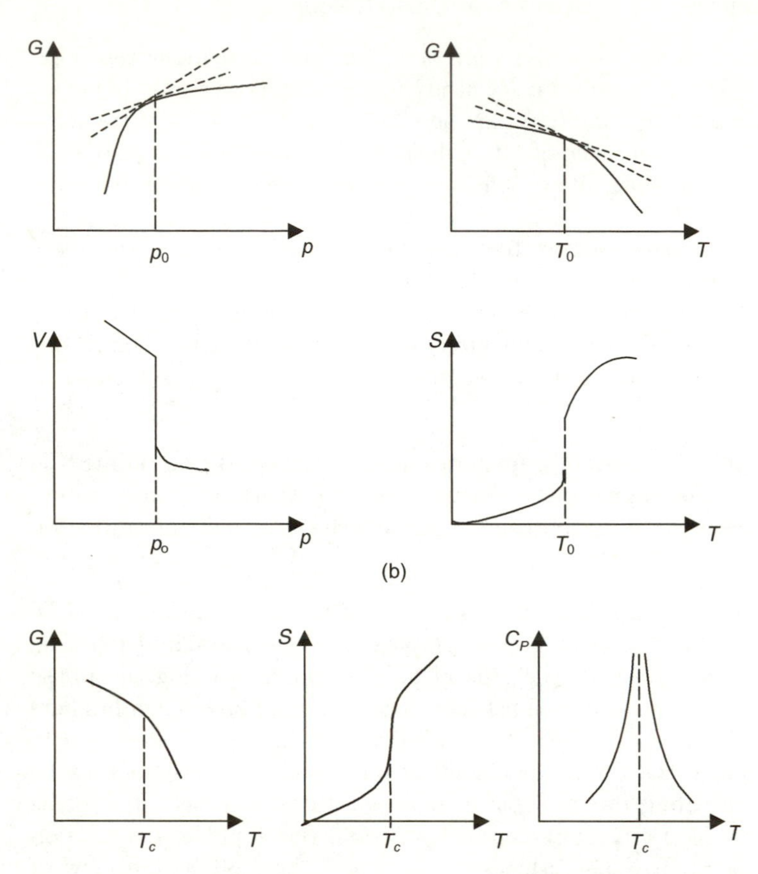

Fig. 1.11. Phase transitions. (**a**) First order transitions. Potentials such as G are continuous in the transition, but the first derivatives and associated quantities (V and S) are discontinuous. **(b)** Second order transitions; the first derivatives of G are continuous, but some second derivatives diverge: C_p, for example, approaches infinity.

This notion of order parameter has an obvious qualitative meaning: when the temperature decreases, the order of the system increases. When a liquid is cooled, it solidifies by passing the solidification point (the crystalline solid is more ordered than the liquid). Moreover, if a ferromagnetic material is cooled below its Curie point, the magnetic order in the system increases (macroscopic magnetization appears, which shows the existence of magnetic order).

For magnetism, the order parameter is the magnetization, while it is the electrical polarization in the case of ferroelectricity. The choice of the order parameter, as we will see later, is not always evident. For example, in the case of superfluidity and superconductivity, the order parameter is the wave function of the superfluid phase and the electrons associated with superconductivity.

Using this notion of order parameter, we can distinguish two **types of transitions**:

- transitions with no order parameter for which the symmetry groups of the two phases are such that none is strictly included in the other: they are always first order (with latent heat) in Ehrenfest's sense;
- transitions for which an order parameter can be defined and for which the symmetry group of the least symmetric phase is a subgroup of the symmetry group of the most symmetric phase. If the order parameter is discontinuous at the transition, it is first order in Ehrenfest's sense; if it is continuous at the transition, it is second order (without latent heat).

More generally, we define **first order transitions** associated with the existence of latent heat on one hand, and all other transitions that can be considered continuous on the other hand. The last category particularly includes multicritical phenomena (transitions higher than second order in Ehrenfest's sense).

1.3 General Principles of Methods of Investigating Phase Transitions

For a material that can undergo a phase transition, three types of questions can be investigated: the conditions of the phase change (temperature, pressure, magnetic field, etc.); the behavior of physical quantities in the vicinity of the transition; the new properties of the material when it undergoes a phase transition and their characteristics.

The Gibbs–Duhem stability criterion provides a simple way of studying the conditions of a phase change since this phenomenon implies disruption of phase stability (for example, evaporation of a liquid phase is observed when it becomes unstable).

Use of this criterion requires calculating the thermodynamic potentials. Moreover, the study of the behavior of physical quantities in the vicinity of a phase transition requires using thermodynamic potentials in one form or another; these quantities thus allow characterizing the new properties of the material. The methods used for these studies are schematically of **two kinds**: those utilizing models representing the system described in a simplified way with **thermodynamic potentials**, and those where the behavior of the material is **"simulated"** by a numerical method.

1.3.1 Calculation of Thermodynamic Potentials and Quantities

Calculation of the thermodynamic potentials and conjugated quantities is based nowadays on classic methods in thermodynamics and statistical mechanics; so there is no need to discuss these principles here.

We note that there is a simple general method for systematically obtaining all of the state functions or thermodynamic potentials describing a system with state variables associated with exact differential forms: this is the method of **Legendre transformations**.

If Y is a function which is only dependent on a single independent variable X (a certain state function)

$$Y = Y(X) \tag{1.9}$$

with

$$P = \frac{\mathrm{d}Y}{\mathrm{d}X} \tag{1.10}$$

The derivative P is obviously the slope of the tangent to curve $Y(X)$. If X is replaced by P as independent variable by the Legendre transformation as follows:

$$\Psi = Y - PX \tag{1.11}$$

then we can show that Ψ is a new state function, that is, a thermodynamic potential. A function of P is formally obtained by calculating X as a function of P in (1.10) and by substituting it in (1.9).

The transformation (1.11) can be generalized by applying it to functions of several variables $Y = Y(X_1, X_2, ...X_n)$.

If $P_i = \mathrm{d}Y/\mathrm{d}X_i$ is the conjugate variable of X_i, function Ψ is defined by the transformation

$$Y = \Psi - \Sigma_i \, P_i X_i \tag{1.12}$$

In general, the choice of variables that can be used to study a system, particularly when it is undergoing a phase transition, imposes the choice of the thermodynamic potential. When working with variables p and T, the free enthalpy G is the good potential; when the system is to be studied with variables T and V, the free energy F is the pertinent potential. We then have the classic relations:

$$\begin{aligned} \mathrm{d}U &= T\mathrm{d}S - p\mathrm{d}V \qquad & \mathrm{d}F &= -S\mathrm{d}T - p\mathrm{d}V \\ \mathrm{d}H &= T\mathrm{d}S + V\mathrm{d}p \qquad & \mathrm{d}G &= S\mathrm{d}T + V\mathrm{d}p \end{aligned} \tag{1.13}$$

For a magnetic system with magnetic induction B and designating the magnetic magnetization by M, we will have the potentials:

$$F = U - TS \quad \text{and} \quad G = U - TS - MB$$

We deduce from these relations, for example:

$$C_v = -T\left(\frac{\partial^2 F}{\partial T^2}\right)_v \quad \text{and} \quad \kappa_T = -\frac{1}{V}\left(\frac{\partial^2 G}{\partial p^2}\right)_T \tag{1.14}$$

We have similar expressions in the case of magnetic systems, and the magnetic susceptibility is then the counterpart of the compressibility for a fluid.

It is evidently essential to have an analytical form of the pertinent thermodynamic potential in order to study a phase transition phenomenon, for example, the free enthalpy G. It is thus necessary to have a model that is a simplified representation of the physical reality and can be used to calculate the thermodynamic potential as a function of the useful state variables.

As an example, take the case of a liquid or solid binary solution with two constituents A and B of respective concentrations x_A and x_B. If G_A^0 and G_B^0 designate the free enthalpies of pure substances A and B, We take $G(x_A, x_B)$ as the free enthalpy of the solution:

$$G = x_A\, G_A^0 + x_B G_B^0 + u\, x_A x_B + RT(x_A \ln x_A + x_B \ln x_B) \tag{1.15}$$

R is the ideal gas constant.

In (1.15), it was assumed that the energy of interaction between constituents A and B in the mixture is proportional to the product of their concentrations (this is the term $u\, x_A x_B$).

We arrive at this form of the energy of interaction if we assume: 1) That the molecules in the solution are positioned next to each other so that each molecule has the same number of near neighbors; this is called a **lattice model**. 2) That the mixture is totally random. 3) That the intermolecular forces are short–range and that only the binary interactions between closest neighbors can be taken into consideration.

The fourth term in (1.15) represents the contribution of the entropy of mixing to $G(x_A, x_B)$.

A solution in which the free enthalpy is given by (1.15) is called a regular solution. Equation (1.15) allows investigating the behavior of the solution, in particular, finding its metastability limit, which is given by the condition

$$\delta^2 G = 0 \tag{1.16}$$

Assuming $x = x_A$ with $x_A + x_B = 1$ and thus $x_B = 1 - x$, the equation is rewritten:

$$G = x\, G_A^0 + (1-x) G_B^0 + u\, x(1-x) + RT[x \ln x + (1-x)\ln(1-x)] \tag{1.17}$$

Condition (1.16) can easily be written:

$$-2u\, x(1-x) + RT = 0 \tag{1.18}$$

This is the equation for a very simple curve which gives the shape of the spinodal. This curve is shown by the dashed line in Fig. 1.12. There is a critical

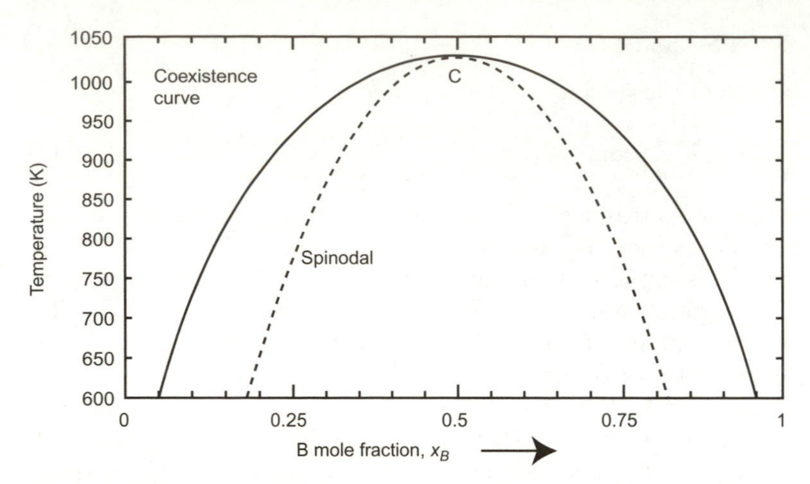

Fig. 1.12. Coexistence line and spinodal for a regular solution. The spinodal is the metastability limit of binary mixture A–B. Mixture A–B becomes unstable inside the spinodal: there is phase separation.

point of coordinates $x = 1/2$ and $T_C = u/2R$ above which constituents A and B form a homogeneous solution regardless of their concentrations.

This critical point, which is simultaneously the highest point of the spinodal and the coexistence curve, is an Ehrenfest second order transition point. The phase transition associated with separation of constituents A and B of the homogeneous phase is called the miscibility gap in the case of a liquid solution.

As for the coexistence curve itself, it can be calculated by writing that the chemical potential of a constituent (A, for example) is the same in the homogeneous phase and in the system after separation. For a regular solution, the following equation is rather easily found for the coexistence curve:

$$\frac{1-x}{x} = \exp\left[\frac{(1-2x)u}{RT}\right] \tag{1.19}$$

The model describing a regular solution has a relatively general extent. When applied to a solid solution in particular, it is called the **Bragg–Williams model**.

Expressions such as (1.17) can be generalized by expanding the excess free enthalpies (that is, the difference between the free enthalpy of the solution or mixture and the free enthalpy of an ideal solution) as a function of the differences in the mole fractions $x_A - x_B$. These expansions in the form of polynomials of the excess free enthalpy G^E

$$G^E = u\,x_A x_B + G_1(x_A - x_B)^2 + G_2(x_A - x_B)^3 + \ldots \tag{1.20}$$

are called the Redlich–Kisten equations.

In general, the correspondence between the thermodynamic quantities and those of statistical mechanics is used. The internal energy U is thus the average energy of a system:

$$U = \langle E \rangle = \Sigma_\alpha E_\alpha e^{-\beta E_\alpha}/Z \tag{1.21}$$

We have a summation on all microscopic states of the energy system E_α. Z is the partition function:

$$Z = \Sigma_\alpha e^{-\beta E_\alpha} \tag{1.22}$$

where $\beta = 1/kT$, k is the Boltzmann constant, and we have:

$$\begin{aligned} &U = -\left(\frac{\partial \log Z}{\partial \beta}\right)_V, \qquad C_V = \left(\frac{\partial U}{\partial T}\right)_V = k\beta^2\left(\frac{\partial^2 \log Z}{\partial \beta^2}\right)_V \\ &F = -kT\log Z, \qquad G = \frac{-1}{\beta}\left(\frac{\partial (V\log Z)}{\partial V}\right)_T \\ &p = -\left(\frac{\partial F}{\partial V}\right)_T = -\frac{1}{\beta}\left(\frac{\partial \log Z}{\partial V}\right)_T \end{aligned} \tag{1.23}$$

Calculation of the thermodynamic quantities is thus reduced to calculation of partition function Z. It is thus necessary to know which form we have assumed for the energy of the system, that is, by the Hamiltonian $\mathcal{H}$. This makes it necessary to select a model representing the system: liquid, gas, metal alloy, ferromagnetic or ferroelectric material, etc.

1.3.2 Equation of State

An isotropic fluid composed of only one molecular species with a fixed number of molecules N can be described by two variables. Variables (p, V, T) are not independent: there is a relation $f(p, V, T)$ between these variables called **equation of state**.

In general, each intensive variable (T, p, magnetic field H, etc.) can be expressed by partial derivatives of functions (thermodynamic potentials) of extensive variables (S, V, N, M, etc.). There are functional relations between variables that generalize the equation $f(p, V, T) = 0$. In the case of a magnetic system, a relation can be found between magnetic field $\boldsymbol{H}$, magnetization $\boldsymbol{M}$, and temperature T. However, the equation of state is most often used in the case of fluids to represent the state of the fluid and predict phase changes.

An equation of state can be obtained by calculating an intensive variable such as pressure p with the expressions for the thermodynamic potentials (Eqs. 1.23, for example) and by searching for the form taken by product pV, for example.

The Taylor expansion of pV, as a function of $1/V$, allows placing the equation of state for a fluid in the form of the **virial expansion**.

We note that the van der Waals equation of state proposed in 1873 is the first equation of state for a real fluid. We will specifically return to these topics in a later chapter.

The continuous increase in the calculating power and speed of computers opened up new prospects for numerical calculation and allowed simulating the behavior of systems composed of a large number of particles (atoms or molecules) in the physics of condensed matter. We can calculate thermodynamic quantities such as the pressure for a finite system and then pass to the **thermodynamic limit** ($N \to \infty$, $V \to \infty$, $V/N = C^{te}$). This is the principle of two **simulation** techniques: **molecular dynamics** and the **Monte Carlo** method, which will be discussed in Chapter 4.

1.3.3 Dynamic Aspects – Fluctuations

A phase transition is not an instantaneous phenomenon since the material generally evolves in the vicinity of the transition. Germs or nuclei of a phase B are formed within the initial phase A (for example, solid microcrystallites are formed within the liquid phase): this is the **nucleation** phenomenon. It is thus necessary to take into account the dynamics of the phase transition. This is a specific problem that will be discussed in Chapter 2. Nucleation phenomena particularly involve the surface tensions and interfacial energies associated with them.

Another aspect of the physics of phase transitions – but one which is not peculiar to them – is the problem of fluctuation of the thermodynamic quantities characterizing one phase of the material (the density of the liquid phase, for example), particularly in the vicinity of the transition. In effect, any system in thermodynamic equilibrium is the site of local variations of the thermodynamic variables around their equilibrium value.

For example, if $n(\boldsymbol{r})$ is the local density of molecules in a fluid with N particles and $< n(\boldsymbol{r}) >$ is the average density with fixed T and p (this is the statistical mean over the entire fluid), $\delta n(\boldsymbol{r}) = n(\boldsymbol{r}) - < n(\boldsymbol{r}) >$ is the fluctuation of the density variable. By definition, we have $< \delta n(\boldsymbol{r}) >= 0$. If the system is uniform, $< n(\boldsymbol{r}) >= n = N/V$.

In the case of fluids, $\delta n(\boldsymbol{r})$ is greater the closer we are to the liquid/gas critical transition point: this property can be demonstrated with the **critical opalescence** phenomenon.

The existence of fluctuations in the vicinity of thermodynamic equilibrium is a completely general phenomenon. Using statistical mechanics, we can demonstrate that there are simple relations between the fluctuations of the thermodynamic quantities and other thermodynamic quantities (Problems 1.1 and 1.2).

Let us now return to the case of a fluid described by its density $n(\boldsymbol{r})$. Assuming that the system is invariant due to translation, the autocorrelation function $H(\boldsymbol{r})$ characterizing the fluctuations is defined:

$$H(\boldsymbol{r}', \boldsymbol{r}'') = H(\boldsymbol{r}' - \boldsymbol{r}'') = 1/N < \delta n(\boldsymbol{r}')\delta n(\boldsymbol{r}'') >= H(\boldsymbol{r}) \tag{1.24}$$

where we set $\boldsymbol{r} = \boldsymbol{r}' - \boldsymbol{r}''$ and where $<>$ is the ensemble average over the system.

If a magnetic solid is involved, a spin–spin correlation function will be introduced between the spins of sites i and j:

$$\Gamma(\boldsymbol{r}_i, \boldsymbol{r}_j) =< (S_i- < S_i >)(S_j- < S_j >) > \tag{1.25}$$

where $<>$ is the ensemble average over the spins. If the system of spins is invariant on translation, we have $< S_i >=< S_j >$ and (1.25) is written:

$$\Gamma(\boldsymbol{r}_i, \boldsymbol{r}_j) = \Gamma(\boldsymbol{r}_i - \boldsymbol{r}_j) \tag{1.26}$$

If we are far from the critical point (Curie point) of the ferromagnetic–paramagnetic transition, the spins are then weakly correlated when we have $\boldsymbol{r} = \boldsymbol{r}_i - \boldsymbol{r}_j \to \infty$: $< S_i S_j >\approx< S_i >< S_j >=< S >^2$ and $\Gamma \to 0$.

On the contrary, in the immediate vicinity of the Curie point, neighboring spins are strongly correlated and $\Gamma \neq 0$.

We are more specifically interested in the case of liquids with local density $n(\boldsymbol{r})$ in total volume V with N molecules (Fig. 1.13). In this volume V, the fluctuations in the total number of molecules are characterized by:

$$< \Delta N^2 >=< \int_V \delta n(\boldsymbol{r})\mathrm{d}\boldsymbol{r} \int_V \delta n(\boldsymbol{r}')\mathrm{d}\boldsymbol{r}' > \tag{1.27}$$

Hence

$$< \Delta N^2 >= N \int_V \mathrm{d}\boldsymbol{r} \int_V \mathrm{d}\boldsymbol{r}' H(\boldsymbol{r} - \boldsymbol{r}') = VN \int_V H(\boldsymbol{r}'')\mathrm{d}\boldsymbol{r}'' \tag{1.28}$$

Using a grand–canonical ensemble (Problem 1.2), we can show that for a fluid:

$$< \Delta N^2 >= N \frac{\kappa_T}{\kappa_T^0} \tag{1.29}$$

where $\kappa_T = -1/V(\partial V/\partial p)_T$ is the isothermal compressibility of the fluid and κ_T^0 is the compressibility for an ideal gas. $< N >$ is the mean number of

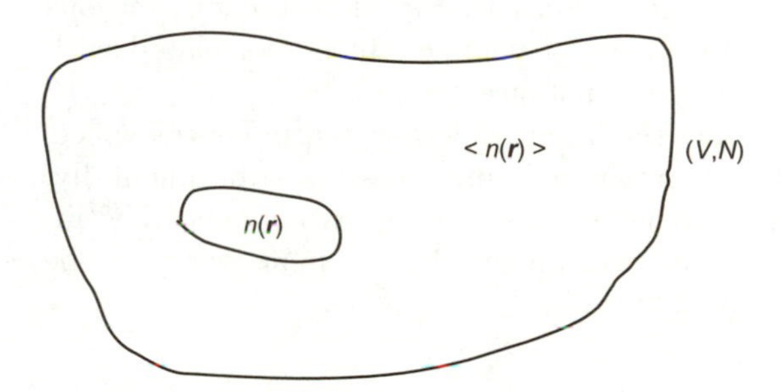

Fig. 1.13. Local density fluctuation in a fluid.

molecules ($< N >= N$ if it is fixed). We thus have:

$$\kappa_T/\kappa_T^0 = V \int_V H(\boldsymbol{r})\mathrm{d}\boldsymbol{r} \tag{1.30}$$

We know that $\kappa_T \to \infty$ at the liquid/gas critical point (this is an experimental finding which can be accounted for by an equation of state such as the van der Waals equation, for example). Integral (1.30) is thus divergent and $H(\boldsymbol{r})$ should have singularity.

Since local density fluctuations are strongly correlated, they can be characterized by correlation length $\xi(T)$, which is a function of the temperature and approaches infinity when $T \to T_C$. We can then take $H(\boldsymbol{r})$ in the simple form, for example:

$$H(r) = C^{st}\frac{e^{-r/\xi}}{r} \tag{1.31}$$

In general, a **structure function** $S(\boldsymbol{k})$ for a material is defined by:

$$S(\boldsymbol{k}) = \frac{1}{N} < n(\boldsymbol{k})n(-\boldsymbol{k}) > \tag{1.32}$$

where $n(\boldsymbol{k})$ is the Fourier transform of $n(\boldsymbol{r})$: $n(\boldsymbol{k}) = \int_V e^{-i\boldsymbol{kr}}\ n(\boldsymbol{r})\ \mathrm{d}\boldsymbol{r}$. For $\boldsymbol{k} \neq 0$, it is possible to show that $S(\boldsymbol{k}) = H(\boldsymbol{k})$, where $H(\boldsymbol{k})$ is the Fourier transform of $H(\boldsymbol{r})$. With (1.31) for $H(\boldsymbol{r})$, $S(\boldsymbol{k})$ assumes the simple asymptotic form:

$$S(\boldsymbol{k}) \propto \frac{1}{k^2 + \xi^{-2}} \tag{1.33}$$

This simple form for $S(\boldsymbol{k})$ is called the Ornstein–Zernike–Debye approximation for the structure function. Near the critical temperature T_C, $\xi(T)$ can be selected in the form $\xi(T) = C^{st}|T - T_C|^{-1/2}$. If $\boldsymbol{k}$ is small, $S(\boldsymbol{k}) \approx \xi^2 \to \infty$.

The correlation function $H(\boldsymbol{k})$ defined by (1.24) or in the form of (1.26) allows characterizing the way in which the different parts of a system are correlated. The **correlation length** $\xi(T)$is a simple spatial translation of this correlation: it expresses the range of fluctuations of a variable characteristic of its state in the material. The larger $\xi(T)$ is, the more the fluctuations in the density of a fluid in one element of volume will be correlated with fluctuations in another element a great distance away.

The structure function expresses the same properties, but in Fourier space. The structure function has the advantage of being accessible experimentally: it can be obtained with a light– or neutron–scattering experiment. The intensity of the light scattered by a medium in which a light wave can be propagated is in effect proportional to $S(\boldsymbol{k})$.

1.4 The Broad Categories of Phase Transitions

We have explained the thermodynamic characteristics of a phase transition which can be used to lay the foundations for a classification of these transitions. Beyond this thermodynamic classification, it is almost impossible to give a satisfactory typology of phase changes because they involve extremely different properties of the substance. Moreover, they are encountered in all systems, whether physical, chemical or biological. Any attempt to classify the transitions thus necessarily has an arbitrary aspect.

Schematically, they can a priori be classified in two broad categories:

- phase transitions for which none of the phases is distinguished by intrinsically different properties: they are characterized by a change of order, structure, or symmetry in the material;
- transitions where one of the phases that appears has a new physical property; it can also be associated with a change in symmetry or structure.

Evaporation and solidification of a liquid, melting and sublimation of a solid, and separation of a mixture of liquids belong in the first category. The same holds for the order/disorder transition in a binary alloy: the atoms of each constituent are randomly distributed on the crystalline sites in the disordered phase at high temperature, while they preferentially occupy certain sites in the ordered phase at low temperature.

Ferromagnetic/paramagnetic, ferroelectric/paraelectric, and superconducting phase transitions belong to the second category.

This classification has the fault of being qualitative since it is not based on any physical criterion which could serve as the basis for a comparison between phase changes.

Phase transitions can also be differentiated with the structure change criterion: transitions **with or without structural change** can be distinguished.

1.4.1 Transitions with a Change in Structure

Many substances in the solid state undergo a phase transition associated with a change in structure: during the transition, the arrangement of the atoms is modified and is associated with a change in the symmetry of the crystal. These are **structural transitions**. This change, or break, in symmetry is characterized by going from a phase with high symmetry to another phase with lower symmetry induced by displacement of atoms in the solid.

A liquid is a phase in which the molecules are a priori disordered: their centers of gravity are randomly distributed in the material. However, there are liquid phases composed of anisotropic molecules that can undergo a phase transition from a totally liquid "disoriented" phase (the molecules are randomly oriented and distributed) to an oriented phase: all of the molecules have their axes oriented in the same direction. These liquid phases are **liquid crystals**, also called **mesomorphic phases**.

From the mechanical point of view, the liquid crystal has the properties of a liquid. However, for reasons of symmetry, if the liquid crystal is deformed, the energy associated with any deformation is a quadratic function of these deformations: this is Hooke's law for solids. The symmetry is broken at the liquid/liquid crystal transition. Solidification of a liquid (liquid/solid transition) is accompanied by a change in structure with the appearance of crystalline symmetry in passing into a crystalline solid state. The solidification of a liquid in the form of a glass, the **glass transition**, is an exception: if there is no organized or long-range-order structure formation. Glasses are disordered materials **beyond equilibrium**.

Structural transitions are often accompanied by a modification of a physical quantity (density, elastic constants, thermal conductivity, etc.) and are sometimes associated with the appearance of a new property (ferroelectricity, for example).

There are essentially two broad categories of structural transitions:

- **order/disorder transitions**;
- **displacive transitions**.

Order–disorder transitions are characteristic of binary metal alloys (for example, Cu–Zn alloy, which is β-brass).

The phase transition can be characterized by an order parameter, as defined in Sect. 1.2.3 by:

$$P = |W_{Cu} - W_{Zn}|/(W_{Cu} + W_{Zn}) \quad (1.34)$$

where W_{Cu} and W_{Zn} are respectively the probabilities of occupation of a site by a copper atom or a zinc atom. For brass, the symmetry of the low-temperature phase is simple cubic; the symmetry of the high-temperature phase is centered cubic and the probabilities of occupation of the sites are equal (50 and 50%).

This order/disorder transition can be associated with a **ferroelectric–paraelectric transition** in some crystals, as we will see.

Displacive transitions are another category of structural transition. They are either characterized by finite displacement of an atom in a crystal lattice from a position of equilibrium (over distances of the order of interatomic distances) or by molecular rotation, which causes lattice distortion.

Although these movements are local, they are nonetheless cooperative: the atoms move collectively, but there is no change in the composition of the material (the atoms do not have the time to diffuse). For example, in perovskites, which are crystalline solids of the general formula ABO_3 ($SrTiO_3$, $BaTiO_3$, $NaNbO_3$, etc.), the phase transition results from rotation of polyhedrons (TiO_6 octahedrons in the case of $SrTiO_3$) in the lattice, which induces a change of symmetry (the symmetry is cubic at high temperature and

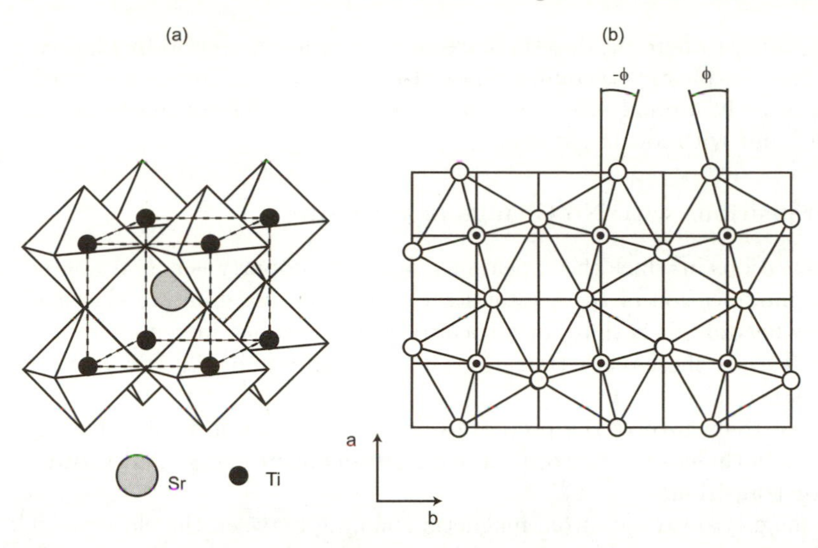

Fig. 1.14. Unit cells of perovskite materials. Their general formula is ABO_3. (**a**) There is a three–dimensional lattice of BO_6 octahedrons. (**b**) The structure was doped with Fe^{3+} ions; the motions of the octahedrons at the transition can be detected by electron paramagnetic resonance. The angle of rotation φ is the order parameter.

tetragonal at low temperature in $SrTiO_3$ with a second order transition at $T_C = 108$ K). The angle of rotation of an octahedron in the lattice is a state variable that can be used to describe the transition and that can be selected as order parameter (Fig. 1.14).

Some of these transitions can displace the centers of gravity of the positive and negative charges of the ions in the lattice; if they no longer coincide, the lattice has a macroscopic electric dipole moment in the absence of a permanent electric field and we then have a ferroelectric material and a ferroelectric/paraelectric transition (this is the case for $BaTiO_3$ and $NaNbO_3$, but not for $SrTiO_3$).

Finally, there are phase transitions in which the structure of the material is modified following motions of molecular rotation or their blocking; they can also modify the size of the crystal lattice. The most typical case is the **liquid/plastic crystal** and **plastic crystal/crystal** phase transition. Plastic crystal phases are intermediate between the liquid state and the crystalline state: they correspond to a solid material with a periodic crystal lattice but where the symmetry of the molecules leaves them one degree of freedom of rotation in the solid state. Carbon tetrachloride, CCl_4, and methane, CH_4, are typical examples of so-called globular molecules that form stable plastic crystal phases (the liquid/plastic crystal transition temperatures are respectively 250.3 K and 90.7 K). The plastic crystal phase is of rhombohedral symmetry in CCl_4; at $-47.4°C$, this phase undergoes a transition into a monoclinic crystalline phase where the degrees of freedom of rotation are totally blocked. The solid in the plastic crystalline phase **flows** easily because the molecules can rotate freely around one axis and be reoriented. These transitions are first order, but with low latent heat.

1.4.2 Transitions with No Change in Structure

Phase transitions in which the appearance of a new property is not correlated with any modification of the crystal structure of the material are classified under this heading; it is thus not associated with a change in symmetry.

We will study these phase transitions in detail later and will only mention their existence here.

Many of them involve the properties of the electrons of a solid. For example, this is the case of **ferromagnetic, superconducting,** and **metal/insulator** transitions.

Ferromagnetism results from magnetic coupling between the electrons in the solid (magnetism is encountered in crystalline solids and metallic glasses).

The superconducting transition is characteristic of the electric properties of some solids: their resistivity vanishes below a certain temperature. This transition is found in metals, cuprates, and some organic compounds and fullerenes doped with alkalis. This transition is second order.

As for the metal–insulator transition, or Mott transition, it is induced in a solid by application of pressure: after the phase transition, the solid has the electrical conduction properties of a metal. It has been demonstrated in silicon and germanium doped with elements from columns III and V of the periodic table (As and Ga, for example), or in solid matrices of rare gases in which metals have been trapped.

Another system in which transitions with no change in symmetry are observed is liquid helium. The liquid phase of helium 4 becomes superfluid below 2.17 K: its viscosity is reduced to zero (it flows without friction in a capillary tube). The superfluid transition has been demonstrated in helium 3 at a temperature of 2 mK. The superfluid transition is second order.

1.4.3 Non–Equilibrium Transitions

Some materials in the liquid state (pure substances or mixtures) undergo a phase transition manifested by modification of the mechanical properties of the system. This transition is induced by blocking of atomic or molecular motions whether the degrees of freedom corresponding to diffusion are frozen or a three-dimensional lattice involving molecules with their solvent is constituted. The new phase formed corresponds to a disordered state: no long-range order of the molecules or atoms. This type of transition occurs at a certain temperature T_g which is not fixed for a given material since it is a function of the heat treatment the system has undergone. The new phase is **not in equilibrium**.

The **glass transition**, that is, the phase change manifested by formation of a glass from a liquid, is the standard example of such a transition. It occurs in substances (elements, organic and inorganic compounds, mixtures) that form liquids whose viscosity becomes very high when they are cooled ($10^7 - 10^{13}$ P). Below their melting point, they first pass into a metastable supercooled state, then they slowly solidify to form a glass. There are very many examples of glassy materials and we will return to them: oxides such as SiO_2, B_2O_3, binary systems such as As–S, As–Se, metal–metalloid alloys such as Fe–B, Pd–Si, organic compounds of the C_2H_5OH type, etc.

Gelation or the sol–gel transition is another type of nonequilibrium transition in a liquid. This transition is seen in solutions of organic or inorganic molecules in a solvent. They can undergo a transition to a more organized phase in certain physical conditions (concentration, temperature, pH): they are constituted of a three-dimensional lattice without periodic distribution of the centers of gravity but with the viscoelastic properties of a solid, and this is the **gel phase**. Gelatin is the standard example of the gel phase, and it is obtained from a solution of denatured collagen in water.

We note the existence of **colloidal phases** which are actually diphasic media in which one of the phases is very finely dispersed in the other one: oil-in-water emulsion, aerosols, foams, etc. The colloidal medium is constituted of particles in unstable equilibrium: the suspension can be destroyed by aggregation of the particles, and this is **flocculation**. The surface properties play an essential role in the mechanisms of formation of colloidal phases.

1.5 The Major Experimental Methods for Investigation of Phase Transitions

The experimental study of a phase transition does not fundamentally differ from the experimental study of a material: the properties and structure of a material and the relations between properties and structure are studied. In a phase transition, it is thus possible to study the properties of a material on either side of the phase transition point, the change in structure at the transition, and the correlations between this change in structure and the evolution of the properties.

As we emphasized, a phase transition is not instantaneous: it is necessary to take into consideration nucleation, a dynamic phenomenon which is also susceptible to an experimental approach.

On one hand, direct experimental methods of investigating the structures and properties of materials and their evolution and on the other hand, indirect methods of investigating the properties can be distinguished.

There is a very wide variety of physical methods for studying the properties of a material: calorimetric measurements to measure the specific heats, differential thermal analysis for detecting fusion, rheological methods for measuring the viscosity and modulus of elasticity, optical methods for measuring birefringence, etc.

The structure of materials and consequently their modification during phase transitions can be investigated with classic microscopic methods, in particular, by electron microscopy, X-ray diffraction, and also by nuclear magnetic resonance (NMR) and neutron scattering. Nucleation phenomena can also be demonstrated and followed with some of these methods, for example, crystallization of a liquid using electron microscopy.

A very large amount of information can also be obtained on material undergoing a phase transition, particularly a second order (or continuous) transition, with particle scattering experiments. These particles can be electrons, photons (visible light or X-rays), phonons (acoustic waves), or neutrons. Neutrons are particularly interesting particles, since they penetrate most materials, which is not true of photons (at least those associated with visible light).

In general, we can show that the intensity of the particles scattered by a material is proportional to the structure function $S(\boldsymbol{k})$ for the material. This is the case for the intensity of the scattered light in a light scattering experiment (Fig. 1.15). The experiment conducted in a fluid in the vicinity of its critical point reveals a very important increase in the scattering intensity in the vicinity of the transition point: this is the **critical opalescence** phenomenon. When $\boldsymbol{k} \to 0$, (it is sufficient to make angle θ approach 0), (1.30) and (1.32) show that compressibility of the fluid can be attained by measuring $S(\boldsymbol{k})$, that is, the intensity of the scattered light (Fig. 1.15).

A scattering experiment can be conducted with neutrons; they are sensitive to the spins of the atoms and are an excellent means of studying magnetic materials, for example. In this case, the scattering intensity (or effective scattering cross-section per unit of volume) is proportional to the structure function.

In NMR, some signals, like the line shape, are also directly correlated with a structure function.

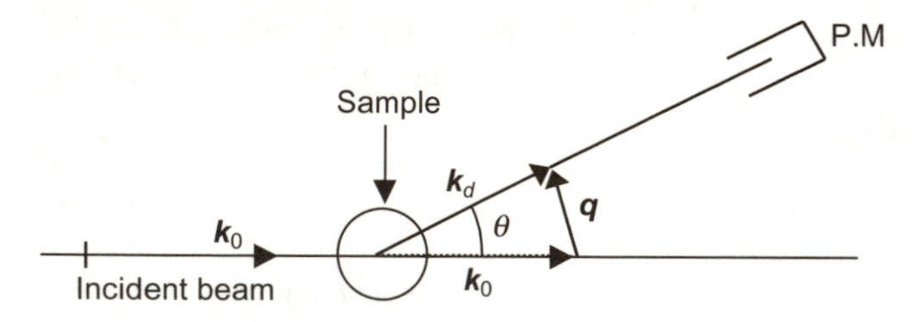

Fig. 1.15. Geometric diagram of a scattering experiment. The incident radiation (photons, neutrons...) corresponds to wave vector $\boldsymbol{k_0}$. Scattered radiation is observed in the direction detected by the photomultiplier, P.M; it corresponds to vector $\boldsymbol{k_d}$. Then $q = k_d - k_0$. In a light scattering experiment, $q \approx 2k_0 \sin\theta/2$. If $\theta \to 0$, then $q \to 0$ and the intensity approaches infinity (1.33).

1.6 The Broad Categories of Applications of Phase Transitions

Table 1.1. Summary Table of Applications of Phase Transitions. The characteristics of the transitions are indicated in the rows and the sectors with potential applications are indicated in the columns. Concrete examples have been cited in some cases.

	Electronics, Data processing	Energy, Electrical engineering	Agriculture, Food industry	Metallurgy, Mechanics	Chemistry, Pharmacy, Process, Engineering	Civil engineering
Latent heat		Thermal engine		Melting metal	Chemical reactor	
2nd-order Transitions	Josephson junctions, Memories	Current transmission			Supercritical	
Multicritical transitions					Oil deposits	
Metastable phases		Superheating of a liquid		Quenching of metals		Earthquakes, Thixotropy Glasses
Conformation of molecules			Gels		Proteins	Clays, Cements
Nucleation, Surface properties		Evaporation of a liquid	Emulsions	Microstructures	Catalysts	Cements

A very large number of phase transition phenomena have important technological applications: in metallurgy, electronics, process engineering, in the food industry, etc. Transitions are also utilized in technical systems (fluids in thermal engines, for example) and in materials. Moreover, numerous natural phenomena originate in a phase transition, evaporation or condensation of water in the atmosphere, for example.

We will explain these specific applications in each chapter in this book. The characteristics of the phase transitions used in large industrial sectors are reported in summary Table 1.1. For example, these include applications (real or potential) of superconductivity (second order transition) in electrical engineering and electronics (production of high magnetic fields, components such as Josephson diodes, for example). Phase transitions are also utilized in the agriculture and food industry and pharmacy (alimentary gels and capsules for encapsulation of drugs, for example).

As for the nucleation phenomenon associated with surface properties, it is found in a very large number of processes and industrial energy systems in civil engineering.

1.7 Historical Aspect: From the Ceramics of Antiquity to Nanotechnologies

The history of the physics of phase transitions is blended with the history of materials technology and more recently with the history of thermodynamics and the physics of condensed matter.

As soon as Man mastered fire, ceramicists and metallurgists were the first to make solid materials from the phase transitions which are the basis of fabrication of ceramics and metals. Ceramics have been known in the Near East since Mesolithic times (eighth millennium before Christ), and they were first produced by baking clay-based paste in an oven. As for bronze and iron, they appeared later, in the second millennium before Christ, probably among the Hittites, as metallurgical operations require mastery of high temperatures (of the order of approximately 2000°C).

The properties of some magnetic materials were also known in antiquity. Thales of Miletus knew more than 2500 years ago that magnetite, or lodestone, attracted iron. The Chinese were undoubtedly the first to study magnetism and the long-distance action of magnets. In Europe, Peregrinus of Maricourt (1269) and W. Gilbert (1600) treated the applications of magnets and developed the first theories of magnetism.

A scientific approach to materials and phase transition problems was only possible when there was a clear concept of the notion of temperature, in fact, only since the 17th century. This notion obviously owes much to the invention of the thermometer, with which Galileo is credited as he invented the air thermometer in 1592. The grand duke of Tuscany Ferdinand II probably made the first operating thermometer, the alcohol thermometer, in 1647; Fahrenheit developed the mercury thermometer between 1708 and 1724. Scientists such as Bacon, Descartes, Galileo, and Boyle believed that heat and its peculiar motions originated in a change of state like melting.

In the 18th century as thermal machine technology emerged with the first steam machines, meteorological phenomena began to be investigated (melting of snow, cooling of hot air by pressure reduction including condensation of water vapor, for example). With the Scottish physicist J. Black, scientists then introduced the important notions of specific heat and latent heat. Thermodynamics gradually established the body of its tenets following the work of Sadi Carnot, then J. R. Mayer, J. Joule, Lord Kelvin, and R. Clausius.

The notions of thermodynamic functions, or state function such as internal energy U and entropy S were formalized and tools were available for systematically studying phase transition phenomena at the end of the 19th century.

The systematic study of phase diagrams, particularly in metallurgy, became possible based on the knowledge of thermodynamic potentials acquired in the work of J. W. Gibbs and P. Duhem.

Moreover, the thermodynamic treatment of surface and interfacial phenomena by J. W. Gibbs opened the way to investigation of the nucleation phenomena which play a large role in phase transitions.

The existence of second order phase transition phenomena was established for the first time in carbon dioxide, CO_2, by T. Andrews in 1869. He discovered the liquid–gas critical point in a light-scattering experiment, the phenomenon of critical opalescence. Three years later, van der Waals provided the first theoretical framework for explaining these results by proposing a state equation for real gases.

Studies of magnetism then followed the research on fluids. Pierre Curie systematically inv estigated high-temperature ferromagnetic properties after J. Hopkinson demonstrated the disappearance of magnetization of iron at a high temperature in 1890. Pierre Curie discovered the law bearing his name that gives the susceptibility as a function of the temperature. Pierre and Jacques Curie had previously established the foundations of a phase transition approach with symmetry principles, thus paving the way to understanding phenomena such as ferroelectricity and piezoelectricity, discovered in 1880.

Pursuing the path opened by P. Curie for studying ferromagnetism, P.Weiss advanced the hypothesis in 1907 that ferromagnetic solids could be characterized by the existence of an internal magnetic field which could account for most of the experimental results of the time: this is the molecular field approach. Paul Langevin arrived at a similar concept and results.

The progress in low-temperature physics at the end of the 19th century permitted liquefying all gases. Liquefaction of helium by Kamerlingh Onnes in 1908 opened the way to the discovery of superconductivity by the same physicist, demonstrated for the first time in mercury in 1911.

At the beginning of the 20th century, physicists had obtained a powerful investigative tool in X-rays, which allowed determining structures and their evolution. The studies of metal alloys also made an important contribution to the physics of phase transitions. G. Tammann was the first to hypothesize the existence of an ordered phase in alloys at low temperature in 1919: this was demonstrated in 1929 by C. H. Johannsen and J. O. Linde in a X-ray diffraction experiment. In 1926, Tammann and O. Heusler observed a specific heat anomaly in a bronze alloy: this was the demonstration of the critical point of the order–disorder transition. Finally, in 1934, L. Bragg and J. Williams introduced the concept of long-range order and with it paved the way to the notion of order parameter.

The analogy between the van der Waals state equation and the equation for ferromagnetics was very quickly perceived: they anticipated the analogous equations for the line of coexistence near the critical point and similar behaviors for quantities such as magnetic susceptibility and compressibility. Nevertheless, the idea of a unified approach to all of these phenomena only arose at the end of the thirties with the work of P. Ehrenfest and L. Landau, who introduced the concept of order parameter. Important theoretical advances were made in the 1920s–1930s utilizing a microscopic approach which made quantum physics and quantum statistics possible. Ising's model (1925), first conceived for treating the case of magnetism and improved by W. Heisenberg, stimulated a very large number of studies generalizing the molecular field approach. It is necessary to note that L. Onsager was able to solve the two-dimensional Ising model for zero-field magnetism without approximation only in 1944.

Experimental methods for investigation of phase transition phenomena were again enriched beginning in 1950: precision thermometry became possible due to the progress in electronics, light scattering, and then later neutron scattering, and nuclear magnetic resonance (NMR) allowed systematically studying all phase transition phenomena in all solid and liquid materials.

The more accurate measurements of the critical points thus allowed showing that behaviors of the type predicted by the molecular field (or "van der Waals") models poorly accounted for the experimental reality.

At the end of the 1960s, stimulated by experimental research, the approach to second order phase transition phenomena was totally renewed. The studies by L. P. Kadanoff, C. Domb, and M. E. Fisher led to the notions of **scaling** and **universality laws**. Borrowing mathematical methods from field theory, K. G. Wilson proposed the **renormalization group** theory that allowed a unified approach to second order phase transitions, in a certain way generalizing van der Waals' ideas.

The technical progress made in heat treatments (utilizing lasers, for example) parallelly permitted obtaining the first metallic glasses and ceramics with increasingly small microstructures. Since the 1960s, the use of new materials (polymers, new alloys, amorphous metallic alloys, semiconductors, ceramics) in many industrial sectors (electronics, aerospace industry, automobile industry, for example) has stimulated research on phase changes in materials. We have progressively passed from microstructures whose size is of the order of the micron to structures on the nanometer scale (one-thousandth of a micron) and since the beginning of the 1990s, possible applications of nanotechnologies have been considered.

Problems

1.1. Fluctuations

The energy of a system with fixed T and V is designated by E and its equilibrium value, which is a statistical mean, is designated by $< E >$. Fluctuations in the energy are characterized by $< \delta E^2 >=< (E- < E >)^2 >$. Show that $< \delta E^2 >= C_v/k\beta^2$ with $\beta = 1/kT$ and C_v the specific heat.

1.2. Fluctuations and Compressibility

Consider fluctuations in the number of particles N in a fluid of volume V around its equilibrium value $< N >$. They can be defined by the relation:

$< \Delta N^2 >=< (N- < N >)^2 >$. Using a grand–canonical distribution and the relations between p, μ and the grand–canonical partition function, set up the relation $< \Delta N^2 >=< N > \kappa_T/\kappa_T^0$; κ_T is the compressibility of the fluid and κ_T^0 is the compressibility of an ideal gas.

1.3. Free Enthalpy of a Binary Mixture

The free enthalpy of a binary mixture is defined by (1.15) which is represented by a $G(x)$ diagram, where x is the concentration of one of the constituents. If G_A and G_B designate the molar free enthalpies of the constituents in the mixture, show that the slope of the tangent at any point in the diagram is given by $G_B - G_A$.

1.4. Fluctuations and Scattering of Light

Light scattering is a very widely used technique for studying phase transitions in transparent fluids and solids (Fig. 1.15). The total scattering intensity at time t corresponding to vector $\boldsymbol{k}$ is:

$$I(\boldsymbol{k}, t) =< |E(\boldsymbol{k}, t)|^2 >$$

where $E(\boldsymbol{k}, t)$ is the electric field of the scattered wave, which is associated with fluctuation of the dielectric constant ϵ of the hypothetically isotropic medium:

$$E(\boldsymbol{k}, t) \propto E_0 \mathrm{e}^{-\mathrm{i}\omega_0 t} \delta\epsilon(\boldsymbol{k}, t))$$

ω_0 is the frequency of the incident wave, and $\delta\epsilon(\boldsymbol{k}, t)$ is the Fourier transform of fluctuation ϵ. The spectral density is defined by:

$$I(\boldsymbol{k}, \omega) = \int_{-\infty}^{+\infty} e^{i\omega t} < E(\boldsymbol{k}, t+\tau) E^*(\boldsymbol{k}, t) > \mathrm{d}\tau$$

1. Calculate fluctuations $\delta\epsilon$ with constant p and derive $I(\boldsymbol{k}, \omega)$ associated with temperature fluctuations δT.
2. Assuming that δT obeys the heat equation, derive $I(\boldsymbol{k}, \omega)$.
3. What is the shape of the spectrum obtained?

1.5. Density Fluctuations in a Heterogeneous Medium

Local variations $\delta F(\boldsymbol{r})$ in the free energy are associated with local density fluctuations $\delta\rho(\boldsymbol{r})$ in volume V by the simple relation:

$$\delta F = \frac{a}{2}\delta\rho^2 + \frac{b}{2}(\Delta\delta\rho)^2$$

1. Calculate the total variation ΔF_k associated with wave vector fluctuations $\boldsymbol{k}$.
2. Calculate the probability of a fluctuation $\delta\rho_k$.
3. Derive the mean value of $\delta\rho_k^2$.
4. The medium is a fluid. What is the physical meaning of a? What is derived from it at the liquid/gas critical point?

2 Dynamics of Phase Transitions

2.1 A Large Variety of Mechanisms

The thermodynamic description of phase transitions given in the preceding chapter says nothing about the evolution of a phase in time nor on the kinetics of the transition phenomenon. For example, the phase diagrams (Sect. 1.2.2) only give information on the conditions of the existence of phases (as a function of pressure, temperature, concentrations of constituents, etc.) and not on the time required for passing from one phase to another when the thermodynamic parameters of a system are changed (the temperature, for example). We know that a phase change is not instantaneous and that it is a dynamic phenomenon; each one has its own kinetics.

The dynamic description of a phase transition implies that the mechanisms of the transformation are known in advance. They are very varied. They most frequently involve the formation of microstructures which are the "nuclei" or "germs" of the new phase that appears during the transformation: this is the **nucleation** phenomenon. For example, nucleation is encountered in a liquid during crystallization when crystallites, the nuclei of the solid phase, are formed. In this case, nucleation is only possible if the atoms or molecules constituting the material can diffuse within it to aggregate and thus form the nuclei of the new phase. Diffusion is then the key mechanism that guides the nucleation phenomenon itself and determines its kinetics. The interfacial properties, particularly the interfacial energy, also play an important role in nucleation; the appearance of the nucleus of a new phase in effect implies the creation of an interface between two phases.

Situations are also encountered in which the transition is not triggered by transport of macroscopic matter by diffusion, but by local fluctuations in the concentration of the constituents. These transitions, found in solid or liquid solutions, occur out of thermodynamic equilibrium: they correspond to spinodal decomposition. This does not involve the initial creation of an interface, and there is no nucleation strictly speaking. The dynamics of spinodal decomposition is thus different from the dynamics of nucleation.

Another situation corresponds to the one encountered in gels. The gelation transition or sol–gel transition (Sect. 1.4.3) is induced by establishment of links between polymer chains in solution (and between solution and polymer): a three-dimensional network is formed from a polymer solution. Gelation is a dynamic phenomenon whose kinetics is a function of the kinetics of establishment of these links.

In the particular case of gelatin, the sol–gel transition is itself triggered by a change in the conformation of the macromolecular chains in the gelatin: they undergo a helix–coil transition which has its own dynamics.

Finally, a last type of transition involves local but collective movements of atoms or molecular groups within the material that trigger the transition. This is primarily the case of displacive transitions (Sect. 1.4.1) encountered in perovskites and in metallurgy, for example.

2.2 Nucleation

2.2.1 The Diffusion Phenomenon – Fick's Law

Diffusion phenomena, that is, movement of atoms or molecules within a gas, a liquid, or a solid over a more or less long distance, determine the dynamics of a large number of phase transitions and make nucleation possible.

We note that diffusion is a phenomenon that permits a physical system to attain a state of equilibrium from an initial nonequilibrium state. There is thus diffusion of heat between the high- and low-temperature parts of a solid (by a conduction mechanism) and diffusion of matter in a solution when there is a concentration difference in it. The spatial and temporal variations in the temperature and density of particles in a system obey equations which have the same form.

Nucleation in a pure phase or a phase with only one constituent, called homogeneous nucleation, involves self-diffusion (diffusion of identical particles).

Designating the number of particles per unit of volume by $n(\boldsymbol{r}, t)$ and the flow of atoms or molecules by $\boldsymbol{J}$, that is, the number of particles crossing a unit of surface area per unit of time in concentration gradient ∇n, we then have the following equation, which is Fick's first law for diffusion:

$$\boldsymbol{J} = -D \nabla n \tag{2.1}$$

where D is the diffusion coefficient. The $-$ sign accounts for the fact that the flow and concentration gradient are of opposite signs. If the phase is pure, D is the self-diffusion coefficient. Taking into account the continuity equation:

$$\frac{\partial n}{\partial t} + \nabla \cdot \boldsymbol{J} = 0 \tag{2.2}$$

we have the general equation for three-dimensional diffusion:

$$\frac{\partial n}{\partial t} = D \nabla^2 n \tag{2.3}$$

Equation (2.3) is **Fick's second law** for diffusion. It does not have a simple solution particularly for a three-dimensional system. The following solution is obtained from (2.3) for a one-dimensional system and with Fourier transformation:

$$n(x,t) = \frac{N}{(4\pi Dt)^{1/2}} \exp\{-\frac{(x-x_0)^2}{4Dt}\} \tag{2.4}$$

N is the total number of particles in an infinite volume parallel to the x axis and unit section perpendicular to this axis, where all particles are concentrated in x_0 for $t = 0$. The mean square of movement of the particles is written with (2.4):

$$< (x-x_0)^2 >= \frac{1}{N} \int_{-\infty}^{+\infty} (x-x_0)^2 n(x,t)\, \mathrm{d}x = 2Dt \tag{2.5}$$

Note that for particles of radius r moving in a fluid of viscosity η, diffusion coefficient D is given by the Stokes–Einstein law.

$$D = \frac{kT}{6\pi r\eta} \tag{2.6}$$

For a three-dimensional isotropic medium, (2.4) is generalized in the form

$$n(r,t) = N(4\pi Dt)^{-3/2} \exp\Big(-\frac{r^2}{4Dt}\Big) \tag{2.7}$$

For particles of radius $r = 0.2\mu$ dissolved in water at ambient temperature, we find that their average displacement at the end of 30 sec is 8μ. This is the phenomenon of Brownian motion.

2.2.2 Diffusion Coefficient and Activation Energy

Diffusion of matter implies movement of atoms or molecules within a solid or a fluid which must have sufficient energy to move particles from their equilibrium position.

The existence of a vacant site, a vacancy, facilitates diffusion in solids. In a liquid, the Frenkel theory is based on the hypothesis of the existence of holes in the liquid which would facilitate molecular motions within the liquid.

Diffusion requires the particles to cross a potential energy barrier so that they can move within a phase (Fig. 2.1).

If ΔG_0 designates the variation in free enthalpy per particle necessary for crossing the potential barrier, probability P_0 that the particles will have sufficient energy to do this is given by:

$$P_0 = C^{st} \exp\Big(-\frac{\Delta G_0}{kT}\Big) \tag{2.8}$$

Since $\Delta G_0 = \Delta H_0 - T\Delta S_0$, (2.8) can be rewritten as:

$$P_0 = C^{st} \exp\Big(-\frac{\Delta H_0}{kT} + \frac{\Delta S_0}{k}\Big) = C^{st} \exp\Big(-\frac{\Delta H_0}{kT}\Big) \tag{2.9}$$

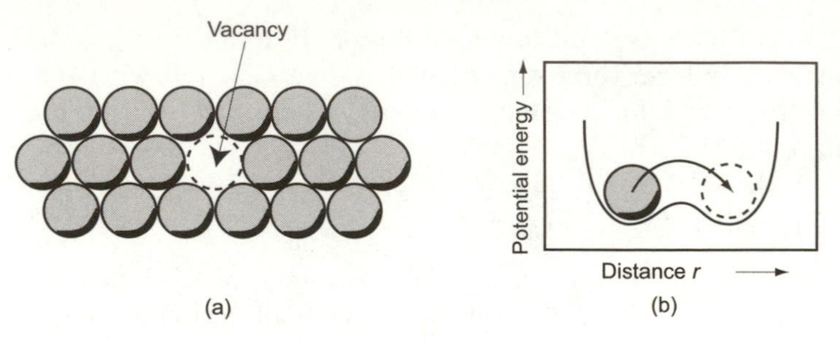

Fig. 2.1. Vacancy in a crystal lattice. The vacancy (**a**) must overcome a potential barrier (**b**) to move, associated with variation of free enthalpy ΔG_0.

In the case of a solid, the particles diffuse by occupying the vacancies present in the material. If ΔH_L is the enthalpy of formation of these vacancies, the probability of the existence P_L of a vacancy in the vicinity of a particle is a Boltzmann function of the type:

$$P_L = C^{st} \exp\Big(-\frac{\Delta H_L}{kT}\Big) \tag{2.10}$$

The probability P_D associated with this diffusion mechanism is thus:

$$P_D = C^{st} P_0 \; P_L = C^{st} \exp\Big(-\frac{\Delta H_0 + \Delta H_L}{kT}\Big) \tag{2.11}$$

P_D can be interpreted as the frequency of "jumps" of a particle from one equilibrium position to another.

The diffusion coefficient D of a particle is inversely proportional to the "lifetime" (that is, the residence time) in an equilibrium position. D is thus also proportional to the frequency of jumps P_D from one vacancy to another. We will then have:

$$D = D_0 \exp\Big(-\frac{\Delta H_0 + \Delta H_L}{kT}\Big) \tag{2.12}$$

This relation can be generalized and D can be written as:

$$D = D_0 \exp\Big(-\frac{\Delta W}{kT}\Big) \tag{2.13}$$

W is the activation energy associated with the diffusion mechanism. Its value is a function of the type of mechanism and thus the nature of the phase.

2.2.3 Nucleation of a New Phase

A phase transition is often initiated by a nucleation process, manifested by the appearance of germs or nuclei of the new phase. This process is made possible by diffusion of atoms or molecules that aggregate to form the nuclei of the new phase (formation of drops of liquid, for example). The size of these nuclei increases in time and at the end of the process (condensation of vapor

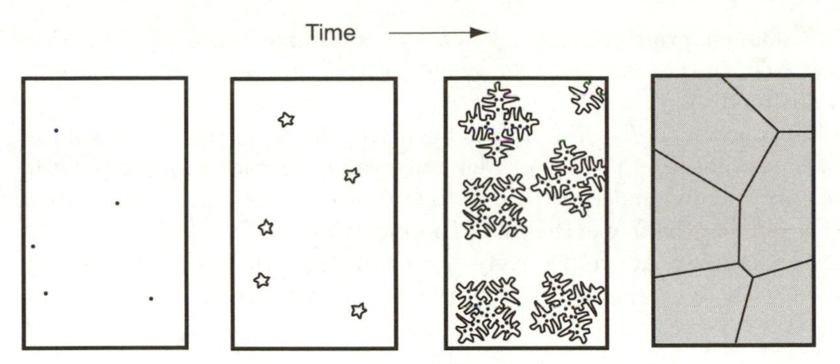

Fig. 2.2. Formation of microcrystallites in a liquid. The solid nuclei or germs are microcrystallites which initiate solidification. They take the form of dendrites in a metal.

or solidification of a liquid, for example), a new homogeneous phase is formed (Fig. 2.2).

From the thermodynamic point of view, a change in the free enthalpy ΔG_n is associated with nucleation and with these two processes:

- passage of atoms or molecules from one phase to another or from one state to another (for example, from the liquid phase to solid nuclei), which corresponds to the variation in free enthalpy ΔG_V.
- creation of an interface between the nuclei of the new phase and the initial phase, associated with a change in free enthalpy ΔG_S.

We thus have:

$$\Delta G_n = \Delta G_V + \Delta G_S \tag{2.14}$$

If Δg_V designates the free enthalpy of formation of the new phase per unit of volume, where V is the volume of the new phase, we can write: $\Delta G_V = V\Delta g_V$. We will assume that the nucleation process is initiated with spherical nuclei of radius r. We will give the principle of the Volmer model for nucleation, specifically applying it to the liquid/solid transition (crystallization) in this chapter. The case of nucleation of a liquid will be treated in Chap. 4.

If γ designates the surface tension (for example, between solid and liquid when a solid/liquid interface with microcrystallites is formed), we have:

$$\Delta G_V = \frac{4\pi}{3} r^3 \Delta g_V \qquad \Delta G_S = 4\pi r^2 \gamma \tag{2.15}$$

and

$$\Delta G_n = \frac{4\pi}{3} r^3 \Delta g_V + 4\pi r^2 \gamma \tag{2.16}$$

Δg_V is positive if $T > T_0$, where T_0 is the temperature of the transition (for example, the solidification temperature). Under these conditions, applying the Gibbs–Duhem principle, the new phase is not stable. Since γ is a positive quantity, ΔG_n is then positive: the nuclei of the new phase are not stable and do not tend to grow.

On the contrary, if $T < T_0$, Δg_V is negative, the new phase (the solid in the case of solidification) is stable. There are values of r for which $\Delta G_n < 0$, and the corresponding nuclei will tend to stabilize the new phase since their formation reduces the free enthalpy of the material.

If Δh_V and Δs_V are respectively the changes in enthalpy and entropy associated with the formation of a unit volume of the new phase, we can write:

$$\Delta g_V = \Delta h_V - T\Delta s_V \tag{2.17}$$

At the solidification temperature T_m, we can postulate $\Delta h_V = -L_f$, where L_f is the latent heat of fusion of the material per unit of volume (Δh_V is negative) and $\Delta s_V = -L_f/T_m$.

In general, we can write:

$$\Delta h_V = -L_f - \int_T^{T_m} \Delta C_p \mathrm{d}T' \tag{2.18}$$

and

$$\Delta s_V = -\frac{L_f}{T_m} - \int_T^{T_m} \frac{\Delta C_p}{T'} \mathrm{d}T' \tag{2.19}$$

where ΔC_p is the difference in specific heat between the liquid and solid phases. Hence:

$$\Delta g_V = -\frac{L_f \Delta T}{T_m} - \int_T^{T_m} \Delta C_p \; \mathrm{d}T' + T\int_T^{T_m} \frac{\Delta C_p}{T'} \mathrm{d}T' \tag{2.20}$$

We assumed $\Delta T = T_m - T$, and this quantity is a measure of the degree of supercooling of the liquid.

In a first approximation, we can hypothesize that $\Delta C_p = 0$. We then have:

$$\Delta g_V = -L_f \frac{\Delta T}{T_m} \tag{2.21}$$

This approximation is totally justified in the case of metals but is much less justified for polymers. In the last case, the integrals in (2.20) can be calculated by assuming that ΔC_p is constant. We then obtain:

$$\Delta G = \Delta H_f \frac{\Delta T}{T_m} - \frac{1}{2}\Delta C_p^f \frac{\Delta T^2}{T_m - \Delta T} \tag{2.22}$$

We took the difference in the specific heats at the melting point ΔC_p^f as the value of ΔC_p. For the remainder of the calculations, we will assume that $\Delta C_p = 0$ and we will thus neglect the second-order terms in ΔT. Equation (2.16) is rewritten:

$$\Delta G_n = -\frac{4\pi}{3} r^3 L_f \frac{\Delta T}{T_m} + 4\pi r^2 \gamma \tag{2.23}$$

When $T < T_m$ ($\Delta T > 0$), ΔG_n has a maximum for $r = r^*$. For $r > r^*$, the formation of nuclei of increasing size results in stabilization of the solid phase since ΔG_n decreases and even becomes negative.

Radius r^* is called the critical radius (it does not correspond to a critical phenomenon, that is, a second order phase transition in the sense defined previously).

We easily find from (2.23):

$$r^* = \frac{2\gamma}{L_f} \frac{T_m}{\Delta T} \qquad \Delta G_n^* = \frac{16\pi}{3} \frac{\gamma^3}{L_f^2} \frac{T_m^2}{\Delta T^2} \tag{2.24}$$

Graphs of Fig. 2.3 illustrate the situation. $\Delta G_n(r^*) = \Delta G_n^*$ is equivalent to the potential barrier that must be lowered for nuclei of critical size r^* to be formed more easily. We see in (2.24) that the more important supercooling is, the more this barrier is lowered. The Volmer thermodynamic model shows that a complete phase transition by nucleation is only possible if the **liquid is supercooled.**

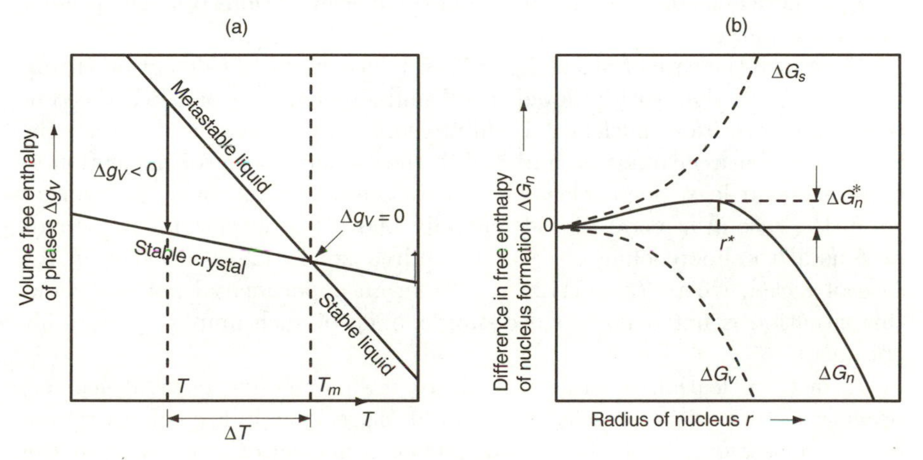

Fig. 2.3. Nucleation of crystallites. (**a**) Variation of Δg_V vs. temperature, $\Delta g_V = 0$ at T_m. (**b**) Variation of ΔG_n, free enthalpy of nucleus formation, vs. radius of nucleus.

The preceding calculation is not sufficient to account for nucleation. Regardless of their radius r, nuclei are not formed in isolation from their environment in the liquid. The probability that a nucleus of radius r will be formed at temperature T must be calculated. In the same manner, the variation of configurational entropy ΔS_n associated with the formation of n_r nuclei of radius r in the material (this is the mixing entropy in classical thermodynamics) must be calculated.

Assume that there are n_r nuclei mixed with n_0 atoms or molecules of liquid; we then have:

$$\Delta S_n = k \ln \frac{(n_0 + n_r)!}{n_0!\ n_r!} \tag{2.25}$$

The total variation of free enthalpy associated with formation of nuclei is written:

$$\Delta G = n_n \Delta G_n - kT \ln \frac{(n_0 + n_r)!}{n_0!\ n_r!} \tag{2.26}$$

or:

$$\Delta G = n_r \Delta G_n - kT[(n_0 + n_r)\ln(n_0 + n_r) - n_0 \ln n_0 - n_r \ln n_r] \tag{2.27}$$

The value of n_r corresponding to equilibrium satisfies the relation

$$\frac{\partial \Delta G}{\partial n_r} = 0 \quad \text{that is} \quad \frac{n_r}{n_0} = \exp\left(-\frac{\Delta G_n}{kT}\right) \tag{2.28}$$

We assumed that $n_0 \gg n_r$. We obtain a classic Maxwell–Boltzmann distribution function. Equation (2.28) gives the nucleus size distribution function. It can be applied to nuclei of critical size r^*.

A numerical example gives an idea of the concentrations of nuclei present in a liquid phase.

If we take the case of nickel $T_m = 1725$ K, a numerical calculation (Problem 2.5) shows that with a liquid metal and supercooling of 10 K, the concentration of critical nuclei n_r^* is infinitesimal (much smaller than 1): the nucleation process cannot be initiated in these conditions. For nucleation to be possible, at least one nucleus per cm^3 of the new phase must be formed.

In the case of a metal (T_m is generally high), this situation corresponds in general to supercooling of several hundred kelvin ($\Delta T = 350$ K in the case of nickel, where $T_m = 1725$ K). From an experimental point of view, this situation is not realistic since supercooling of such amplitude is rarely obtained.

In fact, nucleation is initiated with lower supercooling (several degrees) because the impurities present in the liquid phase, even in low concentration, induce nucleation: this is called heterogeneous nucleation. In a solid, it can be induced on impurities as well as in defects such as dislocations, causing thus melting.

In a situation where nucleation is heterogeneous, the intervention of new interfaces must be taken into consideration: those between the solid nucleus of the new phase and the impurity on one hand, and between the liquid phase and the impurity on the other hand (Fig. 2.4). If ϑ is the wetting angle between the solid nucleus, assumed to be a spherical mass, and the surface of the impurity (a catalyst, for example), one can show that the critical radius r^* is unchanged, but that on the other hand, the expression of the potential barrier ΔG_n^* given by (2.24) is replaced by:

$$\Delta G_n^* = \frac{16\pi}{3} \frac{\gamma^3 T_m^2}{L_f^2 \, \Delta T^2} f(\vartheta) \tag{2.29}$$

where

$$f(\vartheta) = \frac{1}{4}(2 + \cos\vartheta)(1 - \cos\vartheta)^2 \tag{2.30}$$

If $\vartheta = 180°$, the geometric factor is less than 1 and the nucleation barrier is lowered, resulting in an increase in the number of nuclei n_r formed (for $\vartheta = 60°$, $f(\vartheta) = 0.16$, and for $\vartheta = 30°$, $f(\vartheta) = 1.3 \times 10^{-2}$).

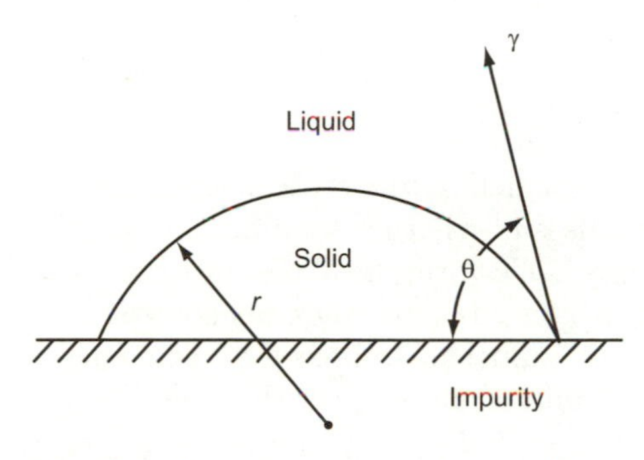

Fig. 2.4. Heterogeneous nucleation.The solid formed is in contact with an impurity, and γ is the surface tension between the liquid and the solid nucleus being formed.

There are few ideal situations such as those just described: in the case of a crystalline solid formed from a liquid which can be a binary, ternary mixture, etc., there is almost never a monocrystalline system corresponding to a perfect crystal. In general, this is a polycrystalline solid material formed from a very large number of small crystals or grains attached to each other by what is called grain boundaries (Fig. 2.5). In the first place, the grain boundaries establish a bond between two crystals, and the atoms are thus much less ordered in this region of the material where the interatomic bonds are much weaker; thus there is a surface energy characteristic of the grain boundary. In the second place, since the structure of the grain boundary is much less dense, it favors more important diffusion of atoms, but also dissolution and binding of impurity atoms whose concentration can be locally important.

The properties of a material, particularly the mechanical properties, are a strong function of its polycrystalline structure and thus the density of the grain boundaries. Microcrystals of metals, alloys, or metalloids can be grown for specific applications (turbine blades, semiconductors, for example), but these are expensive operations.

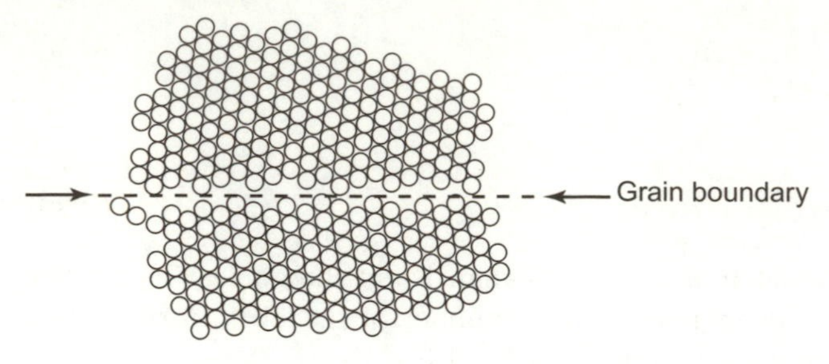

Fig. 2.5. Structure of a grain boundary. In the dashed region, the atoms are less ordered, the structure is less dense, and it favors diffusion the atoms in the solid.

2.2.4 Nucleation Rate

The formation of nuclei, particularly those reaching the critical size r^*, is the stage before the phase transition, regardless of whether solidification of a liquid, condensation of a vapor in liquid or solid form, a solid–solid transition in an alloy, or melting of a solid, which is a more complex phenomenon, is involved. This stage alone is obviously not sufficient to produce a new phase: the nuclei formed must grow within the initial phase. Growth occurs by aggregation of atoms or molecules of the phase undergoing transformation into critical nuclei. This is only possible if the atoms or molecules diffuse within this phase and collide with the critical radii.

A nucleation speed or rate is defined by the relation:

$$I = \frac{\mathrm{d}N}{\mathrm{d}t} = \nu \, n_s \, n^* \tag{2.31}$$

where $N(t)$ is the number of nuclei formed in the system at time t. ν is the frequency of collision of atoms or molecules with critical nuclei, n^* is the number of nuclei of critical size r^* given by (2.28) and (2.24), and n_s is the number of atoms or molecules on the surface of a nucleus. It thus expresses the rate at which critical nuclei are transformed and grow by addition of particles of the initial phase; the nuclei formed from critical nuclei are stable.

The frequency of collisions is obviously proportional to the probability P_0 that the particles can cross the potential barrier ΔG_0 opposing diffusion (2.8). We thus have:

$$\nu = \nu_0 \exp\left(-\frac{\Delta G_0}{kT}\right) \tag{2.32}$$

ΔG_0 is the activation energy involved in the diffusion process and in diffusion coefficient D which we defined (2.12 and 2.13).

In the case of nucleation of a liquid phase, we thus have:

$$I = \nu_0 \, n_s \, n_0 \exp\left(-\frac{\Delta G_n^*}{kT}\right) \exp\left(-\frac{\Delta G_0}{kT}\right) \tag{2.33}$$

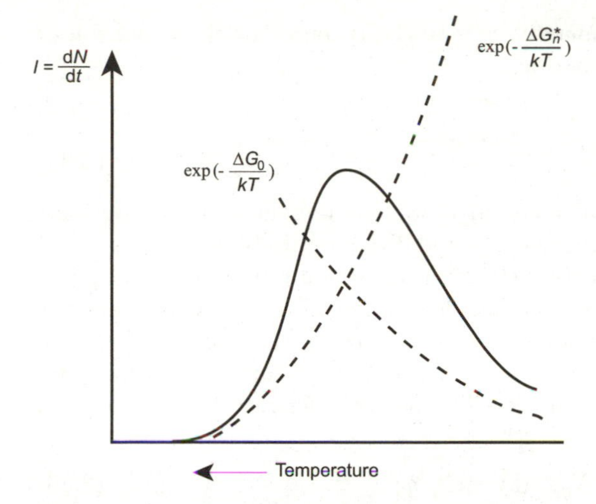

Fig. 2.6. Nucleation rate. The nucleation rate I is the product of two exponential functions which do not vary in the same direction with the temperature. I passes through a maximum. If rapid quenching is conducted, this maximum can be surpassed and the transition can be blocked in this way.

The two exponential terms vary in a different way with the temperature. The first one increases when the temperature decreases below the melting point T_m since $\Delta G_n *$ is a decreasing function of ΔT (2.24). On the other hand, if we assume that ΔG_0 does not vary with the temperature, the second exponential term decreases. The product of these two functions will thus pass through a maximum T_m in the temperature region below T_m. The graph of variation of the nucleation rate I as a function of the temperature is shown in Fig. 2.6.

We thus see that if a liquid is very rapidly cooled (quenched), a temperature region beyond the maximum nucleation rate can be attained and the phase transition can in principle be blocked. This problem will be discussed with respect to the glass transition.

The preceding considerations can be transposed to the case of solid/solid phase transitions initiated by nucleation of microcrystals of one solid phase in the other.

In the case of transformation of the body-centered cubic (b.c.c.) phase of iron to the face-centered cubic phase (f.c.c.), the transition is initiated at 914°C and the nucleation rate passes through a maximum at 700°C.

The problem of the nucleation kinetics can be treated precisely with the Volmer and Weber model of bimolecular reactions (1926): it is assumed that the nuclei of the new phase (microcrystals, for example) undergo a series of transformations similar to those of molecular species during chemical reactions. If E_{n-1}, E_n, and E_{n+1} respectively designate nuclei with $n-1$, n, and $n+1$ atoms or molecules and E_1 designates the individual particle, we can assume that everything "happens as if" we could represent the evolution of the nuclei by the following reactions:

$$E_{n-1} + E_1 \leftrightarrow E_n \tag{2.34}$$

$$E_n + E_1 \leftrightarrow E_{n+1} \tag{2.35}$$

These "reactions" are equivalent in a way to polymerization reactions, where E_1 would be a monomer and E_n and E_{n+1} would be polymers.

If we more specifically consider the transformation represented by (2.35), we have a first process during which nucleus E_n is transformed into nucleus E_{n+1} with reaction rate $q_0 A_n$ and a second process in the opposite direction where nucleus E_{n+1} loses a particle with rate $q_{n+1} A_{n+1}$, and A_n and A_{n+1} are the areas of the corresponding nuclei. The net transfer rate $I_n(t)$ associated with (2.35) is written:

$$I_n(t) = N_n(t)\; A_n\; q_0 - N_{n+1}(t)\; A_{n+1}\; q_{n+1} \tag{2.36}$$

$N_n(t)$ is the number of nuclei with n particles at time t. At equilibrium, we have $I_n(t) = 0$ (the phase transition does not occur). We thus have the relation:

$$N_n^e\; A_n\; q_0 = N_{n+1}^e\; A_{n+1}\; q_{n+1} \tag{2.37}$$

The equilibrium values are given by (2.28) (radius r of an assumed spherical nucleus is proportional to $n^{1/3}$, $A_n \propto n^{2/3}$).

We can thus rewrite:

$$I_n(t) = N_n^e\; q_0\; A_n \left\{ \frac{N_n(t)}{N_n^e} - \frac{N_{n+1}(t)}{N_{n+1}^e} \right\} \tag{2.38}$$

We propose the simplifying hypothesis that there is no reaction of destruction of nuclei with $n+1$ particles if $n > n^*$ in (2.36), where n^* corresponds to critical nuclei of radius r^*. We will then take $N_n(t) = 0$ for $n > n^*$ and $N_n(t) = N_n^e$ for $n < n^*$. The nucleation rate is then written for the stationary state:

$$I^S = N_n^*\; A_n^*\; q_0 = A_n^*\; q_0 \exp\Big(-\frac{\Delta G_n(r^*)}{kT}\Big) \tag{2.39}$$

One finds again an equation similar to (2.33).

If we begin with an initial condensed phase, a liquid, for example, equilibrium is not attained rapidly and it is thus necessary to take into account some transient states. These states are important in numerous first order phase transformations such as nucleation of a liquid in a vapor, crystallization of liquid or supercooled phases, devitrification (crystallization of a glass), nucleation of a new crystalline phase within a crystalline solid, etc.

If we consider more specifically the transformation rate of solid nuclei E_n with n particles $\partial N_n/\partial t$, it is written:

$$\frac{\partial N_n}{\partial t} = I_{n+1}(t) - I_n(t) \tag{2.40}$$

Equation (2.38) can be rewritten in differential form:

$$I_n(t) = -N_n^e \; q_0 \; A_n \frac{\partial}{\partial n}\left[\frac{N_n(t)}{N_n^e}\right] \tag{2.41}$$

Knowing that $N_n^e = N_0 \; \exp\left(-\frac{\Delta G_n(n)}{kT}\right)$ and postulating $D_n = q_0 \; A_n$, we have:

$$I_n(t) = -D_n \frac{\partial N_n(t)}{\partial n} - D_n \frac{N_n(t)}{kT} \frac{\partial \Delta G_n(n)}{\partial n} \tag{2.42}$$

And thus:

$$\frac{\partial N_n(t)}{\partial t} = \frac{\partial}{\partial n}\left(D_n \frac{\partial N_n}{\partial n}\right) + \frac{1}{kT}\frac{\partial}{\partial n}\left(D_n N_n \frac{\partial \Delta G_n(n)}{\partial n}\right) \tag{2.43}$$

This equation, the **Zeldovitch–Frenkel** equation, is the equivalent of the Fokker–Planck equation which describes the evolution of a Markov process in time. It is not surprising that the concentration of nuclei with n particles obeys an equation of this type, since a state with n particles will only be a function of a state with $n-1$ particles.

In (2.43), D_n is equivalent to a diffusion coefficient, but is a function of n. The first term is purely diffusional and in a way represents the kinetic barrier that hinders nucleation, while the second represents the thermodynamic barrier that must be overcome to create nuclei (it is a function of ΔG_n, the activation term).

We note that in a first approximation, the second term can be neglected (particularly for $n \cong n^*$ since $\partial G_n/\partial n = 0$ for $n = n^*$). Equation (2.43) is then simplified:

$$\frac{\partial N_n(t)}{\partial t} = D \frac{\partial^2 N_n(t)}{\partial n^2} \tag{2.44}$$

assuming that D is not dependent on n. We then have an equation in the form of Fick's law (2.3). If t is small, we will then have a special solution:

$$N_n(t) \propto \left(\frac{D}{\pi t}\right)^{1/2} \exp\left(-\frac{n^2}{4Dt}\right) \tag{2.45}$$

For nuclei of critical size n^*, the number only becomes significant at the end of characteristic time $\tau = n^{*2}/4D^*$, where D^* is the diffusion rate for the critical nuclei.

The search for analytical solutions of (2.42) and (2.43) is complicated. In general, the nucleation rate is calculated for the critical nucleus n^*. D. Kashiew gives the following expression for $I_n^*(t)$, for example:

$$I_n^*(t) = I^S\left[1 + 2\sum_{m=1}^{\infty}(-1)^m \mathrm{e}^{-m^2 t/\tau}\right] \tag{2.46}$$

where I^S is the nucleation rate corresponding to equilibrium given by (2.39), for example; $\tau \approx n^*/\Delta G_n$ is the characteristic time.

Another simple expression is given by several investigators (Kantrowitz–Wakeshima–Chakraverty–Feder):

$$I_n^*(t) = I^S[1 - \mathrm{e}^{-t/\tau}] \tag{2.47}$$

It is thus possible to numerically solve the evolution equations for N_n. The first calculations were performed by Turnbull for aluminum at relatively low values of n^* These calculations satisfactorily account for the experimental results.

The variation in time in the number of crystallites formed in a glass, lithium disilicate (Li_2O–$2SiO_2$), is shown in Fig. 2.7, measured by optical or electron microscopy.

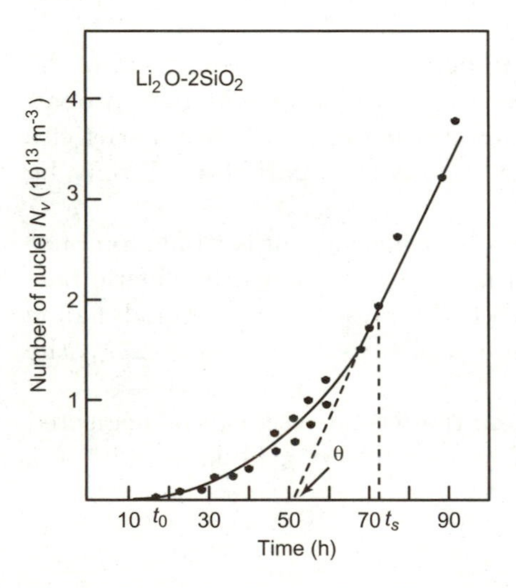

Fig. 2.7. Nucleation in glass. The number of nuclei formed from time t_0, onset of nucleation; t_s corresponds to the beginning of equilibrium. ϑ is the induction time extrapolated from the long-term linear behavior (V. M. Fokin, AM. M. Kalinia, and V. N. Filipovich, J. Cryst. Growth, **52**, 115 (1981), with permission from Elsevier Science).

Glass crystallizes after annealing. The total number of nuclei formed is $N = \int_0^t I(t)\ \mathrm{d}t = I\ t$ at equilibrium with $I(t) = I^S$. For short times, there is a transient period when N varies exponentially. For long times, N varies linearly like $(t - \theta)$, and θ is called the induction time.

More generally, the phase transformation rates with nucleation are represented with diagrams (**time–temperature–transformation or T–T–T**).

These diagrams give the time required for obtaining transformation of a certain fraction (1%, 5%, ... 99%) of the initial phase at a given temperature. An example of a diagram corresponding to a solid/solid transformation in iron is shown in Fig. 2.8. In this diagram, the time required for transforming a given fraction of the solid is minimum at 700°C, which indicates that the nucleation rate is maximum at this temperature. These diagrams are generally difficult to obtain experimentally, particularly in the case of metals where the kinetics is rapid.

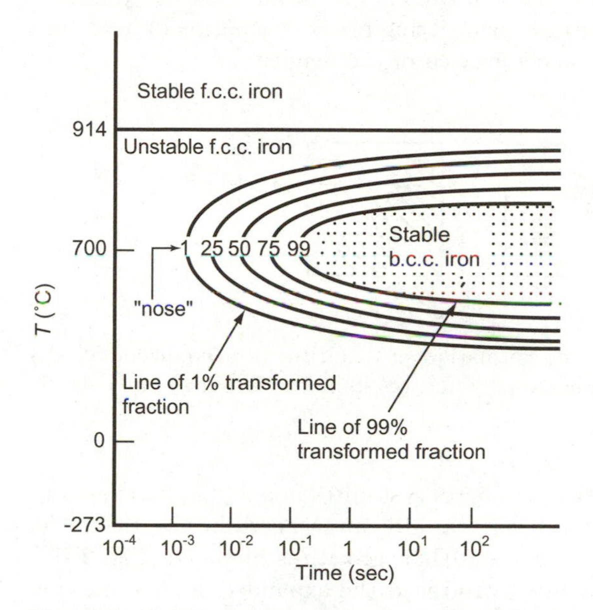

Fig. 2.8. Solid–solid transformation. This diagram shows the f.c.c. phase transition of iron ($T > 914°C$) into b.c.c. The curves indicate the time necessary for converting 1%, 25%, ... 99% of the initial fcc phase at a given temperature. This time is minimum for 700°C, the temperature corresponding to the nucleation rate maximum (M. F. Ashby and D. R. H. Jones, *Materials: Properties and Applications*, Vol. 2, Dunod (1991), p. 73).

2.2.5 Global Phase Transformation – Avrami Model

In most materials, when diffusion phenomena occur, phase transformations take place due to two successive processes:

- formation of germs or nuclei of the new phase: nucleation;
- growth of nuclei from the initial sites.

The nuclei of the new phase grow at the cost of the initial phase (Fig. 2.2), but their growth within a liquid can be anisotropic; microstructures of very different geometric shapes can be formed. In liquid metals, nucleation leads to the formation of dendrites, while in eutectic mixtures, lamellar structures are formed. Molten polymers crystallize from spherical grains, **spherulites**.

The formation of **dendrites** (Fig. 2.9a) results from a purely thermal phenomenon: during its growth, a crystalline nucleus releases its latent heat of crystallization, which tends to increase its temperature – a negative temperature gradient is established in front of the crystal surface favoring diffusion of heat – a protuberance is formed, permitting faster evacuation of heat, and it thus tends to evolve by inducing growth of the dendrite.

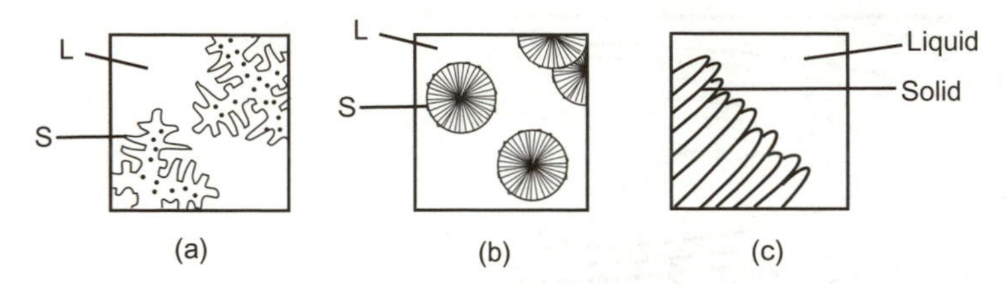

Fig. 2.9. Growth of crystals. (**a**) Formation of dendrites in a liquid metal. (**b**) Spherulitic nuclei growing in a liquid polymer. (**c**) Lamellae forming in an eutectic mixture.

When individual nuclei (that is, microcrystallites in a liquid, for example) grow, steric hindrance takes place very rapidly; after a certain size, the nuclei tend to occupy contiguous volumes and their growth is hindered (Fig. 2.10). To account for this effect, Avrami introduced the expanded volume fraction of nuclei $\phi_x = V_e^{\beta}/V_0$, where V_e^{β} is the volume occupied by the nuclei of new phase β if they can grow without being hindered by the presence of other nuclei; V_0 is the total volume of the phases.

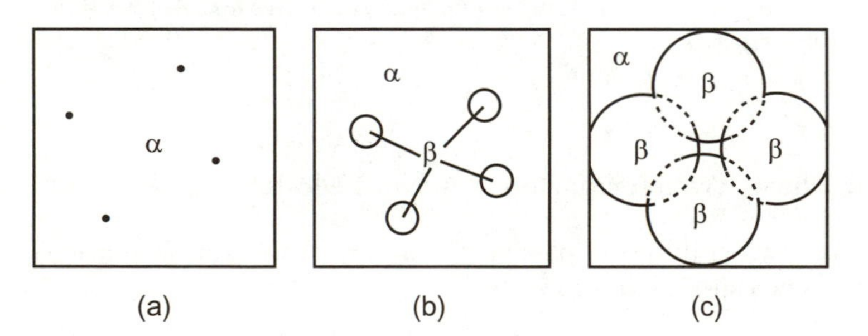

Fig. 2.10. Phase growth diagram. (**a**) Initial rapid nucleation of phase β within initial phase α. (**b**) Growth of phase β nuclei. (**c**) Expanded volume of nuclei β: their growth is sterically hindered.

Theoretically, therefore, ϕ_x could become greater than unity and each nucleus could in principle grow indefinitely in the volume it totally occupied at the end of its growth process. Such a situation is in fact impossible due to steric hindrances. A new parameter is thus introduced: **the real volume**

fraction representing the portion of the volume which is crystallized and which takes into account steric hindrance: $\phi = V^{\beta}/V_0$, where V^{β} is the real volume occupied by nuclei of new phase β.

Avrami hypothesizes that when the "expanded volume" of nuclei $\mathrm{d}V_e^{\beta}$ increases, only the untransformed part of the system (initial phase α) contributes to the variation in the real volume of the nuclei $\mathrm{d}V^{\beta}$. We can thus write:

$$(1 - V^{\beta}/V_0)\ \mathrm{d}V_e^{\beta} = \mathrm{d}V^{\beta} \tag{2.48}$$

or by integration of (2.48):

$$\log(1 - V^{\beta}/V_0) = -V_e^{\beta}/V_0 \tag{2.49}$$

and

$$\phi = 1 - \exp(-\phi_x) \tag{2.50}$$

This expression is the Avrami equation.

In (2.50), ϕ and ϕ_x are functions of the time t and temperature T. We have two extreme situations:

- if $\phi_x \gg 1, \phi \approx 1$, the phase transition is total;
- if $\phi_x \approx 0, \phi \approx 0$, the nucleation and the transition are beginning.

The Avrami equation illustrates the exponential character of nucleation. When the growth of a hypothetically spherical nucleus is controlled by constant mass transfer to the interface, the radius $r(t)$ of the nucleus increases linearly with the growth rate G.

$$\frac{\mathrm{d}r}{\mathrm{d}t} = G \tag{2.51}$$

If nucleation, that is, initial formation of the nucleus, took place at time τ from a critical nucleus r^* (Sect. 2.2.3), we have:

$$r - r^* = G\ (t - \tau) \tag{2.52}$$

For $r \gg r^*$ and $\tau = 0$, the volume of the nucleus will be:

$$V_e^{\beta} = \frac{4\pi}{3}\ G^3\ t^3 = K_3\ t^3 \tag{2.53}$$

If we assume that there are N_n nuclei in the system and that their nucleation occurs at the same time $\tau = 0$, the volume fraction ϕ_x occupied by these nuclei will be:

$$\phi_x = N_n V_e^{\beta}/V_0 \tag{2.54}$$

If $I(\tau)$ designates the nucleation rate, $I\ \mathrm{d}\tau$ nuclei will be formed during $\mathrm{d}\tau$; they will grow and their volume V_e^{β} at time t will be: $V_e^{\beta} = K_3\ (t - \tau)^3$ $(r \gg r^*)$. Hence:

$$\phi_x = \frac{1}{V_0}\int_0^t V_e^{\beta}\ I\ \mathrm{d}\tau = \frac{K_3}{V_0}\int_0^t (t - \tau)^3\ I(\tau)\mathrm{d}\tau \tag{2.55}$$

If we assume that $I =$ Const. and is thus independent of the time, we obtain by integration of (2.55):

$$\phi_x = K_3 \frac{I\, t^4}{4V_0} \tag{2.56}$$

giving

$$\phi = 1 - \exp\Big(-K_3 \frac{I\, t^4}{4V_0}\Big) \tag{2.57}$$

However if the nucleation rate is a function of time $I \propto C^{st} t^{\alpha}$, we see from (2.55) that ϕ can be written as

$$\phi \propto 1 - \exp(-K\, t^n) \tag{2.58}$$

where K is a constant which is a function of the temperature and also of the the geometry via G. This equation is also called the Avrami equation:

- If I is a decreasing function of time ($\alpha < 0$), then $n < 4$;
- If I is an increasing function of time ($\alpha > 0$), then $n > 4$.

The exponent n is called the Avrami exponent and is characteristic of the phase growth mode.

If the nucleus is not spherical and if the nucleation process is controlled by the interface, the exponent is modified. If nucleation occurs through the surface of a cylinder of radius r, its volume will increase like r^2 and we will have $V_e^{\beta} = K_2 t^2$. If nucleation rate I is constant, $n = 3$; if it is a rapidly decreasing function of time as t^{-1}, $n = 2$. In the case of a plate where growth occurs on the faces, $V_e^{\beta} = K_1 t$ and $n = 2$ for a constant nucleation rate.

Another growth process is where growth occurs by diffusion of atoms or molecules of the initial phase to nucleation sites. In this case, the diffusion process controls growth of the nuclei. Their size increases as $t^{1/2}$ and their volume increases as $t^{3/2}$.

For spherical nuclei with a constant nucleation rate, we easily find that $n = 5/2$ and if growth is a rapidly decreasing function of time, $I \propto t^{-1}$ and $n = 3/2$.

The values of the Avrami exponent for different types of growth and geometry are summarized in Table 2.1.

An Avrami exponent which will be characteristic of the growth mode of a material can be found by measuring the untransformed (or transformed) fraction of a material undergoing a phase transition, crystallization, for example. As $(1 - \phi)$ is the untransformed fraction, we obtain from (2.58):

$$\log[\log(1 - \phi)] \propto -n \log t$$

Here n was measured in mixtures of polymers in which the validity of the Avrami equation was verified. In the case of propylene, the experimental

Table 2.1. Avrami exponents n for various geometries and nucleation modes.

	Geometries	Type of nucleation	n
Interface –	Plane	rapidly exhausted	1
controlled	Cylinder	rapidly exhausted	2
growth	Sphere	rapidly exhausted	3
	Sphere	constant rate	4
Diffusion	Plane	rapidly exhausted	1/2
controlled	Cylinder	rapidly exhausted	1
growth	Sphere	rapidly exhausted	3/2
	Sphere	constant rate	5/2

measurements[1] give $2.45 < n < 2.74$; this exponent corresponds to nucleation in the form of spherulites with a diffusion mode.

The recrystallized fraction of the alloy $NiZr_2$ as a function of time is shown in Fig. 2.11. Crystallization from an amorphous phase (a metallic glass) occurs from nuclei whose growth is controlled by the interface. It corresponds to Avrami exponent $n = 4$.

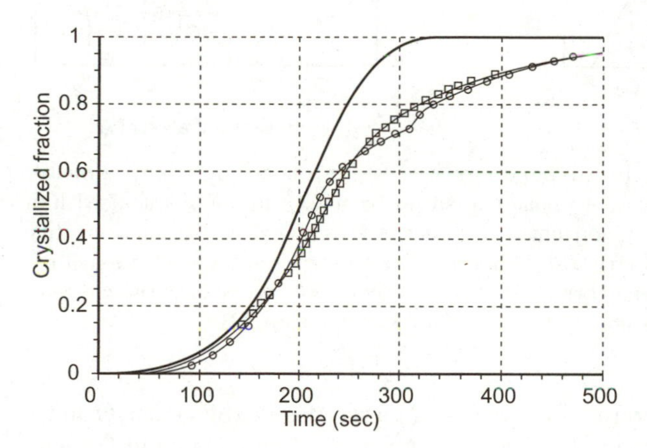

Fig. 2.11. Recrystallization of $NiZr_2$. The fraction crystallized from a metallic glass is determined from the Bragg scattering intensities. The solid curve corresponds to the Avrami equation with exponent $n = 4$ (S. Brauer et al., Phys. Rev. B, **45**, 7704 (1992), copyright Am. Inst. Phys.).

[1] Kim, Polym. Eng. Sci., **33**, 1445 (1993)

2.3 Spinodal Decomposition

The Gibbs–Duhem criterion (Sect. 1.2) led us to define the concept of metastable state and spinodal, which is the curve demarcating the metastability limit. Outside of the spinodal, the physical system becomes unstable. Gibbs was the first to describe this type of phase behavior by proposing a clear distinction between the regions of metastable and unstable equilibrium.

Such a situation is shown in Fig. 2.12 for binary alloy A–B in which an un-mixing phenomenon take place. The liquid crystallizes by a nucleation mechanism followed by growth of nuclei (microcrystallites) when quenching from temperature T_i to point a (between the coexistence line and the spinodal) is conducted. On the other hand, if the quenching process ends at point b beyond the spinodal, the unstable liquid solidifies according to a mechanism called spinodal decomposition: spontaneous fluctuations in concentration cause its transition to a stable solid state without it being necessary to cross the energy barrier corresponding to creation of an interface in order to form nuclei.

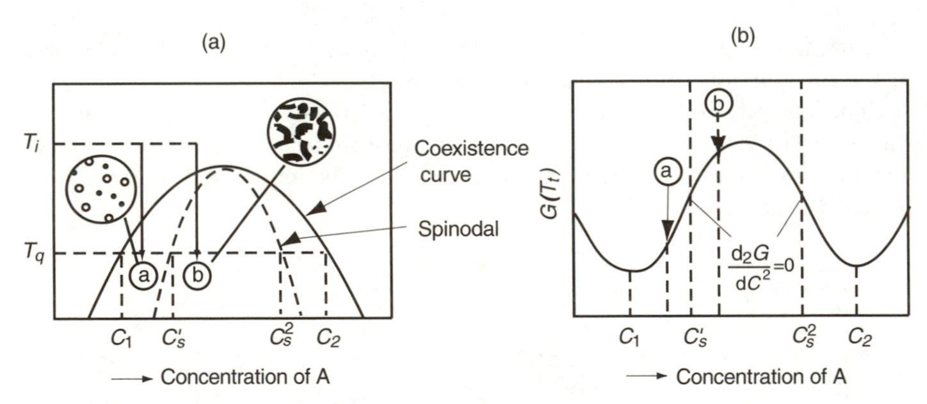

Fig. 2.12. Decomposition in a binary mixture. (**a**) In quenching from temperature T_i, the system undergoes a phase transition by nucleation at point a; within the spinodal at point b, local concentration fluctuations trigger the transition; this is spinodal decomposition. (**b**) $G(T_q)$ diagram for the free enthalpy at quenching temperature T_t. The spinodal corresponds to $\delta^2 G/\delta C^2 = 0$, where C is the concentration of one of the constituents, A, for example, in mixture AB.

Spinodal decomposition is observed in mineral glasses (SiO_2–Na_2O mixtures, as in Pyrex glass), metal alloys, and mixtures of polymers and liquids; in the last case, it is more difficult to reveal.

2.3.1 Thermodynamics of Spinodal Decomposition

We will describe spinodal decomposition in a solid or liquid binary alloy A–B. The concentration of constituent A in the alloy $c = n_A/(n_A + n_B)$ will be designated by c, where n_A and n_B are the numbers of atoms A and B.

We will assume that we have crossed the spinodal and are in the thermodynamic instability region but that the mixture is still homogeneous. It is the site of local and spontaneous fluctuations in the concentration c which are the source of the inhomogeneity within the mixture; the concentration c becomes a local function $c(r)$.

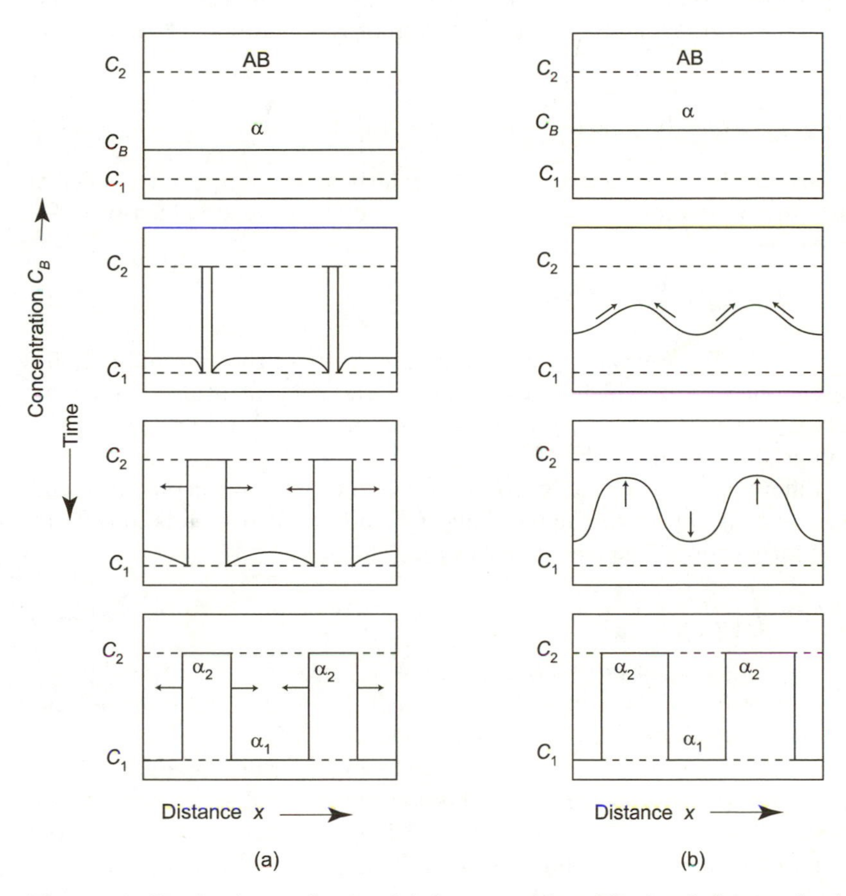

Fig. 2.13. Nucleation and spinodal decomposition. The local changes in the concentration C of constituent A in binary mixture AB. Two phases α_1 and α_2 are formed within the initial homogeneous mixture α of composition C_B. (**a**) by classic nucleation: primitive nuclei are growing with time. (**b**) by spinodal decomposition: the concentration fluctuations increase in the mixture. The final concentrations are C_1 and C_2 (Fig. 2.12).

Let us compare the situation of the mixture in the region where it is unstable on one hand (Fig. 2.13b) and where it is in a metastable state on the other hand (Fig. 2.13a). We find that in the first case, the concentration will vary uniformly and continuously in space and time until the new phase has formed. In the second case, there are important concentration discontinuities corresponding to the appearance of nuclei of the new phase at the beginning of the transformation. In a manner of speaking, there are homogeneous concentration fluctuations in the case of spinodal decomposition and in contrast, heterogeneous fluctuations in the case of the transition by nucleation (Fig. 2.12).

If c_0 is the average concentration in system AB (for example, for constituent A), the free energy per unit of volume $f(c)$ around c_0 can be expanded:

$$f(c) = f(c_0) + (c - c_0)\left(\frac{\partial f}{\partial c}\right)_{c_0} + \frac{1}{2}(c - c_0)^2\left(\frac{\partial^2 f}{\partial c^2}\right)_{c_0} + \dots \tag{2.59}$$

We assume that the fluctuations are of relatively low amplitude, which is obviously an approximation. We will then write the total free energy F for the system as:

$$F = \int \left[f(c) + \frac{1}{2}K\ |\nabla c|^2\right]\mathrm{d}V \tag{2.60}$$

The term $K\ |\nabla c|^2$ accounts for the inhomogeneity of the system in (2.60); it is an energetic term which is a function of the concentration gradient in the system and is in a way the equivalent of the interfacial energy for the fluctuations. It is assumed that $K > 0$.

The difference in free energy ΔF between the inhomogeneous part and the homogeneous part of the material and which thus corresponds to the free energy of formation of the new phase is written as:

$$\Delta F = \int \left[\frac{1}{2}\left(\frac{\partial^2 f}{\partial c^2}\right)_{c_0}(c - c_0)^2 + \frac{1}{2}K|\nabla c|^2\right]\mathrm{d}V \tag{2.61}$$

The first–order term in (2.60) is eliminated, since we know that by definition of the average concentration, $\int(c - c_0)\ \mathrm{d}V = 0$. Terms beyond the second order were neglected in the expansion.

According to (2.61), the homogeneous solution will remain stable if $(\partial^2 f/\partial c^2) > 0$ because in this case, the concentration fluctuations will always be associated with $\Delta F > 0$: they will tend to diminish. Actually, the probability of a concentration fluctuation δc is proportional to $\exp(-\Delta F/kT)$ and it will decrease if δc increases.

On the contrary, if the region of (T, c)–plane corresponding to an unstable thermodynamic state and thus the condition $(\partial^2 f/\partial c^2) < 0$, is involved (and inside the spinodal), i.e. when the amplitude of the fluctuations is large enough, then the first term dominates in integral (2.61). In this case, the concentration fluctuations reduce the free energy and $\Delta F < 0$; the concentration fluctuations then tend to increase and the inhomogeneous phase is the most stable (Fig. 2.13b): there is spinodal decomposition. The situation here differs from the situation where the transition is initiated by nucleation from a sudden concentration fluctuation (concentration peak) followed by growth of the initial nuclei.

At this stage, it is useful to expand the concentration fluctuations in a Fourier series $\delta c(\boldsymbol{r},t) = c(\boldsymbol{r},t) - c_0$. We must assume that the concentration is not only a local variable but that it is also a function of time. Hence we introduce:

$$\delta c(\boldsymbol{k},t) = \int \delta c(\boldsymbol{r},t) \mathrm{e}^{\mathrm{i}\boldsymbol{kr}} \mathrm{d}\boldsymbol{r} \quad \text{with} \quad \delta c(\boldsymbol{r},t) = \int \delta c(\boldsymbol{k},t) \mathrm{e}^{-\mathrm{i}\boldsymbol{kr}} \mathrm{d}\boldsymbol{k} \tag{2.62}$$

To simplify the description, let us assume that we are dealing with a one-dimensional material with fluctuations having the same amplitude A in the form $\delta c_k = A \cos(kx)$. For the moment, we will neglect their dependence on time.

We can write $\Delta F = \int \Delta F_{\boldsymbol{k}} \, \mathrm{d}\boldsymbol{k}$. By integration of (2.61) over the entire volume of the material, we obtain:

$$\Delta F_{\boldsymbol{k}} = \frac{1}{4} V A^2 \left[\left(\frac{\partial^2 f}{\partial c^2} \right)_{c_0} + K \, k^2 \right] \tag{2.63}$$

(integration over the entire volume of $\cos^2(kx)$ gives the factor $1/2$).

We immediatly see in (2.63) that if we are in a region of thermodynamic instability $(\partial^2 f/\partial c^2) < 0$, fluctuations of short wavelength (high k) will tend to increase the free energy and thus stabilize the homogeneous mixture. On the contrary, fluctuations of large wavelength (low k) stabilize the inhomogeneous system and favor the phase transition corresponding to decomposition. We say there is demixing if $\Delta F_k < 0$, i.e.:

$$\left(\frac{\partial^2 f}{\partial c^2} \right)_{c_0} + K \, k^2 < 0 \quad \text{or} \quad k^2 < k_c^2$$

with

$$k_c^2 = -\frac{1}{K} \left(\frac{\partial^2 f}{\partial c^2} \right)_{c_0} \tag{2.64}$$

k_c is a cut off wave vector. It follows that:

- for $k > k_c$, the homogeneous solution is stable;
- for $k < k_c$, the inhomogeneous solution is stable: there is **spinodal decomposition**.

Setting $\lambda_c = 2\pi/k_c$, there will thus be spinodal decomposition for $\lambda > \lambda_c$.

The kinetics of transformation of a binary alloy can be obtained with a diffusion equation. This calculation was performed by J. W. Cahn.

Note that the chemical potential μ of a component of an alloy is $\mu = \partial F/\partial c$. The potential gradient (equivalent to the "thermodynamic force") creates a concentration current $\boldsymbol{J}$ in the solution, $\boldsymbol{J} = -M_0 \nabla \mu$, where M_0

is a transfer coefficient assumed to be constant (it is a function of T). For a binary system, currents $\boldsymbol{J}_A$ and $\boldsymbol{J}_B$ are defined by the relation:

$$\boldsymbol{J}_A = -\boldsymbol{J}_B = M_o \,\nabla(\mu_A - \mu_B) \tag{2.65}$$

When there is no chemical reaction, we have a conservation equation:

$$\frac{\partial c(\boldsymbol{r},t)}{\partial t} + \nabla \boldsymbol{J} = 0 \tag{2.66}$$

If n_A and n_B are the numbers of atoms A and B in the alloy with $N = n_A + n_B$, we can write:

$$\mu_A - \mu_B = \frac{\partial F}{\partial n_A} - \frac{\partial F}{\partial n_B} = \frac{\partial F}{\partial c}\frac{\partial c}{\partial n_A} - \frac{\partial F}{\partial c}\frac{\partial c}{\partial n_B} = \frac{\partial F}{\partial c}\Big(\frac{\partial c}{\partial n_A} - \frac{\partial c}{\partial n_B}\Big) \tag{2.67}$$

that is:

$$\mu_A - \mu_B = \frac{1}{N}\frac{\partial F}{\partial c} \tag{2.68}$$

and thus:

$$\frac{\partial c(\boldsymbol{r},t)}{\partial t} = \frac{M_0}{N}\,\nabla^2\Big(\frac{\partial F}{\partial c}\Big) = M\,\nabla^2\Big(\frac{\partial F}{\partial c}\Big) \tag{2.69}$$

with $M = M_0/N$, where $c(\boldsymbol{r},t)$ is a variable describing an inhomogeneous material, we must take the functional derivative of F with respect to c in (2.69), i.e., from (2.60):

$$\delta F = \int \Big[\frac{\partial f}{\partial c}\,\delta c + \frac{1}{2}\,K\,\nabla_r \delta c\,\nabla_r c + \frac{1}{2}\,K\,\nabla_r c\,\nabla_r \delta c\Big]\,\mathrm{d}\boldsymbol{r} \tag{2.70}$$

Integrating by parts ($\int \nabla_r \delta c\,\nabla_r c = \delta c\,\nabla_r c - \int \delta c\,\nabla^2 c$) and assuming that $\nabla_r c$ is zero on the integration surface, we have contribution $-K \int \nabla^2 c\,\delta c\,\mathrm{d}V$ of this term to δF which is in the form of:

$$\delta F = \int \Big[\frac{\partial f}{\partial c} - K\,\nabla^2 c\Big]\delta c\,\mathrm{d}\boldsymbol{r} \tag{2.71}$$

i.e.:

$$\frac{\partial F}{\partial c} = \frac{\partial f}{\partial c} - K\,\nabla^2 c \tag{2.72}$$

hence:

$$\frac{\partial \delta c(\boldsymbol{r},t)}{\partial t} = M\,\nabla^2\Big[\frac{\partial f}{\partial c} - K\,\nabla^2 c(\boldsymbol{r},t)\Big] \tag{2.73}$$

This is the **Cahn–Hilliard** equation for the evolution of the concentration. It is in the form of a nonlinear diffusion equation.

As we are only interested in concentration fluctuations $\delta c(\boldsymbol{r}, \mathrm{t})$, using (2.59), we can linearize (2.73):

$$\frac{\partial \delta c(\boldsymbol{r},t)}{\partial t} = M\,\nabla^2\Big[\Big(\frac{\partial^2 f}{\partial c^2}\Big)_{c_0} - K\,\nabla^2\Big]\delta c(\boldsymbol{r},t) \tag{2.74}$$

Again introducing the Fourier transforms for $\delta c(\boldsymbol{r},t)$, i.e.:

$$\delta c(\boldsymbol{k},t) = \int \delta c(\boldsymbol{r},t)\,\mathrm{e}^{\mathrm{i}\boldsymbol{kr}}\mathrm{d}\boldsymbol{r} \tag{2.75}$$

Equation (2.74) can be solved by assuming that there is an exponential relaxation mechanism of fluctuations of the type:

$$\delta c(\boldsymbol{k},t) = \delta c(\boldsymbol{k},0)\exp(-\omega_k t) \tag{2.76}$$

where ω_k is the relaxation or damping coefficient.

Inserting (2.75) and (2.76) in (2.74), we immediately deduce by identification:

$$\omega_k = M\ k^2\Big[\Big(\frac{\partial^2 f}{\partial c^2}\Big)_{c_0} + K\ k^2\Big] \tag{2.77}$$

Introducing wave vector k_c, ω_k is rewritten as:

$$\omega_k = M\ K\ k^2[k^2 - k_c^2] \tag{2.78}$$

In the region where the material is thermodynamically unstable, for $k < k_c$, $\omega_k < 0$, the corresponding concentration fluctuations increase with time: they are no longer damped. As in the preceding simplified description (one-dimensional material) we find that: concentration fluctuations of large wavelength intensity with time and cause spinodal decomposition of the binary alloy.

Introducing the structure function $S(\boldsymbol{k},t)$ (see its definition Sect. 1.3.3) $S(\boldsymbol{k},t) =< \delta c(\boldsymbol{k},t)\delta c(-\boldsymbol{k},t) >$ associated with these fluctuations, it will thus have the simple form of an exponential relaxation function:

$$S(\boldsymbol{k},t) = S(\boldsymbol{k},0)\exp(-2\omega_k t) \tag{2.79}$$

The damping coefficient ω_k is the product of two factors:

- a term of Mk^2 in common factor resulting from diffusion in the alloy;
- a term of $K(k^2 - k_c^2)$ that reflects the existence of concentration inhomogeneities and favors spinodal decomposition for small wave vectors.

ω_k is minimum for $k_m = k_c\sqrt{2}$. This value corresponds to a maximum in the rate of increase of concentration fluctuations in the alloy.

Introducing $D_0 = -M(\partial^2 f/\partial c^2)$, ω_k can be rewritten as

$$\omega_k = D_0 k^2\Big[\frac{k^2}{k_c^2} - 1\Big] \tag{2.80}$$

We can graphically represent $-\omega_k/k^2$; it is a linear function of k^2/k_c^2. It was experimentally found that the dynamics of the spinodal decomposition process deviates from this linear behavior.

Let us briefly return to the example of the one-dimensional material discussed at the beginning of the section. We can write the complete Fourier expansion of the concentration fluctuations:

$$\delta c(x,t) = \sum_k \exp(-\omega_k t)(A_k \cos kx + B_k \sin kx) \tag{2.81}$$

only retaining the term $k = k_m$ in the summation, we will then have:

$$\int \delta c^2 \mathrm{d}V = \frac{1}{2} V \exp(-2\omega_{km})(A_{km}^2 + B_{km}^2) \tag{2.82}$$

This expression shows that "everything happens as if" we were dealing with a solid material with two concentrations A_{km} and B_{km} of the constituents and there is indeed phase separation and demixing in the binary mixture.

2.3.2 Experimental Demonstration – Limitation of the Model

Spinodal decomposition has been observed in a large number of materials with very different methods. The phenomenon can be seen in mixtures of liquids, polymers, in glasses, metal alloys, gels, and ceramics. In mixtures of liquids, the concentration fluctuations are rapid except in the vicinity of the critical demixing point (the spinodal is tangential to the coexistence curve at the critical point), and spinodal decomposition due to light scattering (methanol–cyclohexane mixture, for example) can also be observed with interconnected micro-structures which give a spongy consistence to the material.

Spinodal decomposition can be observed by electron microscopy in steels (Fig. 2.14) and glasses (SiO_2 – Na_2O mixture). In steel, during aging the wavelength of the fluctuations in regions rich in carbon increases with time and modulated microstructures formed by spinodal decomposition which modify the mechanical properties of steel are observed. In quenched glass, a true interconnection of the phases formed is observed, with interconnected micro-structures which give a spongy consistence to the material.

The dynamics of spinodal decomposition in metal alloys can also be followed by neutron scattering (Fig. 2.15) and the scattering intensity passes through a maximum and increases in time. Similar experiments were conducted with light scattering in mixtures of polymers (1,4–polybutadiene) and microemulsions (water–decane with a surfactant and agarose gels).

The light intensity proportional to $S(\boldsymbol{k}, t)$ increases in time, but the variation of $-\omega_k/k^2$ is not a linear function of k^2/k_c^2.

In fact, although it satisfactorily accounts for the general evolution of spinodal decomposition, the Cahn model is based on a series of approximations whose validity is debatable after the initial stage of the process. It assumes that there are no random concentration fluctuations of thermal origin in the material. These fluctuations can be accounted for by introducing a random term $R_T(t)$ in the right term of (2.73) (Cook convention). Another approximation is the linear character of the model: coupling of the concentration $c(\boldsymbol{r}, t)$ with other variables, for example, the structural relaxation of the material, is neglected. Finally, the model does not take into account the contribution of terms like $\nabla^2 c^2$ and ∇c^4 in the free energy.

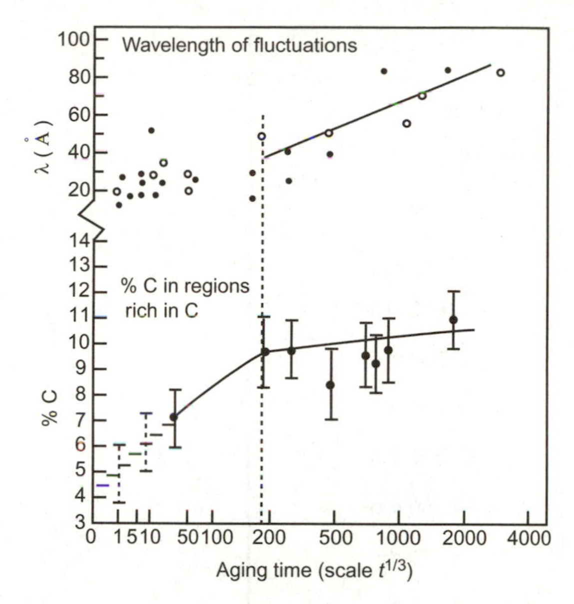

Fig. 2.14. Spinodal decomposition in steel. This phenomenon is observed here in a martensite (iron–nickel–carbon alloy). The wavelength of the concentration fluctuations in regions rich in carbon increases: the structure of the solids is modulated locally, and the concentration asymptotically approaches 11%. The observations were made by transmission electron microscopy (T.E.M.) and by ion microscopy (AP/FIM) [Taylor and Smith, Progress and prospects in metallurgical research, *Advancing Materials Research*, National Academy Press (1987), p. 76].

In the case of polymers, the situation is even more complex, and P. G. de Gennes showed that the interpenetration of polymer chains must be taken into consideration in the expression for the free energy and (2.65), which expresses the concentration current, hence must consequently be modified. It is nonlocal in character due to the connectability of the polymer chains.

Introducing the coefficient $\Lambda(\boldsymbol{r}, \boldsymbol{r}')$ which accounts for interdiffusion of the polymer chains, we will then write:

$$\boldsymbol{J}(\boldsymbol{r}, t) = -\int \Lambda(\boldsymbol{r} - \boldsymbol{r}')\nabla\mu(\boldsymbol{r}', t)\mathrm{d}\boldsymbol{r}' \tag{2.83}$$

In a monophasic region, the damping coefficient ω_k is proportional to k^6.

Finally, we can model spinodal decomposition with methods of numerical simulation (molecular dynamics or Monte Carlo methods). K. Binder investigated spinodal decomposition of a binary alloy with a Monte Carlo method, representing the alloy by a two-dimensional network and assuming the existence of repulsive interactions between next and the nearest neighbors.

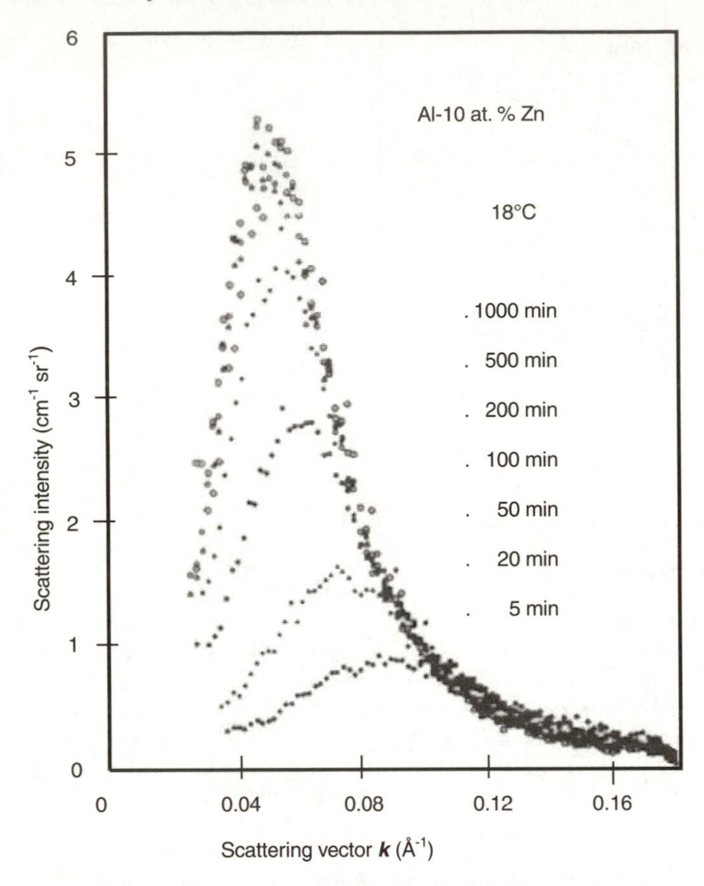

Fig. 2.15. Neutron scattering in an alloy. Spinodal decomposition is demonstrated by neutron scattering in Al–Zn alloy. The scattering intensity due to concentration fluctuations is proportional to $S(\boldsymbol{k}, t)$, increases with the time after quenching, and passes through a maximum (the effective scattering area proportional to it is on the ordinate here) (S. Komura et al., Phys. Rev., **B31**, 1283 (1985), copyright Am. Inst. Phys.).

2.4 Structural transition

The typology of phase transitions introduced here (Sect. 1.41) contains a specific category of transitions; those accompanied by a change in structure. Some of these changes in structure occur without macroscopic diffusion of matter in solids (solid–solid transformation). They are initiated by local motions of atoms or molecular groups which can distort the lattice in the high-temperature phase to form structures of lower-symmetry at lower temperatures.

2.4.1 Dynamics of a Structural Transition – The Soft Mode

As just mentioned, diffusionless transformations result from atomic motions. These movements around equilibrium positions do not occur instantaneously at temperature T_0, the phase transition temperature, and they are actually initiated at a temperature higher (or lower) than the transition temperature. In a way, this pretransition or pretransformation phenomenon is the equivalent of the previously described nucleation process. The atoms or molecular groups that move from their initial equilibrium position are subject to a restoring force whose intensity decreases in approaching the phase transition temperature T_0.

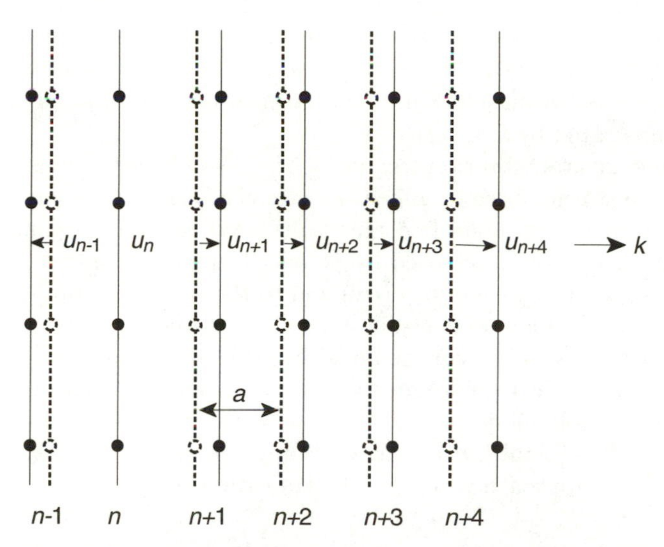

Fig. 2.16. Atoms moving in a lattice. The atoms are located in planes and move with respect to their equilibrium position (dotted lines). U_{n-1}, U_n, and U_{n+1} are movements in different planes; a is the distance between planes.

The phenomenon can be described by considering the motion of atoms located in planes (n) in a privileged direction (Fig. 2.16). If u_i is the motion of an atom of mass m, it is subject to restoring force F_i which can be written as:

$$F_n = \sum_m c_m(u_{n+m} - u_n) \tag{2.84}$$

c_m is a constant that can be calculated from the interaction potential between the atoms in the lattice (the harmonic potential is the simplest choice). We can then write the equation of motion as:

$$m\frac{\mathrm{d}^2 u_n}{\mathrm{d}t^2} = F_n = \sum_m c_m(u_{n+m} - u_n) \tag{2.85}$$

Solutions of this equation are generally sought in the form of $u = e^{i(n+m)ka+i\omega t}$, where k is the vibration wave vector and a is the interatomic distance. Frequencies $\omega = \omega_r + \omega_i$ are the normal lattice vibration modes (real and imaginary parts).

The equilibrium will be stable if all motions of atoms in the lattice $u_n(t)$ are damped with respect to their initial situation. A simple condition for stability of the equilibrium of a crystal lattice phase is derived: $\omega_i > 0$.

This condition is the equivalent of the Gibbs–Duhem condition of thermodynamic stability. The boundary condition for the stability of equilibrium of a crystalline phase will thus be written as: $\omega_i = 0$. This condition will be satisfied if there is a normal mode ω_0 whose frequency is reduced to zero at temperature T_0 in the vibration spectrum of the system (that is, the phonon spectrum). At this temperature, the phase becomes unstable and the solid undergoes a phase transition. We say that the corresponding vibration is a **soft mode**. This characteristic of certain structural phase transitions ($\omega_0(T_0) = 0$) was theoretically demonstrated by Cochran.

Disappearance of the natural vibration frequency of the solid corresponds to disappearance of the restoring force to which the atoms are exposed: their motions are no longer damped, the solid becomes "soft", that is, unstable for this mode, and stability can only be restored by passing to a new phase.

Vibrations associated with a soft mode correspond to the transverse optical modes in the vibration spectrum of the solid, and their frequencies are in the infrared. These soft modes have been demonstrated by Raman scattering, in $SrTiO_3$, for example, which undergoes a structural phase transition at 110 K. In this case, the soft mode is associated with the rotational motion of oxygen atoms in the crystal lattice (they occupy the vertex of TiO_6 octahedrons). Its frequency approaches zero at the transition temperature.

2.4.2 Martensitic Transformation

When the amplitude of the crystal lattice cell distortions related to atomic motions is large, in a solid/solid transformation in a crystalline system, the associated deformation energies become important. In this case, they dominate the nature and kinetics of the phase transition: these are martensitic transformations. The changes in the lattice volume and the deformation energy in a transformation of this type are so important that they suggest a nucleation mechanism. Nevertheless the martensitic transformations are diffusionless transitions. This transformation was named after martensite.

Iron can form a solid solution with carbon at high temperature (the γ phase or austenite) which is a variety of steel (f.c.c. symmetry): it remains stable up to a temperature of 727°C for a 0.8 wt. % carbon content (Fig. 2.17) corresponding to an eutectic point. If austenite is cooled very rapidly (at a quenching rate of the order of 10^5 °C sec^{-1}) at this concentration, it is no longer stable: it is transformed into a new metastable phase,

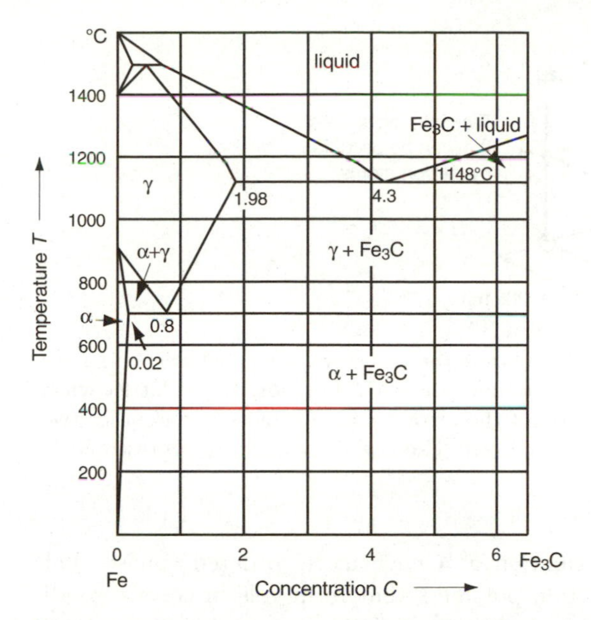

Fig. 2.17. Phase diagram of steel. The different phases of Fe–C alloy are shown as a function of the concentration by weight of carbon. Fe_3C corresponds to cementite. Martensite is formed from the γ phase (austenite) by quenching to 0.8% concentration; it is in a metastable state.

martensite, which is a solid solution of iron and carbon (at ordinary temperature, α–iron in body-centered cubic phase or ferrite can only dissolve 0.035% carbon). Martensite has a b.c.c. structure globally corresponding to the structure of ferrite, but it is locally deformed by the excess carbon remaining in the lattice which gives rise to a local structure of centered tetragonal symmetry by distorting it (Fig. 2.18). Austenite is transformed into martensite by a nucleation–growth process that can be very rapid, and it can propagate in the material at a rate of the order of 10^5 cm · sec $^{-1}$, that is, at the speed of sound; the kinetics is very fast.

This transformation takes place in numerous metal alloys (Fe–Co, Cu–Al–Ni, etc.). It is first order and very often reversible with thermal hysteresis. The martensitic transformation modifies the mechanical properties of an alloy: this is the phenomenon of hardening of steel (Sect. 9.2.2).

2.5 Fractals – Percolation

2.5.1 Fractal Structures

Some transformations in a material are initiated from a state of total disorder (molecules of a polymer in solution, small particles suspended in a liquid, for

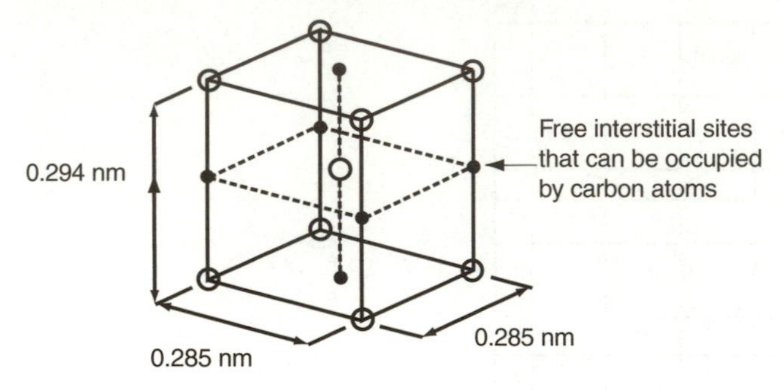

Fig. 2.18. Structure of martensite. During the transition after fast quenching at a 0.8% carbon content, carbon atoms occupy interstitial positions in the lattice which expands in one direction (direction of the arrow) and contracts in the other two. Locally there is a centered tetragonal cell. The carbon atoms are represented by the black points.

example) and lead to the formation of a new, more "ordered" phase. This is the case of solutions of certain polymers that form gels in certain conditions of concentration and temperature (gelatin, for example): the **sol–gel transition** or **gelation** causes the formation of a three-dimensional network.

Colloidal systems are another example of a phase constituted of mineral or organic particles which aggregate to form larger particles (their diameter can be between 1 and 1000 nm).

Indian ink (a suspension of carbon black with a natural polymer, gum Arabic) and milk are examples of colloidal suspensions. Colloidal systems remain heterogeneous disordered phases and for this reason, formation of a colloidal aggregate is a somewhat special phase transition.

Nucleation of dendrites (Sect. 2.2.4), gelation, and formation of a colloidal phase are certainly different processes, but they have a common feature: they are variants of what is called **fractal geometry**.

The concepts of fractal geometry were developed and systematized by B. Mandelbrot. It can be used to describe objects that are invariant under a change in scale: for example, when the scale used to measure them is modified or when the resolution of the observation instrument is changed.

Many structures or objects can be described with the concept of fractal geometry, which has numerous applications in materials science (surface, porosity, aggregate, fracture, phase transformation problems, etc.).

The notion of fractal geometry is based on generalization of the notion of scaling dimension and law. Assume that we are measuring a segment of length L with length unit l. The result of the measurement $L(l)$ is none other than the ratio $\Lambda(l) = L/l$. If we now measure the same segment with length unit l/N, the result of the new measurement will be $\Lambda'(l) = L/l/N = N\Lambda(l)$. The same operation can be formed for a square area $L \times L$ measured with a unit area $l \times l$, where the measured area $S(l)$ is $S(l) = L^2/l^2$. If we take $(l/N)^2$

for the unit of area, we will have a new measurement $S'(l/N) = N^2 S(l)$. For volume V, we will finally have $V(l/N) = N^3 V(l)$.

In the three preceding cases, the fractal dimension of the object or structure is the exponent appearing in the ratio between the result of the measurement of the object and the renormalization factor of the unit of length (in the preceding example, for a segment, square, cube). If D_f is this factor, we will call it the fractal dimension. Two structures whose length ratio is equal to N will thus have a mass relation of $m = N^{Df}$, for example; their **fractal dimension** D_f will thus be $\log m / \log N$.

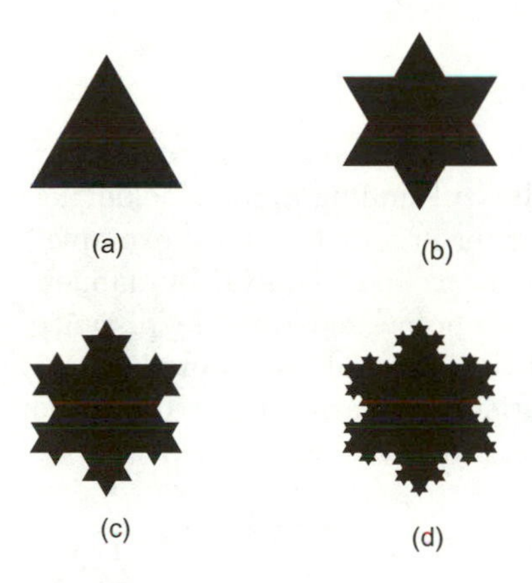

Fig. 2.19a–d. Construction of a fractal structure. A structure of the "snowflake" type is obtained by stages by multiplying the resolution at each stage by a factor of 4. The fractal dimension is log 4/log 3 = 1.26 (Koch construction).

A typical example of such fractal geometry, the Koch construction of a "snowflake", is shown in Fig. 2.19. With a resolution corresponding to length l (1 cm, for example), the initial object is an equilateral triangle. If the resolution is increased by a factor of 4, we have a 12–sided shape (Fig. 2.19b). In each stage where the unit of measurement is reduced by a factor of 3, we multiply the number of segments that can be demonstrated in the structure by a factor of 4. Applying the preceding definition, we will then have $D_f = \log 4 / \log 3 = 1.26$. We also replicated the initial figure during the operation by constructing a structure that can be reproduced indefinitely by similarity. The structure is invariant with a change in scale; it has self-similarity with dimensional factor D_f, which is the fractal dimension. This behavior of the structure in a change of scale is called **self-affinity**.

In general, the natural structures in a material do not resemble the relatively simple structures constructed with the procedure just described and illustrated in Fig. 2.19. They are actually structures exhibiting statistical self-similarity or self-affinity.

For example, this is the situation encountered in a heap of sand in which there is no uniform stacking of grains similar to the periodic distribution of atoms in a crystal lattice, but identical grains are statistically found in moving within the heap. Compare with the periodic structure of a crystal where there is invariance due to translation of this scale structure. This is also the situation in a gel where the same basic structure, the initial polymer which has been dissolved, is randomly found in the material (collagen in the case of a gelatin gel, for example).

These fractals corresponding to a statistical situation can be described with a density–density autocorrelation function $G(r)$ (Sect. 1.3.3) defined here with the expression:

$$G(\boldsymbol{r}) =< \rho(\boldsymbol{r}+\boldsymbol{r}')\rho(\boldsymbol{r}') > \tag{2.86}$$

$\rho(\boldsymbol{r})$ is the local density at point $\boldsymbol{r}$ and $< >$ is a statistical average taken over the entire space. $G(\boldsymbol{r})$ expresses the probability of finding a point belonging to the structure in $\boldsymbol{r}+\boldsymbol{r}'$ (the presence of matter in a material, for example) knowing that there was one of the same nature at $\boldsymbol{r}'$. In an equivalent manner, we can say that $G(\boldsymbol{r})$ is the probability that two points separated by distance $\boldsymbol{r}$ will belong to the structure (to a phase of the material, for example).

For a fractal structure exhibiting this statistical invariance due to a scale change, $G(\boldsymbol{r})$ decreases monotonically with r, but more slowly than with an exponential trend. $G(\boldsymbol{r})$ has the form:

$$G(\boldsymbol{r}) = r^{-\alpha} \tag{2.87}$$

If a scale change by a factor N is made, $G(r)$ should be invariant, $G(Nr) \approx G(r)$ with a multiplying factor which is none other than $N^{-\alpha}$ (assuming the material is isotropic). We can show that in Euclidean space of dimension d, $\alpha = d - D_f$.

The preceding statements can be applied to colloidal aggregates. Small particles of colloidal gold of uniform individual radius $R \approx 7.5$ nm form large aggregates which can be seen with the electron microscope. The number N of gold particles in an aggregate can be counted by varying observation scale l. Plotting $\log N$ as a function of $\log l$ in a log–log graph, we find a linear law: the slope of the line is the fractal exponent of the structure (here $D_f = 1.75$). For soot particles, $D_f = 1.18$. In general, the fractal exponent is between 1 and 2.

The dynamics of growth of colloidal aggregates is also of interest: the average particle size obeys a scaling law as a function of time, indicating the fractal character of the structure. The increase in the average radius of the aggregates as a function of time can be measured by light scattering.

The intensity of the scattered light is proportional to the Fourier transform of function $G(r)$, that is, structure function $S(\boldsymbol{k})$ (Sect. 1.3.3). Assuming that the mechanism of growth of aggregates is controlled by diffusion of particles in the medium, it is possible to show that the intensity is a linear function of the diffusion coefficient, thus $1/R$.

In studying the variation in the scattering intensity in time, one determines the behavior of R and the scaling law $\approx t^{0.56}$ was found. This result is in agreement with a kinetic theory of growth which provides that this exponent for growth is equal to $1/D_f$ ($D_f = 1.57$).

The kinetic theory of aggregation is based on the Smoluchowski model. It represents the increase in the concentration $c_k(t)$ of an aggregate $k (k = 1, 2, 3...)$ which is the number of particles in this aggregate, by the equation:

$$\frac{\mathrm{d}c_k}{\mathrm{d}t} = \frac{1}{2} \sum_i \sum_j K_{ij} c_i c_j - c_k \sum_j K_{ij} c_i \tag{2.88}$$

where K_{ij} are kinetic constants. The evolution of an aggregate k in time results simultaneously from the formation of a new nucleus k by combination with nuclei j. Equation (2.88) has no simple general solution. In some cases, there is an asymptotic solution corresponding to the formation of an infinite nucleus: a **gel**.

Finally, fractal structures can be observed in dendrites formed during crystallization of a metal or a binary mixture; they have strong similarities to colloidal aggregates. The analysis of digitalized images of dendritic structures formed in thin films of Zn and $NbGe_2$ gives fractal exponents of 1.66 and 1.69, respectively (Fig. 2.20).

The mechanism of growth is controlled by diffusion in the medium, and the density of the particles $c(\boldsymbol{r}, t)$ forming the dendrites obeys the equation:

Fig. 2.20. Growth of a dendrite. The Zn dendrite grows according to a fractal structure. It is obtained by electrodeposition of metallic Zn at the interface between an aqueous solution of Zn sulfate and n-butyl acetate (M. Matsushito et al., Phys. Rev. Lett., **53**, 286 (1984), copyright Am. Inst. Phys.).

$$\frac{\mathrm{d}c(\boldsymbol{r},t)}{\mathrm{d}t} = D\,\nabla^2 c(\boldsymbol{r},t) \tag{2.89}$$

where D is the diffusion coefficient. This is an equation of the type describing nucleation (2.44).

The growth of aggregates and dendrites can also be simulated by numerical methods of the Monte Carlo type, for example. The fractal exponents are close to 1.7 for two-dimensional aggregates and close to 2.5 for three-dimensional systems.

The description of a material or phase undergoing transformation by fractal geometry gives no information on the mechanisms of the phenomenon. It gives a conceptual picture for understanding and formalizing the transformation; it also allows comparing different phenomena such as the formation of colloidal phases, dendrites, gels, etc.

2.5.2 Percolation and Gelation

Percolation is at the origin of a mechanical phenomenon that occurs when a fluid flows through a porous medium. A hot liquid, for example, opens a path through the network of microchannels within a powder (coffee in the filter of a coffee pot, for example) to continuously flow through (Fig. 2.21). We say that it percolates through the porous medium. The network of open channels that allow diffusion of the fluid constitutes a continuous series of connections from one end of the diffusing medium to the other.

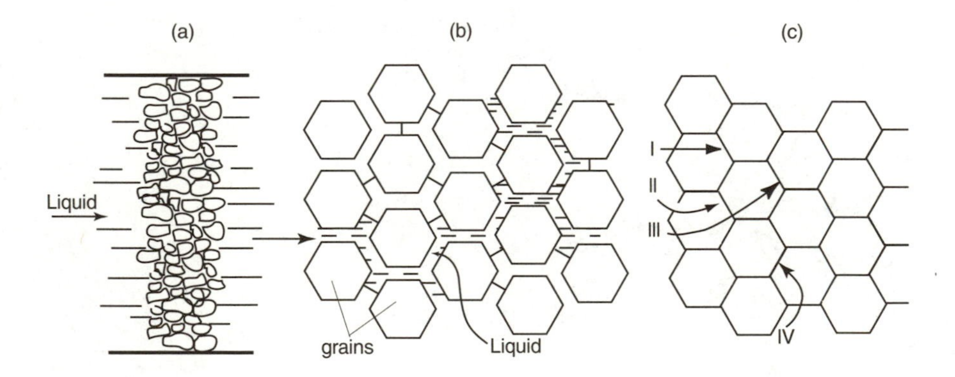

Fig. 2.21. Flow in a porous medium. (**a**) Flow of a liquid through a porous medium: percolation. (**b**) Modeling of the network of pores in the material. This network is represented by channels, some of which are clogged. (**c**) Diagram equivalent to the network modeled in (**b**) showing the connectivity of the network; a solid line represents an open channel and a dashed line indicates a closed channel. This diagram shows "bond percolation". I) Bonds or channels open. II) Channel blocked. III) Nucleus of size $s = 2$. IV) percolation path corresponding to a nucleus with $s = \infty$.

By analogy, percolation describes a phenomenon in which initially independent small structures aggregate or are coupled with each other to form a microstructure or "macro-object", and the bonds between the individual objects can be physical or chemical in nature. A classic example of percolation is the example of metallic particles mixed in an isolating powder: there is a strong increase in the electrical conductivity of the material when the concentration of metal particles is greater than 30%. Beyond this threshold, they have established a sufficient number of bonds or contacts to allow continuous connection in the medium. **Gelation** is similar to percolation. We begin with a solution of macromolecules (gelatin in water, for example), and for a certain concentration and temperature, gelation of the solution is observed; the molecules of the polymer and solvent form a three-dimensional network which is a continuous structure and is in a way the equivalent of an aggregate.

J. M. Hammersley proposed the first mathematical theory of percolation in 1957; it was subsequently applied to different physical situations, to condensed matter in particular by D. Stauffer and P. G. de Gennes. We will later return to treatment of gelation with percolation models and for the moment will only show how the character of the phenomenon is revealed.

Since formation of macroscopic connected masses or aggregates is involved, the material can be modeled in two different ways (Fig. 2.22):

- the material is represented by a network in which all sites are occupied (by a polymer molecule in the case of a gel) and bonds are established randomly between neighboring sites with probability p;
- alternatively, the sites are randomly positioned with probability of occupation p, and the occupied sites are connected if they are close neighbors.

Bond percolation (the analog of percolation of a liquid shown in Fig. 2.21) is involved in the first case, in the second case we deal with, **site percolation**. There is a threshold p_c such that connectivity is complete for $p > p_c$. This is the case in Fig. 2.22b. The totally interconnected network would correspond to the aggregate in Fig. 2.21, for example, or to a gel.

A fractal dimension can be defined for a percolation model. For example, if we take the case of a gel undergoing formation, the polymer and solvent form clusters of increasing size containing a higher and higher number of intermolecular bonds (when the gel is formed, the interconnection is complete). The phenomenon can be modeled by bond percolation. The clusters have a fractal character because they exhibit the same behavior when examined on different scales. If we take a cluster of radius r in the solution during gelation and if we designate the number of bonds established within the mass by $N(r)$, we have $N(r) \approx r^{Df}$, where D_f is the fractal dimension. This situation can be simulated with a Monte Carlo method. For two- and three-dimensional systems, we respectively have $D_f = 1.9$ and $D_f = 2.5$.

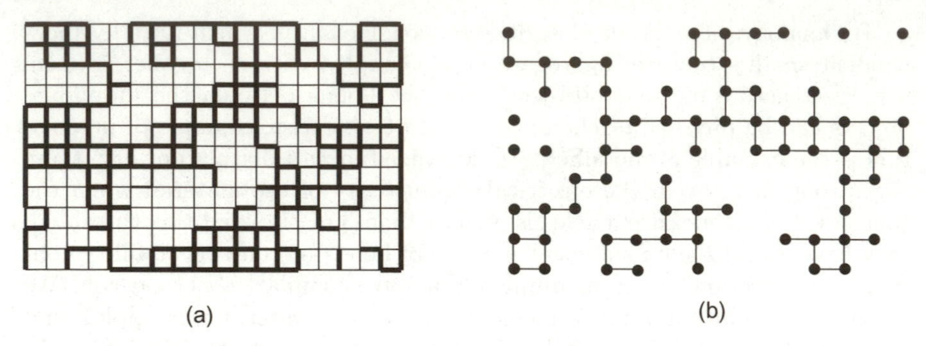

Fig. 2.22. Percolation in a network. (**a**) Bond percolation: all sites in a two-dimensional network are occupied, but they establish a bond with their neighbors with probability p. (**b**) Site percolation: the sites are occupied with probability p, and every occupied site is connected with its neighbors. There is a threshold p_c such that interconnection is total for $p > p_c$ (the case for b).

The percolation threshold p_c is defined by the value of p where a cluster of infinite size is formed in the material. In the vicinity of p_c, the occupied sites form a large interconnected network surrounded by clusters of smaller size.

Assume that a site percolation model is used and that $n_s(p)$ is the number of clusters or nuclei of size s, that is, containing s connected sites. Then $n_s(p)$ is obviously a function of the probability of occupation of a site p.

We then define the average size $< S_m >$ of a cluster by:

$$< S_m >= \sum_s s^2 n_s(p) / \sum_s s\, n_s(p) \tag{2.90}$$

Let ξ be the coherence or correlation length associated with the percolation (Sect. 1.3.3). In the vicinity of p_c, the diameter of a cluster is given by ξ. We can then define fractal dimension D_f by $< S_m >= \xi^{Df}$. We then introduce two critical exponents γ and ν in the vicinity of p_c such that:

$$< S_m > \propto |p - p_c|^{-\gamma} \quad \text{and} \quad \xi \propto |p - p_c|^{-\nu} \tag{2.91}$$

We immediately derive $D_f = \gamma/\nu$.

If $N(r)$ is the number of sites located in radius r around a point in the network, we will then have:

$$N(r) = r^{D'_f} f(r/\xi) \tag{2.92}$$

$f(r/\xi)$ is a dimensionless function.

If P_∞ is the probability that a site in a cluster will be part of the infinite cluster that occupies the available volume of the system, $N(r)$ can also be written, where d is the lattice dimension:

$$N(r) = r^d P_\infty \tag{2.93}$$

Assuming that $P_\infty \propto |p - p_c|^\beta \approx \xi^{-\beta/\nu}$, we find that $f(r/\xi) = (r/\xi)^{\beta/\nu}$ and thus:

$$N(r) = r^{d-\beta/\gamma}\left(\frac{r}{\xi}\right)^{\beta/\gamma} \tag{2.94}$$

Hence $D'_f = d - \beta/\gamma$, which is another expression for the fractal exponent. These exponents have been tabulated by Stauffer.

The correlation length ξ allows discriminating between small- and large-distance scales within the material. At small distances, $r < \xi$, the infinite cluster remains irregular with a large spread in the size of inhomogeneities, and we can consider that this is a self-similar fractal object with fractal dimension D_f (or D'_f).

On the other hand, with a large scale, $r > \xi$, and the cluster can be considered an approximately uniform network whose inhomogeneities can be neglected; its fractal dimension is then the dimension of its space, that is, d. At the limit, if $p = 1$, all sites are occupied and connected, the network is a regular object on all scales, and it actually has no fractal character.

As we will show, certain properties of a gel and the sol–gel transition can be accounted for with a percolation model.

2.6 Dynamics of Phase Transitions and Properties of Materials

When a material undergoes a phase transition, the dynamics of the phenomenon may have an important effect on the substance and properties of the new phase formed. For example, with an initial liquid phase, either a crystalline solid or a glass can be formed by cooling. If the cooling rate is slow, a crystal will generally be produced; on the contrary, if cooling is very fast (quenching), a glass can form. Metallic glasses can be formed from liquid alloys in this way if quenching is very fast (one method used consists in melting a solid alloy by laser pulsing and then rapidly quenching it). When solidification in the form of a glassy phase is conducted by fast quenching, the mechanism of growth of solid nuclei in the mother phase is "short-circuited" in a way: the nuclei of the crystalline phase could have been formed by nucleation, but they did not have the time to grow.

Another example is martensitic transformation in solid iron–carbon solutions; if quenching is fast enough (cooling rate of the order of 10^5 °C$\cdot$ sec^{-1}, a metastable phase is formed: martensite (Sect. 2.42). Martensite, whose carbon content is 0.8%, is very hard but brittle (its resistance to plastic deformation is high). Its brittleness can be reduced by heat treatment (by heating it): carbon atoms then migrate into the solid solution supersaturated with carbon which is then metastable and form Fe_3C precipitates, where the lattice relaxes to the b.c.c. structure of stable equilibrium, α ferrite. The microstructures constituted by carbide precipitates increase the hardness and elastic limit of steel.

The mechanical properties of a material will thus be a very significant function of the kinetics of the liquid–solid or solid–solid transformation.

The kinetics determines to a great extent the nature and size of the **microstructures** which are formed in a solid and can very significantly modify the properties of the material. The microstructures generally describe the microscopic state of a solid on a scale smaller than the size of a crystal, for example, several interatomic distances.

We will return to a complete study of solidification later and will only give an overview of the role played by dynamics in modifying the properties of a solid here.

The microstructures formed during crystallization can be of several natures, as we have seen: dendrites, spherulites, lamellae. If very small microstructures are to be obtained, important nucleation which allows forming a large number of nuclei is required. Homogeneous nucleation (implying a solution free of impurities) favors obtaining small-scale microstructures. The nucleation rate is simultaneously controlled by supercooling and by diffusion in the initial liquid phase.

In general, the existence of small microstructures (microcrystals of small diameter) in a crystalline solid favors elevation of the elasticity limit (which indicates the limit of the forces that must act on the material to permanently deform it). These microstructures are an obstacle to plastic deformation of the solid phase. This phenomenon is observed in metals as well as in crystallized polymers. It is thus advantageous to favor nucleation of a large number of nuclei (and thus a high nucleation rate) and to prevent their growth. Ceramics also have high elastic limits: they are polycrystalline structures. The smaller the diameter of the microstructures in a ceramic, the higher the elastic limit.

An interesting perspective from this point of view is opened by what we call nanomaterials. These materials are microstructures on the nanometer scale. Here, too, we find that in solids with nanostructures, the mechanical properties are very different from those of materials with microstructures, and their ultimate tensile strength is much higher in particular. These materials can be obtained by rapid quenching, which prevents mass crystallization by passing into a glassy phase. We will specifically discuss nanomaterials in Chapter 9.

Problems

2.1. Spinodal Decomposition

Take the expression ΔF of the variation in free energy associated with concentration fluctuations δc in a binary mixture (2.61). Assume that δc obeys the equation:

$$\frac{\partial \delta c}{\partial t} = \tau \frac{\partial \Delta F}{\partial c}$$

τ is the relaxation time.

1. Determine the equation that the Fourier transform $\delta c(k, t)$ obeys.
2. Find a simple solution for this equation.
3. Use it to show that there are fluctuations which are no longer damped in the vicinity of the spinodal.

2.2. Soft Mode

Consider a displacive transition in a solid initiated by displacement q of an ion in a given direction. Let the potential be associated with this displacement:

$$V(q) = V_0 + \frac{1}{2} a\, q^2 + \frac{1}{4} b\, q^4$$

and we assume that $a = \alpha\, (T - T_c)$ and b is constant.

1. Write the equation for the movement of an ion of mass m taking into account its damping γ.
2. In the vicinity of the equilibrium position q_e, assume that $q = q_e + \delta q(t)$. Calculate q_e and write the equation that $\delta q(t)$ obeys.
3. Neglecting m, find the solutions for $\delta q(t)$. What happens when $T \to T_c$?

2.3. Heterogeneous Nucleation

Assume that the conditions are the same as in Fig. 2.4, where a solid spherical nucleus is formed on the surface of an impurity that plays the role of catalyst for nucleation. The solid/liquid, solid/catalyst, and liquid/catalyst surface tensions are designated by γ_{sl}, γ_{sc}, and γ_{lc} and the change in specific free enthalpy is designated by Δg_v.

1. Calculate the areas of the interfaces as a function of r and θ and the corresponding surface free enthalpy.
2. Knowing that the volume of the spherical nucleus is given by:

   $$V = \frac{4\pi}{3} r^3 \left[\frac{(2 + \cos\theta)(1 - \cos\theta)^2}{4}\right] = \frac{4\pi}{3} r^3 f(\theta)$$

   determine the total change in enthalpy associated with nucleation.
3. Determine the critical radius r^*.

2.4. Liquid/Vapor Transition and Interface
Consider the change in liquid/vapor equilibrium in the presence of a spherical interface of radius r.

1. Write the new liquid/gas equilibrium condition. Take the Laplace equation into account.
2. Determine the new pressure at equilibrium when the temperature is fixed and the new temperature at fixed pressure. One designates the liquid/gas surface tension by γ.

2.5. Homogeneous Nucleation
We want to determine the concentration of nuclei in the liquid phase of nickel near the transition temperature.

1. Calculate the concentration of solid nuclei (number of nuclei per cm^3) in the liquid at the solidification temperature T_m for the following radii: $r = 0.5$ nm, 0.7 nm, and 1 nm.
2. Calculate the radius of the nucleus corresponding to a concentration of 1 nucleus/cm^3.
3. Calculate the critical radius r^* for a supercooling of 10 K.
4. Determine the concentration of these nuclei. What conclusion can you draw? $T_m = 1725$ K, $L_f/T_m = 1.4$ J m^{-3} K^{-1}, $\gamma = 0.2$ J m^{-2}, $V_{mol} = 7$ cm^3/mole.

3 Phase Transitions in Liquids and Solids: Solidification and Melting

3.1 Ubiquitous Phenomena

Liquid–solid (solidification) and solid–liquid (melting) phase transitions are certainly among the most widespread in nature and many of them also have very important implications. For example, think of the formation of ice crystals due to solidification of liquid water and the inverse phase transition involved in meteorological phenomena, as well as melting or solidification of metals or metal alloys. We indicated in the preceding chapter that the properties of a solid (in particular, its mechanical and thermal properties) are very significantly a function of the kinetics of solidification and more specifically the size of the microstructures formed during the phase transition.

We should nevertheless emphasize that although melting and solidification are well-known transitions used for a very long time, we are still far from obtaining satisfactory theories for describing them and explaining the mechanisms involved. Although a simple theoretical model like the Lindemann model gives a relatively satisfactory empirical criterion for "predicting" melting of a solid, for example, it does not completely explain the phenomenon.

A dual approach must be used:

- *thermodynamic*: how can the liquid–solid transition and the changes in the properties of the system in the transition (density, specific heat, entropy, etc.) be classified? Knowing the intermolecular forces in the system, can this transition be predicted and its properties in the transition be characterized?
- *kinetic*: what are the microscopic mechanisms that can be used to explain transition phenomena in solids and liquids (melting in particular)? Can their dynamics be characterized?

The thermodynamic approach to solidification also leads to consideration of the situation where certain organic or inorganic materials in the liquid phase pass into a state of metastable equilibrium (supercooling), then solidify to form a glass. In this case, we have a glass transition. We will discuss this important transition, which has a specific character, in another chapter. The kinetics of the transition, particularly the liquid cooling rate, plays a specific role in formation of a glassy phase.

Finally, it is necessary to note that in the case of very special systems such as fullerenes (spherical organic molecules with the formula C_{60}), the very existence of a liquid phase of the material known in either solid form or gaseous form is questioned. Does the liquid phase of fullerenes really exist? This question is still open.

3.2 Characterization of the Phenomena

3.2.1 Thermodynamic Characterization

The different phases of a material are simply described with phase diagrams (Sect. 1.2.2): the equilibrium between a liquid and a solid is represented by a coexistence line in the (p, T)–plane corresponding to the equation:

$$\mu_S = \mu_L \tag{3.1}$$

where μ_S and μ_L are the chemical potentials of the solid and the liquid. This curve is the locus of the melting and solidification points of the material in the (p, T)–plane (Fig. 1.1 and Fig. 3.1).

According to the thermodynamic classification of phase transitions (Sect. 1.2.3), we can say that all liquid–solid transitions are **first order** because they are all associated with a discontinuity in entropy corresponding to a **latent heat**, which is also the change in enthalpy at the transition.

If L_f designates the heat of fusion ($L_f > 0$) at temperature T_m and ΔV is the change in volume in the transition, these quantities do obey the **Clapeyron equation**:

$$L_f = T_m \frac{\mathrm{d}p}{\mathrm{d}T} \Delta V \tag{3.2}$$

Some comments should be made concerning these quantities.

First, note that L_f and T_m vary significantly from one material to another (Table 3.1). L_f is of the order of a kJ/mole for molecular compounds such as O_2 and N_2 (0.444 kJ/mole for O_2 and 0.722 kJ/mole for N_2) and for materials whose atoms or molecular arrangements are spherical (rare gases such as A, Kr, etc., metals such as Na, Cs, etc., organic compounds such as CH_4, CCl_4, etc.). L_f is much higher for materials corresponding to heavier atoms such as iron, silver, or copper, or for germanium and silicon (31.8 kJ/mole for Ge, 46.4 kJ/mole for Si).

On the contrary, we empirically find that ratio L_f/RT_m (R is the ideal gas constant) is of the order of unity for most substances: it is equal to 0.98 for O_2, 1.25 for CH_4, 1.19 for CCl_4, 0.85 for Na, 1.15 for Cu, etc. This empirical rule is the **Richard rule**, which can be justified with statistical mechanics.

Finally, note that $\Delta V > 0$ for the overwhelming majority of solids: the density of the solid is greater than the density of the liquid for a given substance. Water is the most important exception to this rule (the density of ice

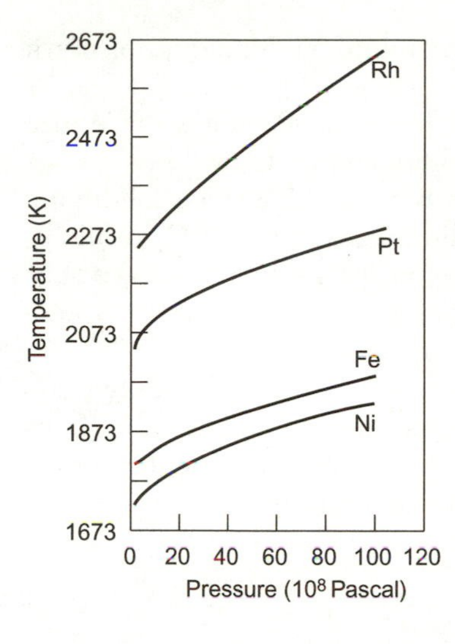

Fig. 3.1. Melting point curve of metals. The melting points of several metals as a function of the pressure are shown in this graph.

Table 3.1. Melting Points and Latent Heats of Fusion.
This table shows the latent heats L_f and melting points T_m of different solids. Note the low scattering of the values of ratio L_f/RT_m, but the high value for water, germanium, silicon, and ethane (Richard's law).

Solid	T_m (K)	L_f (kJ/mole)	L_f/RT_m
O_2	54.4	0.444	0.98
A	83.8	1.18	1.70
CH_4	90.7	0.944	1.25
C_2H_4	104	3.35	3.88
Kr	116	1.64	1.70
Hg	234	2.30	0.98
CCl_4	250	2.74	1.19
H_2O	273	6.00	2.64
Cs	302	2.14	0.85
Na	371	2.61	0.85
Li	454	3.02	0.80
Pb	601	4.86	0.97
Ge	1210	31.8	2.71
Ag	1233	11.3	1.10
Cu	1356	13	1.15
Si	1683	46.4	2.76
Ni	1728	17.6	1.23

is less than the density of the liquid), and germanium, gallium, and bismuth are three others.

The change of ΔV or the density in the liquid/solid transition is also highly dependent on the materials. Liquid sodium contracts by 2% when solidifying ($T_m = 98°C$), while molten NaCl contracts by 25% when solidifying ($T_m = 801°C$). Gallium expands by 3% in solidifying ($T_m = 30°C$), while liquid water expands by 8% in solidifying into ice at 0°C. Finally, we note that 3.2 implies that $\Delta V < 0$ for substances, such as water for example, where $(\mathrm{d}p/\mathrm{d}T) < 0$.

The melting point T_m is a function of the pressure. The American physicist P. W. Bridgman was the first to systematically study the variation in T_m as a function of the pressure at the beginning of the 20th century. Empirical equations correlating T_m with p were obtained, for example:

$$p = a\left[\left(\frac{T_m}{T_t}\right)^c - 1\right] \tag{3.3}$$

where T_t is the temperature of the triple point, T_m is the melting point at pressure p, and a and c are two parameters. This relation was obtained by F. E. Simon and G. Glotzel in 1929; it was verified for high temperatures ($c = 1.4$ for iron, $c = 2.7$ for NaCl, and $c = 9$, with $a < 0$ for ice I).

All of these general comments (presented in Table 3.1) suggest that melting and solidification are "nonuniversal" in a way. With the exception of the empirical and approximate Richard rule, no characteristics common to liquid/solid transitions are found for any organic and mineral substances. It is thus difficult to develop theoretical models with a wide range of validity which can predict the behavior of systems in the transition.

3.2.2 Microscopic Approach

Crystallization of a liquid causes the appearance of a periodic structure within a disordered ensemble of atoms or molecules constituting the liquid phase below its solidification point (we will treat the particular case of glassy phases in another chapter). To use general terminology, we say that crystallization is a transition accompanied by "**symmetry breaking**".

Of course, there is a large variety of crystal lattices characterized by their symmetry, just as a very wide distribution has been found in the values of the thermodynamic quantities characterizing the transition (crystallization temperature and latent heat, for example). This suggests that all of these properties are a function of the nature of the intermolecular forces between the atoms or molecules of the material.

For example, it is possible to predict solidification of systems such as rare gases with simulation methods (molecular dynamics or Monte Carlo). The Lennard–Jones pair potential $V(r)$ representing the interactions between atoms of a rare gas is written:

$$V(r) = 4\epsilon\left[\left(\frac{r_0}{r}\right)^{12} - \left(\frac{r_0}{r}\right)^{6}\right] \tag{3.4}$$

where ε is the potential well depth and r_0 is the interatomic distance for which the potential vanishes; we can then show that the liquid will crystallize in a lattice with face-centered cubic (f.c.c.) symmetry.

In fact, the theoretical models that can be used are very sensitive to the form of the intermolecular potential selected (for example, 3.4): a slight change in the form of this potential can result in a change in symmetry in the crystal lattice predicted by the model. This is particularly the case in metals, whose solid phases have very different symmetries: f.c.c., body-centered cubic (b.c.c.), hexagonal close packed (h.c.p.), etc. Steel, for example, (iron–carbon alloy, (Fig. 1.8)) solidifies at atmospheric pressure and 1598°C in the b.c.c. form, then at 1394°C it passes into a phase of f.c.c. symmetry to return at 912°C to the stable b.c.c. form up to ambient temperature. These two types of crystal structures encountered for the same substance must correspond to forms and values of the free energy that are very close. In the case of metals, the situation is more complex than for materials whose atoms or molecules have a simple spherical shape since the interatomic forces are a function of the volumes occupied by the atoms.

In fact, repulsive interatomic forces play a very important role in solidification (they correspond to the positive term in 3.4). In general, attractive interatomic (or intermolecular) forces (corresponding to the negative term in 3.4) are a very weak function of the orientation and shape of the atoms, but this is not the case for the repulsive forces.

In a liquid phase, the possibilities of arrangement of the atoms and molecules, which are dependent on their shape and orientation, will thus determine the characteristics of the solid phase to a great degree (density and symmetry in particular). They will also influence the value of the heat of solidification L_f and the crystallization temperature T_m. The thermal characteristics of transition points (L_f, T_m) are strongly influenced by the geometry of the atoms or molecules and thus by the interatomic or intermolecular repulsive forces.

These qualitative considerations are supported by some experimental findings. The latent heats of solidification (or fusion) are generally low and less than the latent heats of vaporization: 91.2 kJ/mole for the heat of vaporization and 2.6 kJ/mole for the heat of fusion of sodium; 74 kJ/mole and 1.2 kJ/mole, respectively, for the latent heats of vaporization and solidification of argon, etc. Melting only implies very few "bond" breaks between atoms or molecules and is governed by atomic or molecular rearrangements with rupture of the crystal lattice (or its formation for solidification). If we compare the boiling and melting points of the hydrocarbon series, we find that the boiling points change little when a single C–C bond is replaced by a C=C double bond, or when substitutions are made in the chain (replacing a hydrogen by a methyl, for example). On the contrary, the melting points decrease when these operations are performed. This shows that the attractive forces are altered very little by these substitutions, while the repulsive forces, which are a function of the geometry of the molecules, are significantly altered (for example, the melting point is −95°C for *n*-hexane, C_6H_{14}, while the melting point for 1-hexene, C_6H_{12} decreases to −140°C). Note that the shape and orientation of some molecules can strongly influence the attractive part of their interaction potential and thus determine their properties in the liquid and solid state. This has been verified for covalent molecules such as HCl, HF, etc., and for liquids like water, where the molecules are engaged in hydrogen bonds. For water, the difference between the heat of vaporization and solidification (40 kJ/mole and 6 kJ/mole, respectively) is much less pronounced than for other substances such as argon or sodium, whose atoms are spherical.

To conclude, we can assume that the microscopic approach to the liquid–solid transition based on interaction potentials ((3.4) for example) should account for the phenomena, but it is not possible to find a "universal" form of potentials which would allow predicting the liquid–solid phase transitions for all categories of substances. We will return in this chapter to the methods that allow determining the melting curve of a solid in particular.

3.2.3 Delays in the Transition: Supercooling–Superheating

The description of the states of a body with phase diagrams (Sect. 1.2.2) allows determining the coexistence curves of the solid and liquid phases of a substance which are in stable equilibrium. When one of these lines, such as the melting line, is crossed, there is a phase change. Figure 1.4 shows that the situation can be complicated. It is the projection of the surfaces representing the free enthalpy G of the different phases of a material that can be in two solid forms S_1 and S_2 in the (p, T)-plane (Fig. 1.3).

Beginning with the material in the liquid phase, when its temperature is decreased at constant pressure, reaching point A on the melting curve, the liquid phase should solidify in form S_2 in "normal" conditions. In many systems, however, solidification is not observed at point A and cooling of the liquid phase can be observed up to point B, where it crystallizes in form S_2: we say that the liquid was **supercooled** and was in a state of **metastable equilibrium**. There can also be the more complicated situation where the solid crystallizes at point A in form S_1, which is metastable in these conditions; it then becomes stable at point D.

As we reported, there are also situations for some materials in which the liquid does not solidify in the form of a crystal, but in the form of a glass in passing into the supercooled phase: this is the **glass transition**, which has a very specific character.

The inverse of supercooling can also be encountered: beginning with a material in crystalline form and heating it (segment BA in the diagram in Fig. 1.4), it does not melt at the temperature corresponding to its melting point:

the solid is superheated and is in a metastable thermodynamic state. A solid can achieve superheating (thus preventing melting), but this situation is much rarer than the supercooling observed in a very large number of liquids.

We know how to produce supercooling in numerous liquids; moreover, as shown in Chap. 2, most liquids do not crystallize in the absence of impurities if they enter the supercooled state. The supercooling limit is thus established by homogeneous nucleation of the liquid. The presence of impurities in the liquid phase, which is difficult to avoid, is also another limitation on supercooling since impurities are privileged sites for nucleation of the solid phase.

By analyzing the classic phenomenon of homogeneous nucleation (Sect. 2.2.4) and in particular those based on considerations concerning the nucleation rate, Turnbull showed that the absolute supercooling temperature limit was $T_l = 0.18\,T_m$, where T_m is the melting point of the solid. Turnbull also proposed "short-circuiting" the heterogeneous nucleation induced by the impurities present in the liquid by utilizing finely divided liquids in the form of droplets (typically less than 10 μ in diameter). These droplets can also be coated in a surface layer of another material by immersing them, for example, in an emulsion, which prevents their coalescence; they can also be placed on a substrate. Very important supercooling can be obtained in this way, several hundred K, in numerous molten materials with low melting points (Ga, Hg, Sn, etc.) and high melting points (Fe, Ni, Co, ...). It was possible to exceed the Turnbull limit ($T_l = 0.18T_m$) in small particles of alloys (aluminum–tin, aluminum–silicon) embedded in a solid matrix.

More recently, the studies conducted in microgravity revealed new technical possibilities for studying the limits of supercooling in droplets in the absence of any contact with a surface. Supercooling experiments were particularly conducted in towers in which a high vacuum was created and the drops were left to fall in free fall. An experiment of this type was conducted in Grenoble (B. Vinet et al.) in a free-fall tower (48 m high with a high vacuum, where the pressure inside was $4 \cdot 10^{-10}$ mbar) with drops of tungsten (T_m = 3695 K) and rhenium (T_m = 3459 K). The supercooling obtained with these two metals was 530 K for W and 975 K for Re; it was thus very important. These experiments were conducted with drops 4 mm in diameter and their temperature was measured during the fall by a pyrometric method.

Supercooling has been observed in a wide temperature range in numerous organic and inorganic materials. For example, water could be observed in the supercooled liquid phase in capillary tubes up to approximately 250 K and in small droplets up to 200 K (the appearance of crystallites was followed in these droplets by electron diffraction).

Although supercooling can be observed in many liquids by taking experimental precautions, the lag in melting of a solid, superheating, is on the contrary much more difficult to observe. Superheating of metals and organic and inorganic compounds is nevertheless possible, but in a narrow temperature range; solid helium could thus be brought to the superheated state up to a temperature of 2.3 K above its melting point ($T_m = 2$ K at 30 atm). We will return to the problem of superheating of solids later in this chapter.

We note that melting or crystallization of a material in "normal" conditions of thermodynamic equilibrium between phases is characterized by the relation: $\Delta G = 0$.

For $T > T_m$, the solid phase will be in a metastable state corresponding to superheating. For $T < T_m$, the liquid phase will also be in a metastable state; it is supercooled.

When the metastability limit corresponding to a point on the **spinodal** is reached, the supercooled liquid "should" crystallize and the solid should melt. In certain liquids, the situation is much more complicated due to the existence of a **glass transition**.

A simple criterion can be found for determining this metastability limit:

- the supercooling limit of a liquid is attained when the entropy of the liquid is equal to the entropy of the crystal: below this temperature, we will have the "paradoxical" situation where an ordered system (a crystal) has a higher entropy than the entropy of the same disordered system (a liquid);
- the superheating limit of a crystal is reached when the entropy of the liquid is equal to the entropy of the crystal: above this temperature, there will also be a "paradoxical" situation equivalent to the preceding one.

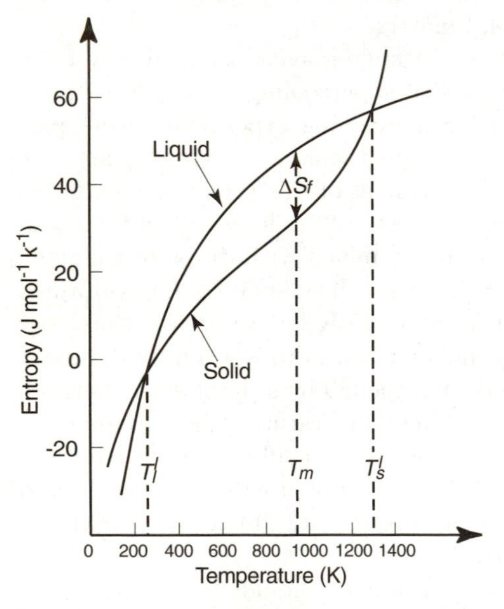

Fig. 3.2. Entropy of liquid and crystallized aluminum. The variation in the entropy is shown as a function of the temperature in the regions of stable and metastable equilibrium. ΔS_f is the entropy of fusion; T_l^l and T_S^l are the limit temperatures of stability (metastability) of liquid and solid (H. J. Fecht and W. L. Johnson, *Nature*, **334**, 50 (1988), copyright MacMillan Magazines Limited).

The limiting superheating and supercooling temperatures T_S^l and T_l^l can be calculated by calculating the corresponding quantities ΔS. This implies that the specific heats of the crystal and liquid are known. An entropy–temperature curve for aluminum which can be plotted from specific heat data is shown in Fig. 3.2. The existing data were extrapolated to the superheated zone and the existence of holes in the crystal lattice were taken into consideration in the calculation. This diagram shows that aluminum could be superheated to a limiting temperature of $T_S^l = 1.38\ T_m$ or 1292 K (in the absence of hole defects, much higher superheating will be obtained with $T_S^1 = 3\ T_m$). Liquid aluminum can be supercooled to $T_l^l = 0.24\ T_m$ or 225 K.

3.2.4 Methods of Observation and Measurement

The solid–liquid transition is a first order transition associated with latent heat. **Differential thermal analysis** (DTA) is a classic experimental technique for determining latent heat and thus the associated phase transition.

The principle of DTA is relatively simple: the temperature of the sample investigated (solid or liquid) is compared with the temperature of another reference sample which does not undergo a phase change in the temperature zone studied and the variation when the temperature of the two materials is raised at a constant rate is recorded (Fig. 3.3). If T_S is the temperature of the material investigated and T_r is the temperature of the reference material and if they are all varied at a constant rate by heating, $T_S - T_r = C^{\text{te}}$. If the material studied undergoes a transition (melting, for example), a temperature peak will be observed and because melting is endothermic, the heat delivered by the heating device is absorbed by this material to melt it at constant temperature while T_r continues to increase. $T_S - T_r$ is detected

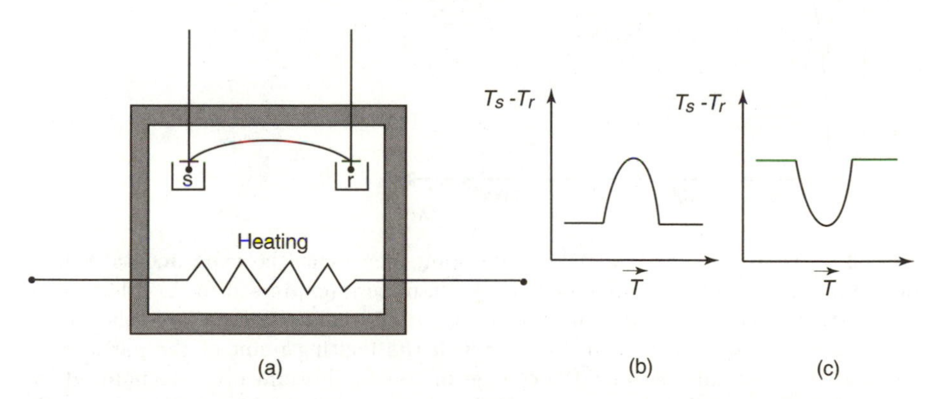

Fig. 3.3. Differential thermal analysis. (**a**) $(T_S - T_r)$ measured as a function of T_S (substance investigated) and T_r (temperature of reference substance) or the furnace temperature. The temperature is varied linearly by heating; (**b**) temperature peak due to an exothermic phenomenon; (**c**) temperature peak due to an endothermic phenomenon.

with a differential method using two oppositely mounted thermocouples. In the case of solidification, since the transformation is exothermic, a temperature peak is observed, and because the material studied solidifies at constant temperature, it stops cooling.

It is not always possible to utilize a method like DTA, which implies direct measurement of the temperature of two samples. A pyrometric method can also be used to measure the temperature of the material investigated by observing the evolution of its brightness in time when it is cooled (or heated). This is the method used to demonstrate supercooling in droplets of molten metals (the brightness is measured with photodiodes).

Melting of a solid phase can also be observed in micro- or nanoparticles (diameter of the order of 100 nm) using "indirect " methods such as electron and X-ray diffraction and electron or infrared microscopy. With this type of technique, the disappearance of the solid phase (and thus melting) is determined by observing the change in the intensity of the spectrum or spectral line in a diffraction experiment. Electron and X-ray diffraction experiments have been conducted on metallic droplets of Pb, Sn, Ag, and In and on bismuth. An experiment of this type conducted on gold droplets up to 20 nm in diameter revealed an event that preceded the bulk-melting (they melted at 600 K, while the melting point of gold is 1337 K (Fig. 3.4). Melting is detected by a change in the intensity of the electron diffraction spectrum.

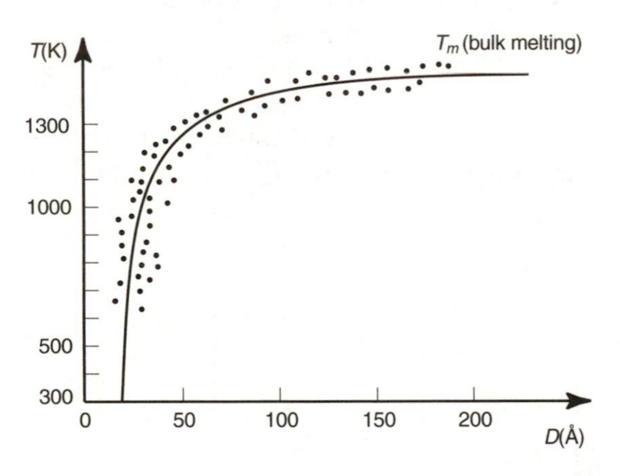

Fig. 3.4. Melting of gold particles. The points represent the experimental values of the melting points for different particle diameters (expressed in Å). Melting of the particles is detected with an electron diffraction experiment. The solid curve is derived from a theoretical model in which the melting point of the particles is calculated taking into account the change in the equilibrium pressure induced by the surface tension (Ph. Buffat and J. P. Borel, *Phys. Rev.*, **A13**, 2287–2297 (1976), copyright Am. Inst. Phys.).

More recent experiments were conducted on CdS nanomaterials between 2.4 and 7.6 nm in diameter: the melting point of CdS in a large volume is 1678 K and the melting point of the smallest particles can decrease to 600 K.

More remarkably, one has shown that the melting point of small aggregates of sodium is very strongly dependent on their size. The melting point of these aggregates (containing 70–200 atoms) can be decreased by almost one-third (120 K) with respect to the melting point of sodium in bulk. Calorimetric measurement showed that the melting point varies periodically with the number of atoms in the aggregates, and this could be due to their greater stability when the electrons form a complete electron layer for certain geometric shapes (icosahedrons and decahedrons, for example). The interest of these observations is far from negligible for applications of nanomaterials when there must not be too great a decrease in the melting point.

Optical methods have also been used to demonstrate solid–liquid transitions. Silver aggregates (approximately 100 atoms) can be doped with molecules of benzene and their UV absorption spectrum can be studied; the spectrum changes on melting of the aggregate, and the melting point can be measured in this way.

For IR-transparent materials, the phase change can be detected by following the evolution of a line in the Raman absorption spectrum. Experiments of this type were conducted at very high pressure (several Mbar) on molecular solids in particles by H. K. Mao. Diamond anvil cells are used for this purpose (Fig. 3.5). Diamond is a material that simultaneously allows transmitting high pressure on the system studied and observing the phase transition by Raman effect. The transformations of solid hydrogen and helium and the transformations of ice have been investigated with this method.

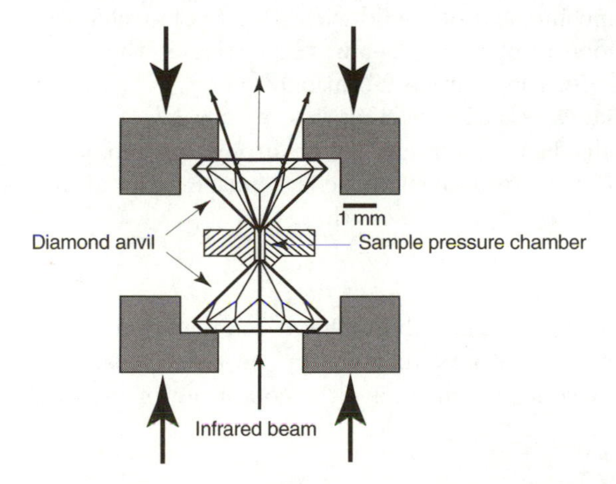

Fig. 3.5. Diagram of a diamond anvil cell. Diamonds allow exercising high pressure (up to approximately 2.5 Mbar) on the material studied and performing optical measurements, by Raman scattering, for example (Laboratory of Physics of Condensed Matter, Université Pierre et Marie Curie, CNRS).

3.3 Melting

In thermodynamic equilibrium, the melting and crystallization points and pressures of a substance are rigorously identical. This would suggest a priori that the theoretical models accounting for these phenomena and used to predict them should both apply to the liquid and solid phases. In reality, the situation is far from being this simple, if only because of intervention of the nucleation mechanism, which plays a particularly important role in the liquid–solid transition close to the crystallization conditions. Moreover, there is currently no satisfactory theory for predicting melting and crystallization phenomena in the most general manner.

3.3.1 The Lindemann Model

The first studies of melting, at least from the beginning of the 20th century, were specially undertaken to find a simple physical criterion for determining the condition of loss of stability and thus melting of a solid. F. A. Lindemann proposed a simple model of melting in 1910 which led to such a criterion. The Lindemann model is based on the empirical finding that when the temperature of a solid is raised, the amplitude of vibration of its atoms around their equilibrium position increases with the temperature. Lindemann hypothesized that when this amplitude of atomic motion attained a certain fraction of the interatomic distance in the lattice, the solid would melt. This model is obviously very simple if not simplistic, and it would seem to apply to monoatomic solids such as metals.

We can begin with the description of a solid using the Einstein model, which allows calculating the specific heat of a solid with statistical mechanics. If the atoms vibrate independently of each other in the lattice at the same frequency ν, assuming $kT > h\nu$, the average vibrational energy $< \varepsilon_x >$ in direction x is given by the Planck equation leading to $< \varepsilon_x >= kT$.

If we designate the amplitude of vibration of the atoms around their equilibrium position by A and the spring constant characterizing the atoms in the harmonic oscillator model by K_ν, we know that:

$$< \epsilon_x >= K_\nu A^2 \tag{3.5}$$

If we assume with Lindemann that the solid melts at temperature T_m so that $A = fa_0$, where a_0 is the interatomic distance in the lattice and f is a constant characteristic of the solid, we then have the condition for melting of the solid:

$$kT_m = K_\nu f^2 a_0^2 \tag{3.6}$$

If Θ_E is the Einstein temperature of the solid, m is the mass of the atoms, and knowing that $h\nu = k\Theta_E$ and $\nu = 1/2\pi(2K_\nu/m)^{1/2}$, the condition given by 3.6 is rewritten as:

$$f^2 = 2\left(\frac{h}{2\pi}\right)^2 \frac{T_m}{k\Theta_E^2 m a_0^2} \tag{3.7}$$

Lindemann hypothesized that factor f should be constant for all solids having the same crystal structure. The data for monoatomic crystalline solids of f.c.c. symmetry are reported in Table 3.2; we find that f is almost constant and an average value of 0.07 can be taken for f.

Table 3.2. Lindemann Parameter.
The interatomic distance, melting T_m, and Einstein Θ_E points for monocrystalline solids are reported. We find that $f \sim 0.07$.

Element	a_0 (10^{-10} m)	T_m (K)	Θ_E(K)	$f 10^{-2}$
Ne	3.16	24	50	6.8
Ar	3.76	84	60	6.3
Xe	4.34	161	40	6.3
Pb	3.50	600	58	8.2
Al	2.86	933	326	6.2
Cu	2.56	1356	240	7.4
Pt	2.77	2044	149	7.7

The Lindemann criterion has the advantage of simplicity and seems to hold for monoatomic crystalline systems. However, it does not provide any fundamental explanation of melting: why does a solid structure suddenly lose its stability at temperature T_m, the melting point? This phenomenon is almost "instantaneous" since melting often occurs over a relatively narrow temperature range in the vicinity of T_m. The Lindemann model obviously simplifies the description of the physical state of a solid significantly by assuming that each atom has the same energy kT_m in melting.

It is nevertheless necessary to emphasize that despite its simple character, the Lindemann criterion is useful because it allows testing modern theories of melting, none of which are satisfactory. M. Born proposed another simple criterion of a similar nature in 1939: melting would be initiated when the shear modulus of the solid is reduced to zero. The solid would then be incapable of withstanding the infinitesimal shear forces which would impose on it important strains: the solid would "flow" and would thus no longer be stable. In fact, abnormal behavior of the moduli of elasticity of a solid in the immediate vicinity of melting has never been experimentally demonstrated. This behavior could be imagined if melting were a second order phase transition, which is not the case. The Born model does not seem to be pertinent, although recent simulations using molecular dynamics methods reveal that the elastic shear modulus decreases rapidly near T_m.

3.3.2 The Role of Defects

Other models have been proposed in attempting to explain melting. Some of them are based on the hypothesis that defects in the crystal structure are in a way nuclei or centers of initiation of melting of the solid. One of these models, due to N. F. Mott and J. K. Mackenzie and proposed in 1957, it assigns a privileged role to **dislocations** in the solid (defects in the periodic alignment of the atoms or molecules in the crystal). According to this model, the dislocation density increases with the temperature and beyond a certain temperature (corresponding to T_m), this density surpasses a critical threshold which breaks the stability of the crystal and thus causes it to melt. This model does not allow predicting T_m with precision, and its predictive value is thus limited.

Another more satisfactory and older theoretical model was proposed by J. Frenkel; it explains melting of a crystal by the existence of holes in the crystal structure. Beyond a critical concentration of holes in its structure, the solid loses its stability and melts. This model is coherent with the Frenkel theory of liquids: it assimilates a liquid with a quasi-crystal lattice constituted of cells that can be occupied by atoms or molecules but where a very large number of cells remain vacant and the atoms move from one cell to another.

The Frenkel and Mott models for melting were recently generalized by H. F. Fecht. He assumes that point defects (holes, dislocations, etc.) are an integral part of the crystal structure and that they destabilize it when the temperature increases. These defects can be introduced by mechanical stresses or by irradiation.

The thermodynamics of phase changes allows simply to demonstrate the role of defects. Continuing the calculation begun in Sect. 2.2.3 (2.20 and seq.), we can show that ΔG, the difference in free enthalpy between the liquid and ideal solid with no defects, is written in good approximation in pure metals as:

$$\Delta G = 7\Delta S_f \frac{\Delta T\, T}{T_m + 6T} \tag{3.8}$$

and in metal alloy liquids which can form glasses as:

$$\Delta G = 2\Delta S_f \frac{\Delta T\, T}{T_m + T} \tag{3.9}$$

where ΔS_f is the entropy of fusion (in general, $\Delta S_f = 1.1k$ per atom for pure metals and $1.5k$ for bimetallic compounds). $\Delta T = T_m - T$ is the degree of liquid supercooling and T_m is the melting point.

If c is the concentration of defects in the solid, the variation ΔG^V of the free enthalpy of the solid associated with these defects is then given by the expression:

$$\Delta G^V = c(\Delta H^V - T\, \Delta S^V) + kT[c \ln c + (1-c)\ln(1-c)] \tag{3.10}$$

where the last term in this expression is the entropy of mixing; ΔH^V and ΔS^V are the changes of enthalpy and entropy associated with the defects.

If the defects are holes, in general $\Delta H^V = 9.28\, kT_m$ per atom or molecule and $\Delta S^V = 2k$.

The changes in ΔG and ΔG^V for different concentrations as a function of T/T_m for a metal alloy that can form a glass in the solid phase are shown in Fig. 3.6. The solid-line curve of $\Delta G(T)$ corresponds to the transition of the supercooled liquid with the solid (at $T = T_m$, $\Delta G = 0$ by definition). The thermodynamic condition for melting of a crystal with holes is no longer written as $\Delta G = 0$ but as:

$$\Delta G^* = \Delta G - \Delta G^V = 0 \tag{3.11}$$

The melting point for each concentration is obtained directly in the diagram: it corresponds to the point of intersection of $\Delta G(T)$ with each $\Delta G^V(c)$ curve. This diagram immediately shows that the melting point is significantly decreased by the presence of point defects in the crystal lattice.

This conclusion is obtained by observing the effects of irradiation (with γ rays in particular) on metals: it decreases the melting point by creating point defects in the crystal structure.

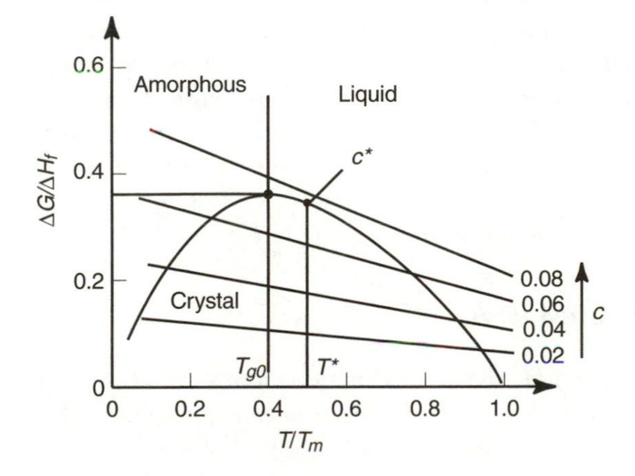

Fig. 3.6. Melting of a solid with defects. The solid curve shows the change in free enthalpy on melting of an ideal solid and the lines show the variations in free enthalpy corresponding to the formation of defects in the solid crystal for different concentrations c. The apex of the curve is the supercooling limit of the liquid. The superheating limit of the solid is at T^*. The ideal crystalline solid was used as reference ($G = 0$). T_m is the melting point in the absence of defects. Below T_{g0} the liquid transforms into an amorphous state which keeps S and H values at T_{g0}; ΔG remains thus constant and is represented by an horizontal line. (H. J. Fecht, Nature, **356**, 133–135 (1992), copyright Macmillan Magazines Limited).

The differences in entropy ΔS and ΔS^* between liquid and crystal on one hand and between liquid and crystal with defects on the other can be calculated with thermodynamic equations:

$$\Delta S = -\frac{\partial \Delta G}{\partial T}; \quad \Delta S^* = -\frac{\partial \Delta G^*}{\partial T} \tag{3.12}$$

The supercooling limit of the liquid is reached when its entropy becomes equal to the entropy of the ideal solid, thus for $\Delta S = 0$. The limiting temperature T_{g0} corresponds to the top of curve ΔG where the tangent is horizontal in Fig. 3.6. This point is the glass transition limit. For $T < T_{g0}$, the liquid will have a lower entropy than the solid. This situation has no physical meaning: this is the **Kauzmann paradox**. To avoid it, the supercooled liquid "has to" solidify by going into the glassy state.

If we now consider the inverse situation, that of a solid with defects, the stability limit of the solid (in fact, the superheating limit) will be reached when the conditions $\Delta G^* = 0$ and $\Delta S^* = 0$ are simultaneously satisfied. In the diagram in Fig. 3.6, this situation corresponds to the temperature T^* and the concentration of defects c^* for which curves ΔG^V and ΔG are tangential. Solids corresponding to $c > c^*$ are no longer stable.

The point with coordinates T^* and c^* correspond to a phase transition where $\Delta G^* = 0$ and $\Delta S^* = 0$ and thus $\Delta H^* = 0$. This would be a transition with no latent heat and thus the equivalent of a second order phase change. The diagram in Fig. 3.6 corresponds to a relatively high critical concentration $c^* = 0.077$ which is difficult to attain experimentally, even by irradiation. For a "normal" solid, instead $c \approx 10^{-5}$ in the vicinity of T_m. This model relies on the a priori assumption that the concentration c of defects in the crystal is frozen in, thus creating a metastable thermodynamic state which can be compared to the liquid state at the same temperature.

3.3.3 Melting and Surface of Materials

Neither the Lindemann criterion nor the preceding thermodynamic model involving the density of defects in the crystal lattice actually explain why a solid suddenly loses its stability and melts, passing into the liquid state. It would seem that all of these models are compatible with the fact that the surface of the solid must play an essential role in triggering melting. In effect, experiments on melting of small particles (particularly those conducted by Ph. Buffat and J. P. Borel on gold particles) all show a very important decrease in their melting point (it can be reduced by approximately 30%). This phenomenon can be simply interpreted by the fact that the amplitude of vibration of the atoms on the surface of a solid is 1.5–2 times higher than in its bulk, and the Lindemann criterion for melting of the solid is *ipso facto* satisfied more rapidly.

An increase in the concentration of defects (holes, for example) in a solid creates a surface (the defects can also be concentrated on the surface of the material) which favors its destabilization.

A series of experimental results tends to support the hypothesis that the surface of the solid plays an essential role in melting. Small particles of silver coated with a layer of gold could be superheated to a temperature of 24 K above their melting point. Solid argon incorporated in an aluminum matrix was also brought to 730 K in the state of superheating (which corresponds to superheating of approximately 650 K); in this case, the persistence of the solid phase is followed using electron diffraction spectrum. Transient superheating of aluminum could even be conducted at approximately 1000 K, but for a time less than 10^{-9} sec. This experiment was conducted on solid aluminum films heated by laser pulses; the transition was followed by picosecond electron diffraction. In this case, superheating corresponds to a transient state rather than a metastable state.

These experiments show on one hand that relatively important superheating can be obtained and on the other hand that the surface plays an important role in the mechanism that results in melting of a solid. Contrary to the situation encountered in small particles, where an "advance" in melting was observed, vibrations of the atoms on the surface of a solid incorporated in a solid matrix are restricted and their amplitude is smaller than in the totally free solid. The Lindemann stability criterion is thus only satisfied at a higher temperature: this leads to observation of superheating over a relatively important range. We can also estimate the mean square of the amplitudes of vibration of atoms from the intensity of the vibrational spectra as well as the corresponding temperature. The temperature is higher for a material incorporated in a matrix than in the free state. We immediately deduce from (3.7) that T_m is higher and superheating can be observed.

The observation of melting on the surface of a lead crystal suggests the intervention of surface effects as factors that trigger melting. In an ionic diffusion experiment on the surface of lead, melting of the metal begins 40 K below T_m: the surface begins to melt and the thickness of the molten layer increases when approaching T_m and reaches some 20 layers at 1 K below T_m. This all happens as if melting was a quasicontinuous phenomenon on the surface and in nearby layers.

The current experimental data and the Lindemann criterion itself thus support the hypothesis that melting is triggered by instability of vibrations on the surface of a solid or at the solid–solid interface.

3.4 Solidification

As for the melting of a solid, there is also no satisfactory theoretical approach for solidification of a liquid which would explain the mechanisms involved and predict the physical conditions (pressure and temperature) of the phase transition, the phase diagrams for a very large number of materials, and the variation of state variables such as the density in melting.

Crystallization of a liquid is a more complex phenomenon especially since it is manifested by **symmetry breaking**: a crystalline solid formed from a liquid has a certain symmetry (body-centered cubic, face-centered cubic, hexagonal, etc.), while an isotropic liquid is invariant by translation. The simplest crystals can be modeled by comparing the atoms of which they consist with **hard-spheres** and by assuming that these spheres were closely packed to constitute the crystal lattice. The "holes" between the contact points of the spheres (Fig. 3.7) are the sites for positioning a second layer of atoms and so forth. There are many compact crystal structures constituted in this way in solids. Thus, 25% of the elements crystallize in f.c.c. form (the most compact) and 20% crystallize in closed packed hexagonal form. Note that only in 1998 was it possible to show, by computer calculation, that the packing of hard-spheres corresponding to f.c.c. symmetry allowed obtaining the average density by filling 74% of the space. Kepler guessed this in 1611.

In the case of metals, which are relatively simple elements, there is a large variety of crystalline forms and it is difficult to predict the type of crystal that will be formed from the liquid phase based on considerations taking into account solely the size or mass of the atoms alone.

Iron, as we have seen, crystallizes at atmospheric pressure first in the body-centered cubic (b.c.c.) form, then it undergoes a transition to a f.c.c. phase, and finally it returns to the b.c.c. phase at lower temperature. The free energies corresponding to these different structures are very close to each other and the situation is even more complicated for molecular materials, organic materials in particular. It is also necessary to take into consideration the nucleation phenomenon that triggers solidification in a thermodynamically unstable state.

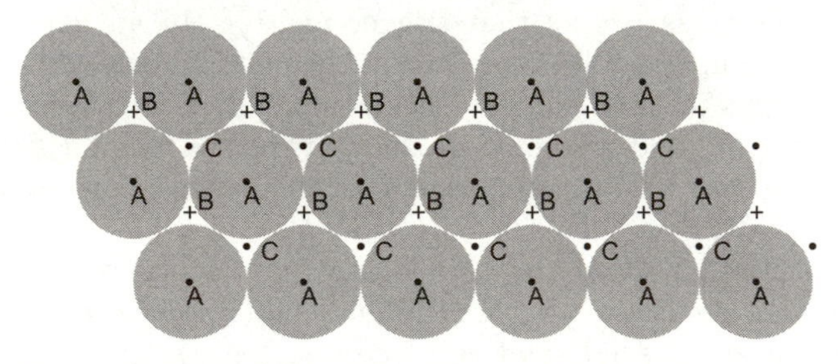

Fig. 3.7. Compact stacking of layers of spheres. A solid phase is constituted in this way. If the second layer is centered on B, the third layer can be centered on A or C; in the first case, there will be compact hexagonal symmetry and in the second case, a f.c.c. structure.

3.4.1 Theoretical Approach to Crystallization with Intermolecular Potentials

The progress made with molecular dynamics and Monte Carlo methods applied to liquids since the beginning of the seventies allowed approaching the liquid–solid transition on this basis. Most of the theoretical treatments are based on calculating the free energy of the system, where the stable phase is the phase with the lowest free energy. At the phase transition, potentials F of the two phases should be equal.

It is of course necessary to introduce an intermolecular potential to calculate this free energy. If Φ is the total potential energy of N atoms or molecules, it will be a function of all their coordinates $\boldsymbol{r}$. If these particles are considered **hard-spheres**, a simple form of the binary interaction potential $u_2(r)$ which prohibits all penetration of the spheres with each other can be selected: $u_2(r) = 0$ if $r > r_0$, $u_2(r) = \infty$ if $r < r_0$ (Fig. 3.8).

A potential of the Lennard–Jones type given by 3.4 can also be taken.

The most recent theoretical approaches for accounting for solidification consist of performing a perturbation calculation with a liquid phase of fixed density. The average density of a solid which is not locally uniform is treated as if it corresponded to the average density of a homogeneous liquid. The free energy of the system is then considered as a function of the density, which is itself a function of $\boldsymbol{r}$: we say that it is a density functional. This method is called the **density functional** approach in statistical mechanics; it was initially introduced by Kohn and Hohenberg to describe the behavior of electrons in a crystal.

The free energies of the liquid and solid phases are calculated. The thermodynamic conditions that permit equality of these potentials determine the solidification point. It is also necessary to introduce the autocorrelation functions of the phase density and the associated S structure functions in the calculation (Sect. 1.3.3). These developments were initially due to Kirkwood, then Haymet, Oxtoby, Ramakrishnan, and Yussouf. We will only give a broad outline of the model here.

For a system of N classical particles (atoms or molecules) of mass m in volume V, the Hamiltonian H_N representing the energy is written:

$$H_N = \sum_{i=1}^{N} \frac{p_i^2}{2m} + V(\boldsymbol{r}_1 \ldots \boldsymbol{r}_N) + \int \rho_m(\boldsymbol{r}) U(\boldsymbol{r}) \mathrm{d}\boldsymbol{r} \tag{3.13}$$

where $\boldsymbol{r_i}$ and $\boldsymbol{p_i}$ are the position and time coordinates of the particles, $U(\boldsymbol{r})$ is an external potential, and

$$\rho_m = \sum_{i=1}^{N} \delta(\boldsymbol{r} - \boldsymbol{r_i}) \tag{3.14}$$

is the macroscopic density.

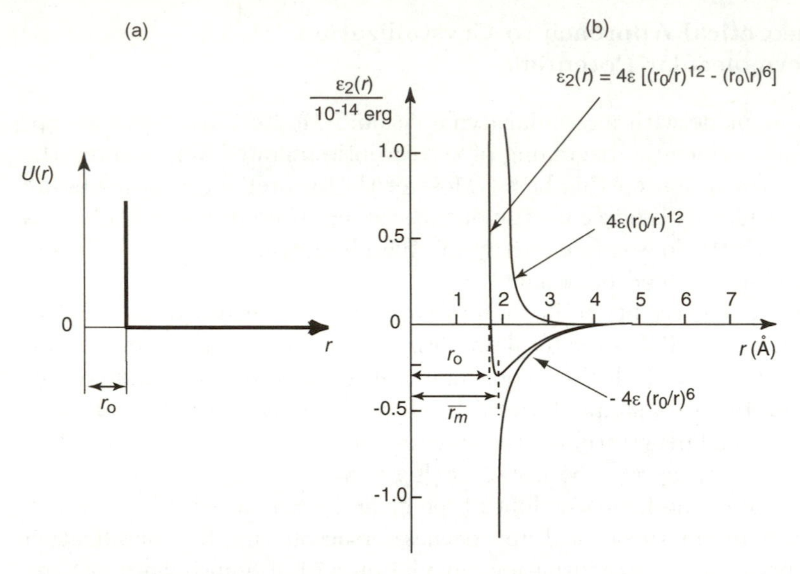

Fig. 3.8. Intermolecular potentials. (**a**) Hard-spheres. For $r < r_0$, the interaction potential between two molecules is infinite: they repel each other. For $r > r_0$, the potential is zero; (**b**) Lennard–Jones potential given by (3.4).

The grand canonical distribution and the associated grand partition function Ξ are introduced. The grand potential J of statistical mechanics is:

$$J = -kT \ln \Xi \tag{3.15}$$

with

$$\Xi = \sum_{N=0}^{\infty} \frac{1}{N!} \int \exp[-\beta(H_N - \mu N)] \prod_{i=1}^{N} \frac{\mathrm{d}\boldsymbol{r_i}\mathrm{d}\boldsymbol{p_i}}{h^3} \tag{3.16}$$

These quantities correspond to those of thermodynamics:

$$J = -pV \ ; \ \frac{pV}{kT} = \ln \Xi = -W \tag{3.17}$$

We will define the density of the inhomogeneous system (solid or pseudoliquid) $\rho(\boldsymbol{r})$ as the average value of $\rho_m(\boldsymbol{r})$ calculated with the grand canonical ensemble, that is

$$\rho(\boldsymbol{r}) = < \rho_m(\boldsymbol{r}) > \tag{3.18}$$

We can then show by calculating the ensemble average that $\rho(\boldsymbol{r})$ is in the form:

$$\rho(\boldsymbol{r}, \mu, T) = \frac{e^{\beta\mu}}{\Lambda^3} \exp C(\boldsymbol{r}, \mu, T) \tag{3.19}$$

μ is the chemical potential and $\Lambda = h/(2\pi mkT)^{1/2}$ is the de Broglie thermal wavelength.

In (3.19), the term $e^{\beta\mu}/\Lambda^3$ corresponds to the approximation of an ideal fluid with no interactions between particles (ideal gas), the term $\exp C(\boldsymbol{r}, \mu, T)$ takes into account all interactions between $2, 3, \ldots n$ particles. $C(\boldsymbol{r}, \mu, T)$ is dimensionless here (it is in kT units).

If μ_L, μ_S and T_L, T_S respectively designate the chemical potentials and temperatures of the liquid and solid phases, we can then write:

$$\rho_L(\mu_L, T_L) = \frac{e^{\beta\mu_L}}{\Lambda^3} \exp C_L(\mu_L, T_L) \tag{3.20}$$

$$\rho_S(\mu_S, T_S) = \frac{e^{\beta\mu_S}}{\Lambda^3} \exp C(\boldsymbol{r}, \mu_S, T_S)$$

since C_L corresponds to a homogeneous system, the liquid, it is not a function of $\boldsymbol{r}$.

$$\ln \frac{\rho_S}{\rho_L} = \beta(\mu_S - \mu_L) + C(\boldsymbol{r}, \mu_S, T_S) - C_L(\mu_L, T_L) \tag{3.21}$$

We necessarily have $\mu_L = \mu_S$ and $T_L = T_S = T$ on the liquid/solid coexistence curve, thus:

$$\ln \frac{\rho_S}{\rho_L}(\boldsymbol{r}, \mu, T) = C(\boldsymbol{r}, \mu, T) - C_L(\mu, T) \tag{3.22}$$

$C(\boldsymbol{r}, \mu, T)$ is a function of the density, and we can thus expand it around its value C_L corresponding to a homogeneous liquid of density ρ_0. It is then written:

$$C(\boldsymbol{r}_1) = C_L + \int C^{(2)}(\boldsymbol{r}_1, \boldsymbol{r}_2)[\rho(\boldsymbol{r}_2) - \rho_0]\mathrm{d}\boldsymbol{r}_2$$
$$+\frac{1}{2}\int C^{(3)}(\boldsymbol{r}_1, \boldsymbol{r}_2, \boldsymbol{r}_3)[\rho(\boldsymbol{r}_2) - \rho_0][\rho(\boldsymbol{r}_3) - \rho_0]\mathrm{d}\boldsymbol{r}_2\mathrm{d}\boldsymbol{r}_3 \tag{3.23}$$

$C^{(2)}(\boldsymbol{r}_1, \boldsymbol{r}_2)$ is the direct correlation function of the liquid. For a homogeneous and isotropic liquid, it is invariant by translation and we can write:

$$C^{(2)}(\boldsymbol{r}_1, \boldsymbol{r}_2) = C^{(2)}(|\boldsymbol{r}_1 - \boldsymbol{r}_2|)$$

$C^{(3)}(\boldsymbol{r}_1, \boldsymbol{r}_2, \boldsymbol{r}_3)$ is the three–particle direct correlation function. The terms $C^{(2)}(r)$, C^3, and so on take into account all interactions between particles and are a function of the structure of the liquid phase.

In the first order, we can write:

$$\ln\left[\frac{\rho(\boldsymbol{r}_1)}{\rho_0}\right] = \int C^{(2)}(|\boldsymbol{r}_1 - \boldsymbol{r}_2|)\ [\rho(\boldsymbol{r}_2) - \rho_0]\mathrm{d}\boldsymbol{r}_2 \tag{3.24}$$

where $\rho_L = \rho_0$ is the density of the homogeneous liquid.

A key result of the theory of the liquid state shows that the Fourier transform

$$c(\boldsymbol{k}) = \rho \int \mathrm{e}^{\mathrm{i}\boldsymbol{kr}} C^{(2)}(\boldsymbol{r}) \mathrm{d}\boldsymbol{r}$$

is directly correlated with the structure function $S(\boldsymbol{k})$ by the **Ornstein–Zernike** equation (Appendix B).

$$c(\boldsymbol{k}) = 1 - \frac{1}{S(\boldsymbol{k})} \tag{3.25}$$

We can obtain $S(\boldsymbol{k})$ and thus $c(\boldsymbol{k})$ with X-ray and neutron–diffraction experiments or with simulation calculations.

Introducing function W given by (3.17) correlated with grand potential J, the different ΔW in the thermodynamic potentials per particle for nonuniform systems W (solid equivalent to a pseudoliquid) on one hand and uniform W_1 on the other hand is written as a limited expansion:

$$\begin{aligned} \Delta W &= W - W_1 = \int [\rho(\boldsymbol{r}_1) - \rho_0] \mathrm{d}\boldsymbol{r}_1 \\ &+ \tfrac{1}{2} \int C^{(2)}(\boldsymbol{r}_1, \boldsymbol{r}_2)[\rho(\boldsymbol{r}_2) - \rho_0][\rho(\boldsymbol{r}_1) + \rho_0] \mathrm{d}\boldsymbol{r}_1 \mathrm{d}\boldsymbol{r}_2 \end{aligned} \tag{3.26}$$

The third-order terms in $C^{(3)}$ were neglected in this expression.

If we assume that the nonuniform phase corresponds to the solid, the density $\rho(\boldsymbol{r})$ will be a function of $\boldsymbol{r}$ and a Fourier expansion can be performed as:

$$\rho(\boldsymbol{r}) = \rho_0 \left[1 + \eta + \sum_n \mu_n \mathrm{e}^{\mathrm{i}\boldsymbol{k}_n \boldsymbol{r}}\right] \tag{3.27}$$

where the wave vectors $\{\boldsymbol{k}_n\}$ correspond to the reciprocal crystal lattice associated with the solid, μ_n are the Fourier components and $\eta = (\rho_S - \rho_0)/\rho_0$ was introduced, where ρ_S is the density of the solid phase and η is the relative variation in density at the liquid–solid transition.

There is a specific crystal structure for each series of vectors $\{\boldsymbol{k}_n\}$. The most stable form corresponds to the minimum of the thermodynamic potentials (for example, $\boldsymbol{k}_n$ will be selected for a f.c.c ., b.c.c. crystal, etc.).

Using (3.18) and (3.24) and knowing that $\rho_0 = \exp(\beta\mu + C_L) / \Lambda^3$ for the homogeneous liquid, we obtain straight away:

$$\rho(\boldsymbol{r}_1) = \rho_0 \mathrm{e}^{c_0 \eta} \exp\left[\sum_n \mu_n c_n \mathrm{e}^{\mathrm{i}\boldsymbol{k}_n \boldsymbol{r}_1}\right] \tag{3.28}$$

where $c_n = \rho_0 \int C(\boldsymbol{r}) \mathrm{e}^{\mathrm{i}\boldsymbol{k}_n \boldsymbol{r}} \mathrm{d}\boldsymbol{r} = c(\boldsymbol{k}_n)$ and $c_0 = \rho \int C(\boldsymbol{r}) \mathrm{d}\boldsymbol{r}$, that is:

$$\ln\left[\frac{\rho(\boldsymbol{r}_1)}{\rho_0}\right] = c_0 \eta + \sum_n c_n \mu_n \mathrm{e}^{\mathrm{i}\boldsymbol{k}_n \boldsymbol{r}_1} \tag{3.29}$$

We also have:

$$\Delta W = (c_0 - 1)\eta + \frac{1}{2}c_0\eta^2 + \frac{1}{2}\sum_n c_n\mu_n^2 \tag{3.30}$$

The third-order contribution in $C^{(3)}$ was neglected.

We can show that $c_0 - 1 = -\rho_0\, kT\, \kappa_T^{-1} < 0$. Note that the values of $\eta = 0$ and $\mu_n = 0$ are always solutions of equations, particularly of (3.30), and they correspond to the liquid state.

In the liquid–solid transition, we should have $\Delta W = 0$; if $\Delta W < 0$, the solid is then the stable phase in applying the Gibbs–Duhem criterion.

To calculate the parameters of the solid phase (η, μ_n) (3.28) must be solved self-consistently.

The equation (3.29) can be rewritten on one hand by integrating the two terms and on the other hand by multiplying each side by $\exp(\mathrm{i}\boldsymbol{k}_j\boldsymbol{r})$ and integrating them, which gives:

$$1 + \eta = \mathrm{e}^{c_0\eta}V^{-1}\int \exp\Big[\sum_n c_n\mu_n \exp(\mathrm{i}\,\boldsymbol{k}_n\boldsymbol{r})\Big]\mathrm{d}\boldsymbol{r}$$

$$\mu_j = \mathrm{e}^{c_0\eta}V^{-1}\int \exp(\mathrm{i}\,\boldsymbol{k}_j\boldsymbol{r})\exp\Big[\sum_n c_n\mu_n \exp(\mathrm{i}\,\boldsymbol{k}_n\boldsymbol{r})\Big]\mathrm{d}\boldsymbol{r} \tag{3.31}$$

To be rounded off these calculation requires that $C(\boldsymbol{r})$ be determined by defining the intermolecular potential form. These calculations can be performed using the interaction potential $V(\boldsymbol{r}_1, \boldsymbol{r}_2, \boldsymbol{r}_3, \ldots \boldsymbol{r}_n) = 1/2\sum_{i,j} u(\boldsymbol{r}_{ij})$ where the binary potentials $u(\boldsymbol{r}_{ij})$ can be potentials of the hard-sphere or Lennard–Jones type.

One may determine $C(\boldsymbol{r})$ can also be determined from the experimentally found structure function $S(\boldsymbol{k})$ (3.25), using X-ray or neutron diffraction, for example (Fig. 3.9).

Alternatively $C(\boldsymbol{r})$ can be calculated using the **Percus–Yevick** method (Appendix B). For a hard-sphere potential, we have:

$$-C(r) = \lambda_1 + 6\xi\lambda_2 r + \frac{1}{2}\xi\lambda_2 r^3 \tag{3.32}$$

with $\lambda_1 = (1+2\xi)(1-4\xi)^4$, $\lambda_2 = -\left(1+\frac{1}{2}\xi\right)^2(1-\xi)^{-4}$ and $\xi = \frac{\pi}{6}\rho r_0^3$. We find c_0 and c_n by Fourier transformation. We also obtain the equation of state

$$p\rho\beta^{-1} = \frac{1+\xi+\xi^2}{(1-\xi)^3} \tag{3.33}$$

On this basis, A. D. J. Haymet determined the parameters for solidification of a hard-sphere liquid in the form of a crystal of (f.c.c.) symmetry. To do this, it is necessary to solve (3.30) and (3.31) with $\Delta W = 0$ for the transition and knowing $C(r)$ and thus c_n via (3.32). We can proceed by iteration, giving a value for the density ρ_0 of the liquid and an in initial set for $(\boldsymbol{k}_n)$ and (μ_n). A good choice for $\boldsymbol{k}_n$ is to take the value of k corresponding to

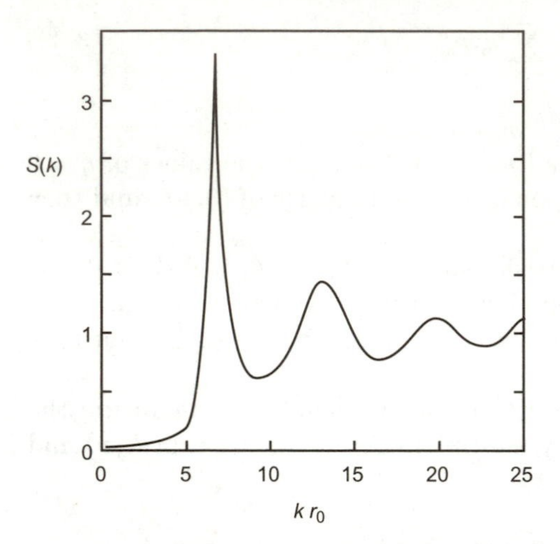

Fig. 3.9. Structure function for a liquid. Structure function $S(k)$ for a liquid calculated with a simulation method using a hard-sphere potential ($u(r) = 0$ for $r > r_0$) is shown. $S(k)$ can be obtained with a X-ray or neutron diffraction experiments (A. D. J. Haymet, J. Chem. Phys., **78**, 4644 (1983), copyright Am. Inst. Phys.).

the wave vector for which structure function $S(k)$ has a maximum peak. The lattice constant a is selected so that the density of the solid $\rho_S = 4a^3$ (for a fcc lattice) agrees with the predicted value of $\rho_S = \rho_0(1+\eta)$.

The results of this calculation are shown in Fig. 3.10a. The solid–liquid transition at equilibrium intervenes when $\Delta W = 0$, which corresponds to $\rho = 0.976$ in the example selected (in r_0^3 units) with a 6% density discontinuity at the transition. A solution is no longer found for the equations (that is, the values of μ_n) for $\rho_0 r_0^3 < 0.93$, which would correspond to the metastability limit of the solid.

This model can be used by taking a Lennard–Jones intermolecular potential. In this case, the reduced variables $T^* = kT/\varepsilon$ and $\rho^* = \rho r_0^3$ are introduced to describe the system. The diagram obtained is shown in Fig. 3.10b.

If we compare this result with the measurements on argon ($r_0 = 0.34$ nm and $\varepsilon/k = 119$ K), we find that the relative predicted density variation in the transition η is too small: it varies from 13 to 11%, while it is 18% in practice.

The Lindemann parameter (3.7) introduced in the model to study melting can be calculated.

By definition:

$$f^2 = a_0^{-2} \int r^2 \rho(r) \mathrm{d}r \tag{3.34}$$

With a hard-sphere potential, we find $0.059 < f < 0.065$, values satisfactorily close to Lindemann's theoretical result of $f = 0.07$.

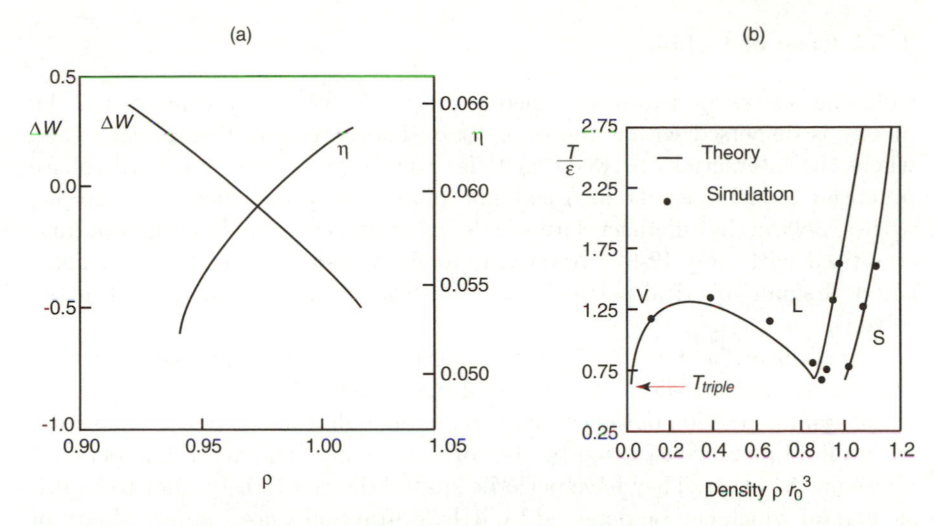

Fig. 3.10. Liquid–solid transition. (**a**) Result of the calculation applying the density functional model to a system of molecules with a hard-sphere potential; η gives the density discontinuity in the liquid–solid transition (in ρr_0^3 units) [A. D. J. Haymet, J. Chem. Phys., **78**, 4646 (1983)]. (**b**) Result of the calculation for a Lennard–Jones intermolecular potential. The points in the diagram represent the results of a simulation calculation (W. A. Curtin and N. W. Aschcroft, Phys. Rev. Lett., **56**, 2777 (1986), copyright Am. Inst. Phys.).

On the other hand, for a Lennard–Jones potential, the values are higher: $0.075 < f < 0.009$. On the whole, the agreement with the Lindemann model for melting is satisfactory.

All of these models based on perturbation calculations by selecting special forms of binary interaction potentials have the merit of predicting certain types of liquid–solid transition. However, they all failed until recently in attempting to predict solidification in the form of a crystal of b.c.c. symmetry. They have the drawback of giving no information on the type of mechanism that triggers crystallization, but they satisfactorily show that two phenomena are competing. We see in (3.30) that when ΔW is to be decreased from a liquid state (the solid state corresponds to $\Delta W < 0$), we can operate on the first two terms, which are negative ($c_0 < 0$ and $\eta > 0$) because they tend to decrease W when the temperature decreases; on the contrary, the third term is unfavorable to solidification if $c_n > 0$. Solidification indeed seems to represent competition between two phenomena: – contraction of the liquid – formation of a periodic structure that triggers the solid phase. Once more, it would seem that water is a special case ($\eta < 0$) which must be treated with another model.

3.4.2 Case of Colloids

Colloidal phases are two-phases media (cf., e.g., Sect. 1.4.3) where one of the phases is dispersed within the other one. The preceding theoretical model where the interactions between particles were represented by a hard-sphere potential describes a colloid. A colloidal phase can be obtained, for example, with a polymethyl methacrylate emulsion produced by polymerization and stabilized with poly-12-hydroxystearic acid. A colloidal emulsion is constituted of small spherical particles less than a micron in size dispersed in the solvent.

Crystallization of a colloidal phase can be observed if its polydispersity is low (the sphere size distribution should not exceed 10% of the average size). If the volume fraction of the colloid is large enough, there is phase separation with a fluid phase constituted by the solvent on one hand and a microcrystal phase on the other. They have periodic spatial distributions similar to a classic crystal which can be detected by light diffraction since the periodicity of the crystal is of the same order of magnitude as the wavelength of visible light. This particularly explains the iridescence observed in a suspension of colloidal silica and in opal crystals (a variety of silica).

Colloidal particles are micro- or nano-sized objects and a form of the interaction potential between the particles that accounts for the physical forces existing in the medium must be selected. For the colloidal system to be transformed into a stable crystal, it is necessary for the electric charges carried by the surface of the particles to be low so that the long-range Coulomb forces are also weak. If this is not the case, the Coulomb repulsive forces between the electric charges will block the appearance of crystalline order in the structure. It has been shown that spherical particles can undergo a disorder/order transition in a fluid for a volume fraction of $\phi \approx 0.50$. This transition was predicted by simulation calculations using a hard-sphere potential: this is the **Kirkwood–Alder transition**. The particles are ordered in space within a fluid (their spacing is of the order of their size). This transition has been observed in polystyrene particles 720 nm in diameter as well as in silica microbeads on which a fluorescent molecule was grafted. Colloidal phases are true models for studying the formation of crystals and glasses.

The simplest potential for describing these interaction forces is:

$$V(r) = \frac{V_0}{r} \exp(-Kr) \tag{3.35}$$

This is the Yukawa potential with screening constant K^{-1} which gives a good description of a charge-stabilized colloidal system. However, it loses its validity if the electric charges are too small or if K^{-1} is too low because particle size effects become very important.

The phase diagrams can be determined with this form of $V(r)$ either by molecular dynamics or by calculation of the free energies of each phase. In this way, it is possible to show that the fluid phase crystallizes in the form of a solid with b.c.c. symmetry if the charge is high or the screening length is large. On the contrary, if K^{-1} is small, a crystal of f.c.c. symmetry is formed; in this case, the potential is long-range and equivalent to a hard-sphere potential.

Charge-stabilized colloidal crystals have very low particle densities and their mechanical stability is also very low: their shear moduli are much lower than those of normal crystals. These colloidal solids can easily be destroyed by simple mechanical stirring and they then undergo a transition to a metastable liquid phase.

These crystallization-destabilization phenomena in colloidal systems can be observed by light scattering. The characteristic crystallization times of charge-stabilized colloidal crystals are of the order of 1 to 10 sec for b.c.c. crystals; on the contrary, they are several weeks for solids of f.c.c. symmetry.

Crystal-like structures can also be obtained in plasma in which colloidal particles constituted of melamine and formaldehyde beads 7 μm in diameter have been incorporated. These **plasma crystals** are a genuine experimental model for studying the solid–liquid transition.

3.4.3 Crystallization and Melting of Polymers

Polymers are giant molecules, i.e. macromolecular, formed of several hundred individual units called monomers. A polymer is in the form of a **linear** or **branched** chain.

Polyethylene, $-\ (CH_2)_n\ -$, is a typical example of a linear polymer. Branched polyethylene, where the polymer chain contains side groups (several groups per hundred carbon atoms in the linear chain), can also be obtained by chemical synthesis. Synthetic materials or natural products (such as proteins, for example), the applications of polymers have been significantly evolving for more than 50 years now.

The existence of crystalline phases in polymers has been observed for a long time. A plastic milk bottle (made of polyethylene) is opaque or translucent while some plastic glasses (made of polystyrene) are transparent. This difference is related to the presence of crystallinity in polyethylene, while polystyrene is not crystalline but amorphous. The size of the crystals formed is a function of the chemical nature, molecular conformation, and microstructures formed in the polymer. In fact, polymers differ from the usual crystal structures due to the semicrystalline character of their solid phases; they are partially in the crystalline state and partially in the amorphous state.

The cooling rate of a molten polymer is an important factor that determines the nature of the solid phase formed: amorphous or crystalline. When the polymer passes from the liquid state to the amorphous state, it undergoes a **glass transition** which occurs at the **glass transition temperature** T_g.

Crystallization of a polymer like polystyrene is a two-stage process:

- first the $(CH_2)_n$ chains all take on a zigzag planar conformation (the C–C bonds all form the same angle of 112°)

- second, the polymer chains are rearranged to form a set of periodically aligned parallel rods (Fig. 3.11). The structure of the crystal can be more complex; for example, this is the case when the carbon chain has a helical conformation and the helices are arranged parallel to each other to form a three-dimensional crystal (a situation encountered with polytetrafluoroethylene $(CF_2)_n$).

In fact, in a semicrystalline polymer, the crystalline and amorphous phases are interpenetrating: the same polymer chain which can contain several thousand carbon atoms that simultaneously belong to amorphous and crystalline regions (Fig. 3.11).

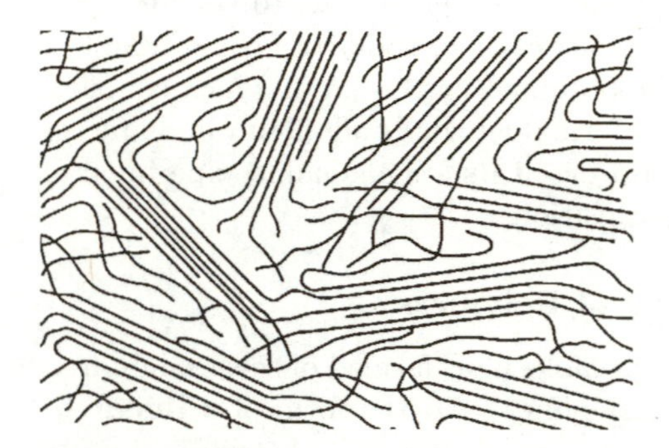

Fig. 3.11. Structure of a solid polymer. The solid consists of a mixture of crystalline and amorphous phases. The polymer chains are aligned parallel to each other in the crystalline phases and are randomly structured in the amorphous phases. The same polymer chain can belong to a crystalline phase and an amorphous phase, thus ensuring cohesion of the solid. Overall, it is a semicrystalline state.

As we previously indicated (Sect. 2.2.5 and Fig. 2.9), a polymer crystallizes from the molten phase from spherical nuclei, **spherulites**.

These structures can be revealed by optical microscopy by observing an unoriented plastic film (polyethylene or polystyrene, for example) between crossed polarizers and analyzers (Fig. 3.12). The diameter of the spherulites, which can vary from one material to another, is of the order of magnitude of the wavelength of visible light. On observation with the electron microscope, the spherulites appear to be constituted of lamellae arranged around a central nucleus. Nucleation of the polymer is generally heterogeneous since it is produced from the impurities present in the molten polymer.

The degree of crystallinity (that is, the concentration of crystallites in the solid phase) is of great practical importance since certain properties (mechanical and optical in particular) are strongly dependent on it. It can be measured by determining the specific phase volume or by X-ray diffraction.

A polymer generally crystallizes in two stages, as in a classic solidification process: – nucleation of crystal nuclei – growth of nuclei. They are generally

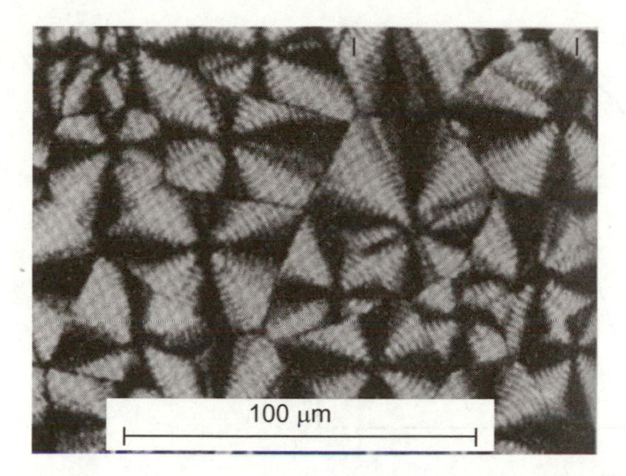

Fig. 3.12. Spherulites in a polymer. These spherulites, which are microcrystalline structures in a polymer film, can be observed by optical microscopy between crossed polarizer and analyzer (scale of 100 μ; L. Monnerie, in *La Juste Argile*, Editions de Physique (1995), p. 228).

spherulites of radius r which is a linear function of time: $\mathrm{d}r/\mathrm{d}t = G$ (2.51). If W_l is the polymer mass remaining in the liquid state at time t and W_0 is its initial mass; then we can show with the **Avrami model** that in general:

$$\frac{W_l}{W_0} = \exp(-Kt^n) \tag{3.36}$$

where n is the **Avrami exponent** (see 2.58).

Equation 3.36 allows us to follow the crystallization of a polymer by measuring the change in its volume in time. The Avrami exponent n can be found by determining the slope of the expression log (W_l/W_0) as a function of $\log t$. We see in Fig. 3.13 that in the case of crystallization of polypropylene, its behavior is given by an Avrami equation, at least in the initial stage, with $2.2 < n < 3$. As we noted previously, this exponent corresponds to nucleation of spherulites with a diffusion mode ($n = 5/2$).

A model like the Avrami model obviously gives no information on the molecular processes involved in the crystallization of a polymer. Any theory of the crystallization of polymers must explain a certain number of phenomena:

- crystals are generally lamellar
- polymer chains are often folded
- the crystal growth rates are a function of the crystallization temperature and molecular weight of the polymers.

The best formulated kinetic theory is the theory of Hoffman and Lauritzen; it is an extension of the Volmer model for nucleation. It departs from the hypothesis of evolution of the process in two steps: nucleation and growth. The crystal grows from the pre-existing surface of the nuclei (spherulites) on

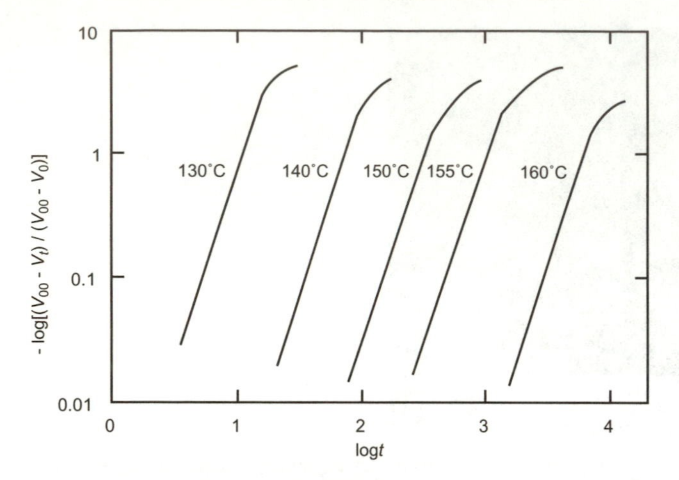

Fig. 3.13. Avrami exponent for crystallization of a polymer. Volumes V_0, V_t, V_∞ of the polymer, polypropylene, measured in the initial (liquid) state at time t (solid and liquid) and $t = \infty$ when it is totally crystallized. The solid phase is formed of spherulites whose nucleation obeys the Avrami law with $2.2 < n < 3$ [Parrini and Carrieri, Makromol. Chem., **62**, 83 (1963)].

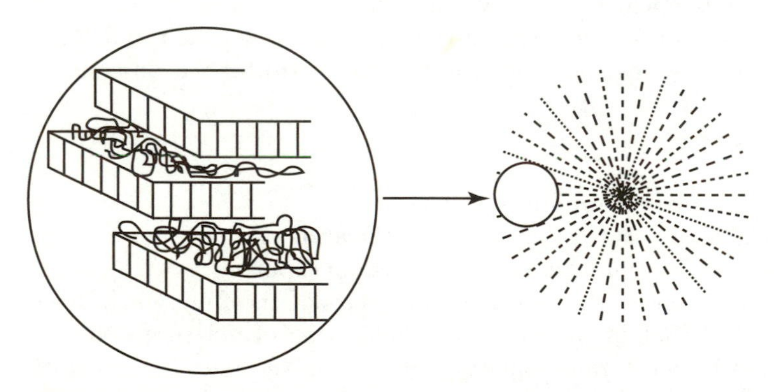

Fig. 3.14. Polymer spherulite. It is formed by interpenetrating organized lamellae and amorphous zones. The chains are folded inside the lamellae (Fig. 3.15) perpendicular to the radial direction of the spherulites.

which the polymer chains aggregate in successive layers (Fig. 3.14). The flexible chains can fold back on themselves; they are positioned perpendicular to the axis of the spherulites.

If Δg_V is the change in free enthalpy per unit of volume due to nucleation of the initial crystal and γ_e and γ_S are respectively the surface energies of the folded polymer chains and of each lamella deposited on a spherulite (Fig. 3.15), we can write the total variation in free enthalpy:

$$\Delta G_n = 2bl\gamma_s + 2nab\gamma_e + n\,abl\Delta g_V \tag{3.37}$$

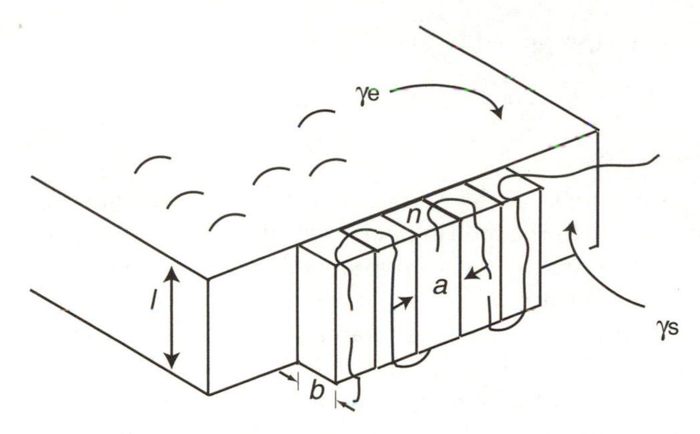

Fig. 3.15. Lamellar growth of a polymer. The polymer chains are deposited on the surface of a spherulite, forming successive lamellar layers [R. J. Young and P. A. Lovell, *Introduction to Polymers*, Chapman and Hall (1995)].

Here n chains are attached to each other over length l , where each strand has a surface cross section ab was postulated.

The expression calculated in Sec. 2.2.3 can be used again for Δg_V. If T_m^0 designates the melting point of a polymer crystal in the absence of surface, T being the observed fusion temperature, and L_f is the heat of fusion and introducing $\Delta T = T_m^0 - T$, $\Delta g_V = -L_f \Delta T / T_m^0$, (3.37) is rewritten as:

$$\Delta G_n = 2bl\gamma_s + 2nab\gamma_e - n\,ablL_f \frac{\Delta T}{T_m^0} \tag{3.38}$$

where n is generally large, hence the first term in (3.38) can be neglected and we then obtain a "critical" value l_0 for which $\Delta G_n = 0$; the crystal is then stable for $l = l_0$:

$$l_0 = \frac{2\gamma_e T_m^0}{L_f \Delta T} \tag{3.39}$$

The thickness of the crystalline lamellae of the polymer is inversely proportional to ΔT, which has been verified experimentally.

Melting of a polymer differs very clearly from melting of other crystals: a single melting point cannot be defined for the same material, as melting takes place over a relatively wide temperature range; melting is a function of the thermal history of the material and particularly the crystallization temperature. These observations particularly complicate the selection of the melting temperature T_m^0 in the preceding crystallization model. In fact, T_m^0 can be estimated by extrapolation by successively measuring the melting points of a crystal solidified at different temperatures and thus at different cooling rates. T_m corresponds to the melting point of an infinitely slowly solidified crystal.

Using a model similar to the preceding one, we can show (Problem 3.4) that:

$$T_m = T_m^0 - 2\gamma_e \frac{T_m^0}{lL_f} \tag{3.40}$$

Since $L_f > 0$, the melting point of the crystal will always be less than T_m^0 and the difference with respect to the melting point T_m^0 of an ideal crystal, in the absence of surface, (for $l = \infty$), is inversely proportional to its thickness, which has been verified experimentally.

In addition to the thermal history of the material, many other factors affect melting of a polymer. Structural factors in particular play an important role. Branching of the polymer chain tends to decrease the melting point T_m: in the case of linear polyethylene, $T_m = 138°C$, while $T_m = 115°C$ for branched polyethylene. These side branches play a role similar to the presence of impurities or defects in the crystal structure, which tend to decrease the melting point. Moreover, relatively important differences are observed in the behavior in melting of a crystallized polymer: – linear polyethylene undergoes neat melting: 70% of the crystallinity disappears over a temperature interval of 3–4°C – branched polyethylene melts over a much wider temperature range: 60% of the crystalline state disappears over an interval of approximately 40°C. Branched polyethylene undergoes a premelting phenomenon 20–30°C below its melting point, which is obviously a serious handicap for the technical use of this material.

Finally, it is necessary to note that some polymers can crystallize under the effect of stresses. Natural rubber only crystallizes slowly at ambient temperature ($T_m \approx 35°C$), but crystallization is very rapid if it is strechted. In this case, the polymer chains tend to align in the direction of the force applied, and this alignment favors growth of crystallites since it tends to reduce the variation in enthalpy in crystallization.

3.5 Crystallization, Melting, and Interface

A liquid–solid phase transition evolves in two stages, as we showed (Sects. 2.2.3 and 2.2.4):

- formation of nuclei of the new phase (nucleation);
- growth of these nuclei.

They require the surface of the nuclei of the new phase: it is necessary to create an interface in order to make the nuclei form, and they grow by addition of atoms or molecules on the interface between the mother phase and the new phase. The example of polymers shows that this process can be complex. The growth rate of the nuclei (crystallites, for example) is a function of the probability that the atoms or molecules will be able to bind on an interface, and the nucleation rate I is given by (2.33). Interfacial phenomena thus play an essential role in crystallization and melting.

3.5.1 Surface Melting

It has been observed for a very long time that liquid films can exist on the surface of a solid at temperatures much lower than its melting point (the first observations were made by Tammann in 1910). These phenomena have been found in a very large variety of materials (rare gases in the solid state, metals and semiconductors, molecular solids) and for temperatures $T \approx 0.9\,T_m$, where T_m is the melting point of the bulk solid. It is also assumed that the existence of a very thin liquid film on the surface of ice makes this solid almost self-lubricated and thus favors sliding of skis and skates. Melting of grain boundaries at the interface between two solids is a similar phenomenon.

Premelting of a solid can be observed by proton backscattering, X-ray, electron, and neutron diffraction experiments. Experiments of this type have been conducted on lead and more recently on ice down to −183°C; they showed that a liquid film of water persists on the surface of ice up to this temperature.

From the thermodynamic point of view, one notes that if the interface between a solid and its vapor is wet by the liquid phase following premelting, the free enthalpy of this interfacial layer will be lower than in the absence of liquid. Thus it tends to remain in thermodynamic equilibrium up to a temperature below the melting point of the solid. We can say that the free enthalpy of the solid can be reduced by "converting" a solid layer into a liquid film. The thickness of the liquid film that allows this new state of equilibrium is the one that minimizes the free enthalpy of the system. The free enthalpy G_f per unit of area for the interface of a solid wet by a liquid film of thickness d is written:

$$G_f(T,p,d) = \rho_l \mu_l(T,p)d + \gamma(d) \tag{3.41}$$

μ_l and ρ_l are the chemical potential and density of the liquid as a whole (neglecting the surface) and $\gamma(d)$ is the surface tension.

For $d = 0$, $\gamma = \gamma_{SV}$ (there is no longer any liquid) and for $d = \infty$, $\gamma = \gamma_{LV} + \gamma_{LS}$ (the solid is totally covered by the liquid). γ_{LS}, γ_{LV}, and γ_{SV} are the liquid/solid, liquid/vapor, and solid/vapor surface tensions. Introducing $\Delta\gamma = \gamma_{LV} + \gamma_{LS} - \gamma_{SV}$, we can represent the interface situation by the equation:

$$\gamma(d) = \Delta\gamma f(d) + \gamma_{SV} \tag{3.42}$$

where the function $f(d)$ characterizing the interfacial energies varies from 0 to 1 when $0 < d < \infty$.

At thermodynamic equilibrium, the chemical potentials μ_f and μ_S of the molten layer and of the solid should be equal :

$$\mu_f(T,p,d) = \frac{1}{\rho_l}\left(\frac{\partial \Delta G_j}{\partial d}\right)_{T,p} = \mu_l(T,p) + \left(\frac{\Delta\gamma}{\rho_l}\right)\left(\frac{\partial f}{\partial d}\right) = \mu_S(T,p) \tag{3.43}$$

(μ_f is obtained by differentiating G_f with respect to $\rho_l d$).

The classic result is obtained: the surface term shifts the liquid–solid equilibrium curve. If (p_0, T_0) designates the normal equilibrium conditions of the liquid and solid phases in the absence of a surface, we can then write an expansion limited to the first order for $\Delta\mu = \mu_S - \mu_l$:

$$\Delta\mu(T,p) = \frac{l_f}{T_0}(T - T_0) - (\frac{1}{\rho_l} - \frac{1}{\rho_s})(p_s - p_0) \tag{3.44}$$

where l_f is the latent heat of fusion and p_s is the new equilibrium pressure in the presence of an interfacial film.

Utilizing the Clapeyron equation and introducing the slopes of the sublimation and melting curves of the material, we determine:

$$\Delta\mu = \alpha l_f \,\frac{T - T_0}{T_0} \tag{3.45}$$

with:

$$\alpha = \left[1 - (\frac{\mathrm{d}p}{\mathrm{d}T})_{S,V} \Big/ (\frac{\mathrm{d}p}{\mathrm{d}T})_{S,L}\right] \tag{3.46}$$

Since the slope of the melting curve is much greater than the sublimation curve, $\alpha \approx 1$, and we can rewrite (3.43) and (3.45):

$$\Delta\mu = l_f \frac{T - T_m}{T_0} = \frac{\Delta\gamma}{\rho_l}(\frac{\partial f}{\partial d}) \tag{3.47}$$

Knowing that the molecular interaction potentials intervening in the interfacial energies are functions that quadratically decrease with the film thickness d, we can take $f(d) = d^2/(d^2 + \sigma^2)$, where σ is a constant whose order of magnitude is that of the diameter of the molecules. Introducing $t = (T_0 - T)/T$ and assuming $d \gg \sigma$, we then obtain:

$$d = \left(-2\sigma^2 \frac{\Delta\gamma}{\rho_l l_f}\right)^{1/3} t^{-1/3} \tag{3.48}$$

If the interactions between molecules are short–range, a law of exponential dependence should be selected for $\partial f/\partial d \propto \exp(-cd)$. We would then obtain $d \propto |\ln t|$.

The two laws of variations of d show that the film thickness tends to be very important in the vicinity of the melting point T_0 and that it decreases according to an exponential or logarithmic law when moving away from it, which is physically understandable.

Specific heat measurements on rare gas films absorbed on a graphite substrate demonstrate the premelting phenomenon with the existence of a liquid layer below the melting point T_0.

3.5.2 Size Effect on Small Particles

Gibbs showed that the melting point of small solid particles immersed in a liquid is decreased. This is the **Gibbs–Thomson effect**, resulting from the contribution of the surface energy to the free enthalpy, which is strongly dependent on the curvature of the interface.

The free enthalpy G of a solid of area A in equilibrium with its mother liquid phase is given by:

$$G = N_S \mu_S + N_l \mu_l + A\gamma_{Sl} \tag{3.49}$$

where μ_S and μ_l are the chemical potentials of the solid and liquid, γ_{Sl} is the solid/liquid surface tension, and N_S and N_l are the number of molecules in the liquid and solid phases. In view of conservation of the total number of molecules ($dN_S + dN_l = 0$) at equilibrium, we will have a minimum of G with respect to the change of the number of particles:

$$\Delta\mu = \mu_S - \mu_l + \gamma_{Sl} \frac{\mathrm{d}A}{\mathrm{d}N_S} = 0 \tag{3.50}$$

$$\Delta T = \left(\frac{1}{r_1} + \frac{1}{r_2}\right) \frac{\gamma_{Sl} T_0}{\rho_S l_f} \tag{3.51}$$

We can perform a limited expansion of $\Delta\mu$ as in (3.44). If r_1 and r_2 are the radii of curvature of the surface $\mathrm{d}A/\mathrm{d}N_S = (1/r_1 + 1/r_2)/\rho_S$, setting $\Delta T = T_0 - T$, we then have for a spherical surface:

$$\Delta T = \frac{2\gamma_{Sl} T_0}{\rho_S l_f r} \tag{3.52}$$

(r is positive for concavities directed toward the solid). Let us assume that a particle has a size greater than the size corresponding to equilibrium and must then melt to regain this situation, but melting requires absorption of heat which will decrease the temperature of the medium and the solid and thus further remove the particle from its equilibrium state. The situation is strictly the opposite for particles whose size is less than the equilibrium size.

Interfacial phenomena are very important in porous media, where they modify the thermodynamic conditions of the liquid/solid transition. The solidification point of a liquid is decreased in a porous medium. $|\Delta T|$ can be calculated with (3.51), but by taking also into account the complexity of the geometry on one hand and wetting of the walls of the pore system by the liquid on the other. These interfacial phenomena also play a very important role in soils, where water is in the liquid state at temperatures below 0°C.

3.5.3 The Special Case of Ice

As we previously emphasized, water is a substance whose liquid and solid phases, **ice**, have "abnormal" properties:

- liquid water has maximum density at 4°C;
- the density of ice is lower than the density of the liquid phase (with $\rho_S =$ 0.920 g cm^{-3} and $\rho_l = 0.997$ g cm^{-3} at 0°C, $\rho_l = 1$ g cm^{-3} at 4°C);
- the specific heat C_V of liquid water is particularly high (with $C_V = 75.24$ J·K^{-1}·mole^{-1} for H_2O, $C_V = 32$ J·K^{-1}· mole^{-1} for liquid sodium, and $C_V = 22.57$ J·K^{-1}·mole^{-1} for benzene);
- the isothermal compressibility of liquid water has a minimum at 46°C;
- The melting curve of ice has a negative slope (Fig. 1.1).

The anomalies in the behavior of the thermodynamic variables of water are due to its very special microscopic structure. The molecules of water H_2O have a V-shape with an angle of 109.5° for H–O–H and they are involved in hydrogen bonds with their neighbors: each molecule can participate in four H bonds and thus form a three–dimensional structure foreshadowing the structure of ice. Ice has a tetrahedral structure in which the oxygen atoms are located in the center of a tetrahedron composed of four other oxygen atoms where they occupy the vertices and are at a distance of 0.276 nm (Fig. 3.16). This characteristic structure of water, not found in molecular compounds with hydrogen bonds such as HF, HCl, C_2H_5OH, etc., was demonstrated by Raman spectroscopy and neutron scattering. The hydrogen bonds of the liquid phase are not stable and their lifetime is of the order of a picosecond.

Hydrogen bonds tend to give an unstable "structure" to the liquid phase of water and can explain its anomalies. We can hypothesize that these bonds tend to form an unstable labile three-dimensional network within the liquid, whose stability tends to increase in the vicinity of 0°C and in the supercooled state. Light scattering experiments conducted by J. Leblond and J. Teixeira

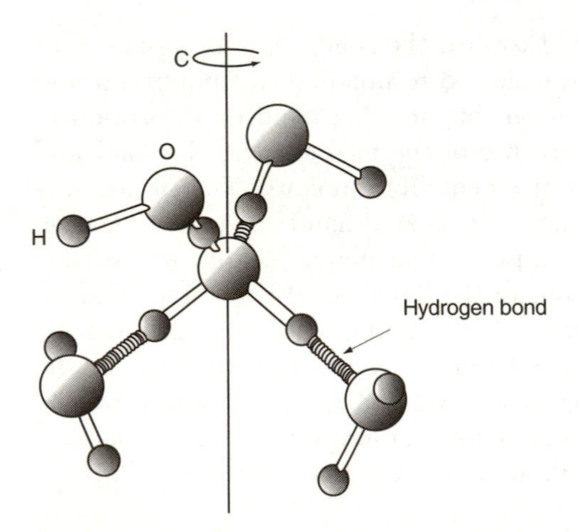

Fig. 3.16. Structure of water. The H_2O molecules form a tetrahedral structure: one molecule occupies the center of a tetrahedron whose vertices are occupied by oxygen atoms. Hydrogen bonds are established between the central molecule and the molecules occupying the four vertices.

revealed a strong increase in the compressibility of supercooled water: as if there were a metastability limit around −45°C corresponding to a point on the **spinodal**; when approaching it, thermodynamic quantities such as the specific heat C_p and compressibility diverge.

The behavior of water near 0°C and in the supercooled phase was interpreted by J. Teixeira and E. Stanley with a **percolation** model (Sect. 2.5.2): the liquid/solid transition is compared to **gelation**, and the hydrogen bonds play the role of bonds in a percolation model. This model allows predicting the anomalies of density, compressibility, and the thermal expansion coefficient.

Other models have been proposed to explain the anomalous properties of water by considering the structures of the liquid. One has assumed the existence of microstructures in the water constituted of cubes whose vertices are occupied by eight molecules or water or by rings formed by four molecules bound by hydrogen bonds (the sublimation energy of ice of 11.7 kcal·mole^{-1} can be reasonably attributed to rupture of two bonds), and the value of the specific heat and behavior of water are found with this model. A third model postulates the existence of several types of populations of oxygen atoms in liquid water: a molecule is surrounded by first neighbors located at the vertex of the tetrahedral structure (Fig. 3.16), while a layer of second neighbors is at a distance of 0.34 nm and 0.45 nm. When the temperature increases, the population of second neighbors has a tetrahedral structure with defects and with five neighbors for the water molecule occupying the center of the tetrahedron (the average coordination number is 4.5).

Ice exists in several crystalline forms (Fig. 1.1): eleven varieties of crystalline ice in all. The most recent form, metastable ice XII, was obtained at a pressure of 2–6 kbar. Calculations seem to show that ice XI would remain stable up to 2000 K at 4 Mbar and that it has compact hexagonal symmetry. Two varieties of amorphous ice can also be obtained:

- by solid–phase deposition from the vapor phase;
- from ice melted at high pressure and solidified.

The first work on amorphous ice was done in 1984 by O. Mishima, L. D. Calvert, and E. Whalley with ice I at 77 K melted at 10·kbar, and supercooled water was resolidified in the form of amorphous ice, which can be studied at atmospheric pressure. Amorphous ice obtained under these conditions has a higher density ($\rho = 1.31$ g cm^{-3} at 77 K and 10 kbar) than when formed by vapour deposition. Reheated to 125 K at 1 bar, the high density amorphous ice is irreversibly transformed into low density ($\rho = 1.17$ g cm^{-3}) amorphous ice. Amorphous ice can also be produced by splat cooling of liquid water ($> 10^6$ K sec^{-1}), for example, by accelerating droplets of water in a supersonic jet and by projecting them on a copper plate at 77 K. The glass transition temperature of water is approximately 136 K.

Based on simulation calculations with molecular dynamics, E. Stanley suggested that the existence of varieties of amorphous ice of different densities and the anomalies found in supercooled water (particularly the important increase in compressibility) would only be a manifestation of the coexistence of two different forms of supercooled water (high- and low-density), where the transition line between these two liquid phases ends in a critical point. This remains to be proven experimentally.

3.6 Very Numerous Applications

A very large number of industrial processes utilize solidification or melting phenomena. Natural phenomena involving a liquid/solid transition, particularly transitions involving water in liquid (including supercooled) or solid (ice or snow) form, which play an important role in meteorology, are also found in nature. In the Earth's crust, melting or crystallization of numerous mineral species also intervenes in many geological phenomena. These questions will be specifically treated in the last chapter.

3.6.1 Melting – Solidification in Metallurgy

Heat treatment of a material plays a key role in materials science because it very broadly establishes the relationship between properties and structure. Quenching of steels is a typical example of heat treatment that produces a very hard metal: the austenite phase passes into the martensite phase, which is a metastable phase supersaturated with carbon (Sect. 2.4.2).

In many metallurgical treatments, a material (metal or alloy) is forced to undergo a phase transition into the metastable state by quenching from high temperature and the solid phase (this is the case for martensite). A similar result is obtained by irradiation.

The cooling rate of the liquid or solid phase of the material is the key parameter of any metallurgical process since it broadly determines the physical properties of the solid, particularly through the size of microstructures. It probably plays an essential role in all quenching operations.

Until the beginning of the sixties, only classic treatments which scarcely allowed exceeding rates of $10^3\,\mathrm{K\,sec^{-1}}$ for surface melting–solidification processes utilizing a laser beam (for fabricating amorphous metallic thin films, for example).

At the two ends of the rate scale, the size of the dendrites formed during solidification varies from 5000 to 200 μ with slow quenching, to 5–0.01 μ with much faster quenching. The metallurgical procedures at average solidification speed (1 to $10^3\,\mathrm{Ks^{-1}}$) lead to formation of dendrites of 5–50 μ and the size can be homogenized by annealing. These are the processes used in pressure molding and surface coating operations.

At the beginning of the sixties, Duwez developed the so-called gun quenching technique. It can be used for rapid quenching of small quantities of liquid alloys by projecting drops of molten metal through a hole with a shock wave. The metal is practically atomized, it cools, and then solidifies in the form of a thin film on contact with a metal plate ("splat cooling" method). This technique allowed obtaining very high cooling rates (from 10^4K sec^{-1} to 10^{10}K sec^{-1}) for the first time. There are many versions of this technique, in particular, all of the so-called jet methods where the molten metal is projected onto a cooled substrate:

- jet obtained by centrifugation;
- jet obtained in the form of plasma.

Cooling and quenching can be obtained by impact of molten metal particles on the surface of a cooled cylinder (Alcoa process, which allows cooling rates of $10^6\text{K} \cdot \text{sec}^{-1}$, (Fig. 3.17a).

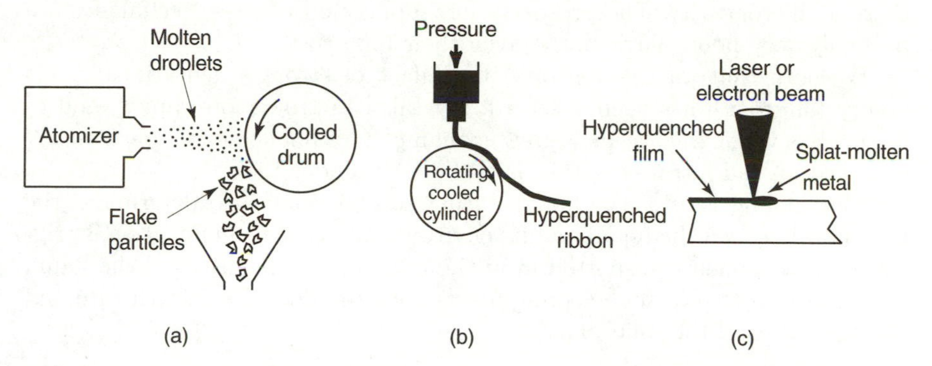

Fig. 3.17. Fast quenching methods. (**a**) Alcoa process: the particle jet is cooled on contact with a drum and solidified flakes are formed. (**b**) Fabrication of a thin film by projection of liquid metal under pressure onto a cooled cylinder; (**c**) Fast melting of the surface of a metal by laser beam or electron impact.

The use of powerful lasers led to new advances in fast quenching methods by eliminating the necessity of totally melting the material before quenching. Local melting of the surface of a material can be induced (over a thickness of 10 to 1000 μm) by irradiating it with a laser beam; it solidifies again almost instantaneously once irradiation stops (Fig. 3.17c). The laser beam has very high energy density (of the order of $10^6\text{W} \cdot \text{cm}^{-2}$) and is concentrated on a spot 0.1 to 1 mm in diameter on the surface (either neodymium–glass pulsed lasers or continuous CO_2 lasers). This method allows attaining quenching rates of 10^6 to 10^{13}K sec^{-1} and is used for fabricating metal films in the amorphous state (fast quenching allows "short-circuiting" the stage of crystallite nucleation). A new method that simultaneously utilizes several powerful laser beams allows hardening metal alloy coatings. The laser beam induces vaporization of surface atoms, creating iron plasma (if steel is being treated) which falls out as "rain" on the surface of the material. All of these methods have the advantage of creating few surface defects.

Contact between the liquid undergoing solidification and an external surface on one hand and gravitation on the other can also be eliminated. This is the goal of the **levitation** experiments conducted on metal drops on which important supercooling could be obtained. Levitation can be obtained by:

- an electromagnetic technique (the metal drop is suspended in an electromagnetic field);
- ultrasound;
- an aerodynamic technique (the drop is suspended in high-speed gas flow).

In Earth's gravitational field, there is always a gravity microgradient within a material. This can be eliminated either by utilizing a tower in which a supercooled drop falls or by conducting an experiment in a spaceship where there is microgravity. The prospects for application of these techniques are probably very poor, particularly given their high cost.

We note that on the contrary, the effect of strong gravitational fields on crystal growth has been observed: crystals can grow more rapidly and to larger sizes when gravitation is greater than g. This phenomenon, particularly observed in lead nitrate crystals, is still poorly understood.

The cooling rate of the liquid during solidification also determines the heat flow between the liquid and its environment. When solidification is triggered, heat is then transported from the apex of the dendrites to the liquid phase. The degree of supercooling determines the dendrite growth rate and the properties of the solid phase.

3.6.2 Molding of Polymers

The possibility of relatively easily melting polymers (their melting points are lower than for most metals) facilitates all operations for molding them from the liquid phase. The two important molding operations are **extrusion** and **injection molding**. In these two techniques, the polymer is brought above its melting point to make it pass into the liquid state.

Extrusion is the method used for continuous molding of polymers; it allows fabricating plastics with a small cross section over a great length (tubes, films, cable sheathing, etc.). The polymer is placed in the extruder in the form of granules or powder and is moved with an Archimedes screw to a heated zone (Fig. 3.18a). The polymer melts under the effect of shear and input of external heat. The extruder head is connected to a die whose shape is a function of the object manufactured (tube, sheathing, etc.). The temperature inside the extruder varies from 150°C to 370°C as a function of the polymers used (PVC, polyurethanes, etc.). The solid plastic obtained is generally semicrystalline.

Injection molding is the most widely used method for polymers. It consists of injecting a molten polymer inside a mold (Fig. 3.18b,c) and can be used to mass produce objects of very varied sizes (toys, ski boots, household articles, vehicle bumpers, etc.). As in extrusion, the polymer is melted in the mold; the melt accumulates in front of the screw, which acts as a piston for injecting the molten polymer in the mold. The item is formed in the mold under pressures that can attain several hundred bar.

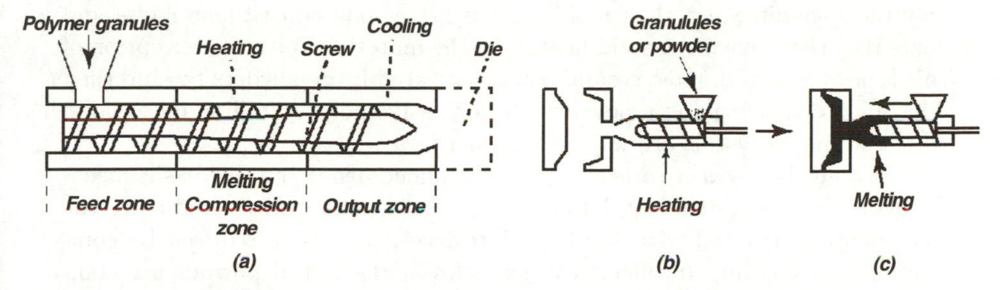

Fig. 3.18. Molding of polymers. (**a**) extrusion method: the polymer is introduced in the form of granules and moved by the screw to a heated zone where it melts. The extruder head has a die and the plastic is drawn at its outlet. (**b, c**) injection molding. The polymer is first heated and pushed by the Archimedes screw; it is then injected in the mold under pressure.

3.6.3 Production of Sintered Ceramics

Ceramics now constitute a vast category of materials with multiple applications. Ceramic parts were initially fabricated by firing a clay paste (clay is called *keramikos* in Greek), a compound of hydrated aluminum silicate and other materials. Ceramics have been used for a long time due to their hardness, refractory properties, and chemical resistance (particularly their corrosion resistance). Their brittleness has nevertheless been a serious handicap to their use. Ceramics are in effect sensitive to small defects and cracks which tend to spread within the material because it is generally formed by **polycrystalline structures**. There are also transparent ceramics which are glassy phases, **glass ceramics**, with low crystallinity.

An attempt was made to remedy this drawback by developing new ceramics, particularly with new textures that improve their mechanical performances (breaking strength). Composite structures have been made in this way:

- ceramic/metal associations (**cermet**).
- disperse phases reinforced with fibers or particles.

The textures consisting of Al_2O_3 (matrix) + ZrO_2 (disperse phase) and Si_3N_4 (matrix) + SiC (fibers) are examples of such composite structures.

All ceramics are manufactured at high temperatures and their melting points are high (close to 2700°C for zirconia ZrO_2, approximately 1500°C for porcelain).

The polycrystalline character of ceramics causes the microstructures inside the material to play an essential role since their size greatly determines the properties. This problem will be discussed in Chapter 9.

Porcelain is manufactured from a fired clay paste. It is hydroplastic, that is, water is immobilized between the clay particles, acting as a lubricant which facilitates shaping and thus molding. On firing, one constituent melts and coats the others, ensuring solidification of the material. With the exception of this type of material, most ceramics are fabricated from powders by sintering. This allows transforming a powder into a massive polycrystalline material in which all microcrystals are welded together in a way.

Sintering involves a series of physical phenomena, but its basic mechanism is decreasing the total free energy of the material by reducing the free energy associated with solid/gas interfaces, which are reduced by compaction and heating. In effect, the particles of the initial powder are compacted under pressure, then brought to a high temperature (a temperature of $T \approx 2/3\ T_m$). Under these conditions, the atoms diffuse from one particle to another, causing the formation of "necks" between the grains and thus reducing the porosity: this is sintering (Fig. 3.19). The atoms essentially diffuse from grain boundaries localized in the necks between particles; the atoms fill the pores, thus reducing the interfaces and the energy associated with them.

Sintering can also be conducted in the presence of a liquid phase: small amounts of additives allow forming a liquid phase when the temperature is raised; diffusion of atoms and formation of grain boundaries and thus sinter-

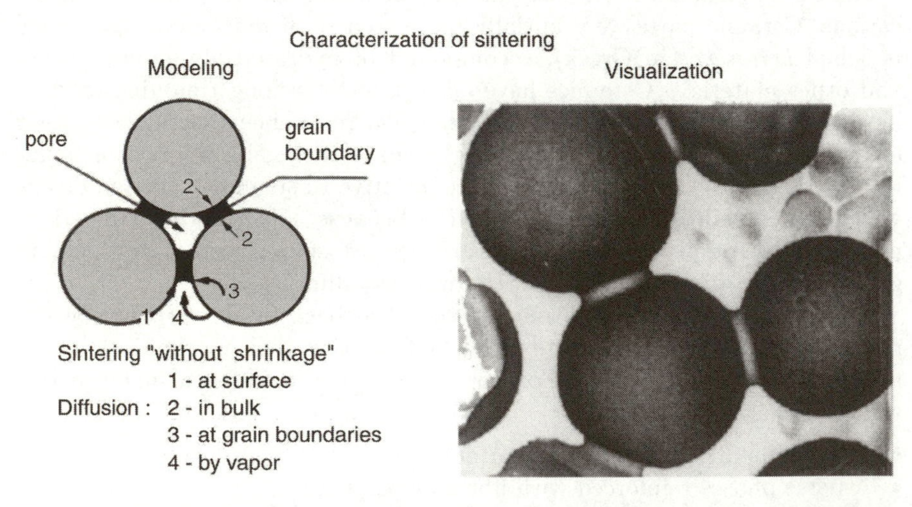

Fig. 3.19. Sintering of a ceramic. The mechanisms involved in sintering of a ceramic manufactured from powder grains are shown. (**a**) Model of the structure; (**b**) electron microscopic image (B.A.M. Anthony, Laboratoire des Hautes Temperatures, Orleans, CNRS).

ing are then favored. If the materials produced are denser, they will be less resistant to high temperatures.

There are other techniques for production of ceramics, for example, chemical vapor deposition (or CVD). Silicon nitride Si_3N_4 is also manufactured by chemical reaction in the vapor phase, where the ceramic is deposited on a substrate.

Ceramics for technical applications have greatly evolved in the last twenty years since they can now withstand high temperatures (1200–1400°C) and can thus be used in the hot parts of engines and gas turbines operating at high temperatures (1300°C). Ceramic materials are also used as heat shields for spacecraft (space shuttles, for example). Zirconium oxide ZrO_2 and silicon nitride Si_3N_4, boron nitride BN, silicon carbide SiC, alloys of the type Al_2O_3–Si_3N_4 (SiAlon) are examples of these thermal and mechanical high-performance technical ceramics (low thermal conductivity and thermal expansion coefficient).

Problems

3.1. Richard and Trouton Rule for Melting

Assume that a liquid phase can be described with a hole model: N atoms or molecules of mass m can occupy the entire volume V of a solid while moving freely, and the presence of unoccupied cells in the structure (holes) facilitates these motions. In the solid phase, on the contrary, the atoms all occupy fixed positions (lattice sites) with no possible permutations. Assume that the interaction potential between atoms is binary and of the hard-sphere type: $u(r_{ij}) = \infty$ if $r_{ij} < r_0$, $u(r_{ij}) = 0$ if $r > r_0$.

1. Calculate the partition functions of the solid and the liquid.
2. Determine the variation in the free energy at the solid/liquid transition and the variation in the enthalpy. Assume that the interaction potentials are identical in both phases. The Stirling equation is used for $N!$
3. Find a simple relation between L_f and T_m.

3.2. Sublimation

Solid–vapor equilibrium is investigated in a system constituted of N atoms. The solid is a crystal in which each site is occupied by one atom. The number of atoms in the solid state is designated by N_s and the number of atoms in the vapor phase is designated by N_g. The Einstein model is used to describe the vibrations of the atoms in the crystal: each atom oscillates in each direction with frequency ν. The energy necessary for extracting one atom from the solid is designated by φ.

1. Write the energy of the solid and its partition function at temperature T.
2. Calculate the partition function of the vapor phase, assuming that it is an ideal gas.

3. Determine the total free energy then N_g at equilibrium. Calculate the pressure p of the gas.

3.3. Melting of a Solid with Defects

The situation is as described in Sec. 3.3.2 for a solid with concentration c of defects (holes) in the vicinity of the stability limit corresponding to critical concentration c^* at temperature T^*. The defects increase the disorder in the crystal structure as if there were a heterogeneous solid/liquid phase associated with local density fluctuations.

1. Calculate Δs^* associated with these fluctuations and the associated fluctuation probability w as a function of c.
2. Determine the variation of w. How does w behave at T^*? What can you deduce from this for the solid?

3.4. Melting of Polymers

Let us take the crystallization model described in Sec. 3.4.3 (Fig. 3.15) and study the melting of a polymer lamella. Let T_m^0 be the melting temperature neglecting surface effects, γ_e the surface energy of a chain folded over a length l and L_f the melting energy per unit volume.

1. Calculate the change in free enthalpy at fusion.
2. Derive the melting temperature T_m.
3. What can be concluded from this?

4 Phase Transitions in Fluids

4.1 The Approach Using Equations of State

In the early history of physics, it was assumed for a very long time that there were three phases, or states, of matter: gas, liquid, and solid. The transition from one state to another occurs at a given temperature and pressure corresponding to the physical state with the lowest free energy F.

In the case of simple liquids, there are three phase transitions: the liquid/vapor transition (boiling and condensation); the liquid/solid transition (melting and solidification); sublimation, which corresponds to a solid/vapor transition. In a mixture of liquids, we also see a unmixing transition: the different liquid phases separate. If several phases are present at the same temperature and pressure, we say that there is **coexistence of phases**. The Gibbs phase rule allows determining the possible number of coexisting phases.

In the case of a fluid (liquid or gas), its physical state can be described with pressure and temperature variables and with the equation of state $p = p(V, T)$ which expresses the pressure as a function of the volume and temperature. This function can be represented by a surface in three-dimensional space (Fig. 4.1). Several cross sections of the surface are shown for different temperatures. The highest temperature, T_6, corresponds to the simplest equation of state, the equation for an ideal gas:

$$pV = nNkT \tag{4.1}$$

N is Avogadro's number, k is the Boltzmann constant ($R = kN$ is the ideal gas constant), and n is the number of moles. The relation between p and V at fixed T given by equation (4.1) represents a hyperbola. In general, the relation between p and V at a given temperature is called an **isotherm**; it is obtained by vertical sections of the surface through planes parallel to axis V. At temperatures below T_3, the isotherms no longer show a behavior of the ideal gas type and they include a rectilinear part. The isotherm corresponding to T_2 contains three segments. The middle one is a straight line which indicates that the system will go from the vapor phase to the liquid phase when the volume is decreased.

By projection of the surface representing the equation of state on the (p, T)-plane, we see that the liquid–gas coexistence line is projected along a line that ends at **critical point** C. At point A, another branch representing

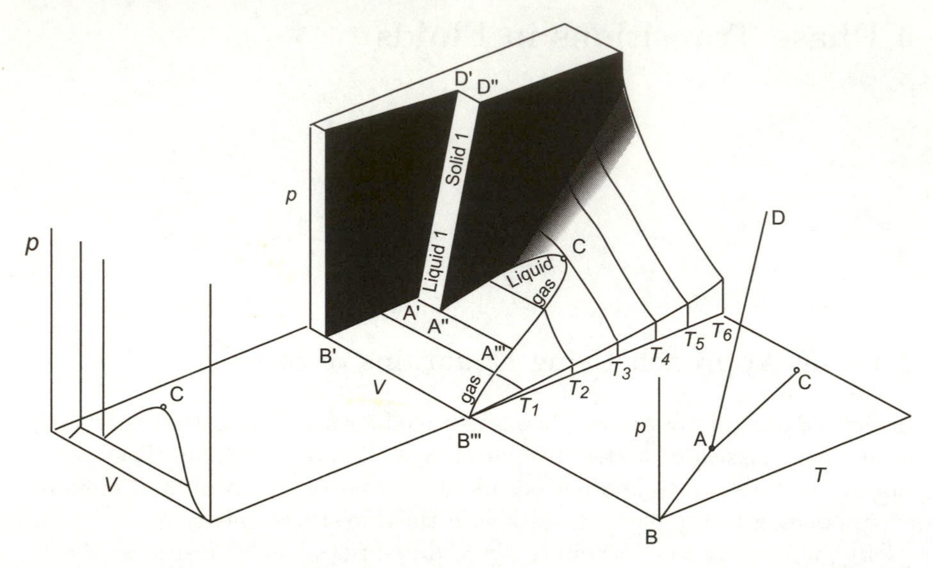

Fig. 4.1. p, V, T diagram for a pure substance. The surfaces represent the equation of state $p = p(V, T)$. Several cross–sections are shown for temperatures T_1, T_2, ... as well as projections of the coexistence curves on planes (p, T) and (p, V).

the projection of the liquid–solid coexistence surface is obtained. Point D is not a terminal point for liquid–solid coexistence (contrary to point C for liquid and gas), and this branch extends to infinity. The pressure and temperature coordinates of point A are fixed, this is the **triple point** used as reference in thermometry. Three densities are observed at point A: gas, liquid, and solid. In most materials, the density of the solid is higher than the density of the liquid. Water is one of the exceptions (Chap. 1).

Several important questions can be raised concerning liquids and solids:

1. Why can two phases coexist at a given pressure and temperature? This is because the free energy is identical for the two phases even though the spatial arrangements of the molecules are different.
2. How does the transition occur in fact? We find the formation of initial "nuclei" of a phase in the form of bubbles or droplets or crystallites and it is thus necessary to explain their formation.
3. Can phase transitions be simulated, with a computer, for example? We will give the principle of the method.

The study of phase transitions in fluids is more complicated if several constituents are present and eventually may show **unmixing**. Finally we have the existence of transitions in non-equilibrium situations where glasses and gels can be formed; they are specifically discussed in Chaps. 5 and 6. Polymerization within a liquid phase is another subject altogether, which we will not directly discuss in this book.

It is clear that phase transitions in fluids, particularly in liquids, have a great number of industrial applications in sectors as varied as the chemical industry, metallurgy, agriculture and the food industry, and petroleum engineering. We will address the relevant issues for those sectors.

4.2 The Liquid–Gas Transition in Simple Liquids

4.2.1 Van der Waals Equation of State

Van der Waals was the first to propose, in 1881, an equation of state that could describe the transition from a gas to a liquid. He modified (4.1), replacing V by $V - Nb$ and p by $p + N^2a/V^2$. The first modification takes into account the fact that the available volume for each molecule must be decreased by the volume already occupied by $N - 1 \cong N$ molecules present in volume V, and referring here to one mole, $n = 1$ (in fact, van der Waals showed that b is four times the volume of the individual molecule). The second modification comes from taking into account the long-range attraction of the other molecules. For the molecules of a liquid confined in a vessel, the attractive forces cancelled out for molecules located inside the volume of the liquid, but those located on the surface of the liquid are subject to forces from the molecules which are below this surface only. These uncompensated forces have a resultant directed towards the inside of the liquid, which gives a supplementary contribution to the pressure.

Hence, the so-called van der Waals equation:

$$\left(p + \frac{N^2a}{V^2}\right)(V - Nb) = NkT \tag{4.2}$$

It contains two constants: b, which is of the order of the volume of the individual molecules, and a, representing the attractive interaction.

This equation has three essential characteristics: it allows representing the isotherms that can be found experimentally, although this representation does not totally conform to experimental reality; from the mathematical point of view, a third-degree equation allows obtaining the liquid/gas transition; it particularly allows obtaining the law of **corresponding states**, as we will show.

The van der Waals equation represents reality only incompletely and from this point of view cannot be compared with other laws of physics such as the Newton or Maxwell equations. In the older engineering literature one finds many empirical equations of state; polynomials with around 20 parameters reflecting sets of many measurements. It is remarkable how well the van der Waals equation, using the critical point to determine a and b, reproduces such a surface, except in the neighborhood to the critical point and at low temperature. In order to obtain the region of coexisting phases the van der

Waals equation had to be augmented by the maens of the Maxwell's rule (Sect. 4.3.2).

4.2.2 The Law of Corresponding States

Equation (4.2) can be rewritten in the form of a cubic equation in V:

$$pV^3 - N(pb + kT)V^2 + aN^2V - abN^3 = 0 \tag{4.3}$$

This equation may have three roots for a given value of pressure p. The three roots coincide when (4.3) is rewritten as:

$$p_c(V - V_c)^3 = 0 \tag{4.4}$$

for $p = p_c$ and $T = T_c$. This is possible if:

$$V_c = 3Nb; \quad p_c = \frac{1}{27}\frac{a}{b^2} \quad \text{and} \quad kT_c = \frac{8}{27}\frac{a}{b}$$

The critical isotherm is a cubic parabola which has a horizontal tangent at the point (p_c, V_c), which is also an inflection point.

Eliminating constants a and b, which are characteristic of a fluid, the van der Waals equation then assumes the simple form:

$$p/p_c = \frac{8(T/T_c)}{3(V/V_c) - 1} - \frac{3}{(V/V_c)^2} \tag{4.5}$$

Introducing the **reduced variables**, $p_r = p/p_c$, $T_r = T/T_c$, $V_r = V/V_c$ defined by reference to the critical coordinates p_c, T_c, V_c, (4.5) is then written:

$$p_r = \frac{8T_r}{3V_r - 1} - \frac{3}{V_r^2} \tag{4.6}$$

This equation can be interpreted simply as: the equation of state is the same for all fluids if the critical parameters are used to define the units. This is the situation shown in Fig. 4.2. This is called the **law of corresponding states**. This law was used by van der Waals to predict the critical temperature of helium before it was even possible to attain this low temperature (T_c = 5.2 K, p_c = 2.75 bar), while the Dutch physicist Kamerlingh Onnes had already plotted the isotherms above the critical point. Kamerlingh Onnes used them to estimate the critical temperature by comparing the isotherms of H_2 and He in 1907.

The law of corresponding states does not uniquely apply to the van der Waals equation; it is a general law that can be derived by statistical physics, as de Boer and Michels did in 1938. It is assumed that the molecular interaction potential has the form $\varepsilon(r) = \varepsilon_0 f(r/r_0)$. Here ε_0 is the energy and r_0 is the length characteristic of the molecules in the medium; r_0 is the radius of attraction, for example, in a hard-sphere model or the distance at which the

Lennard–Jones potential passes through zero (Fig. 3.8). If we calculate the partition function corresponding to a system of molecules with a potential such as $\varepsilon(r)$, it is possible to show that it is written in the form of an integral which is only dependent on two dimensionless parameters $T^* = kT/\varepsilon_0$ and $V^* = V/(r_0^3 N)$. The pressure is then easily expressed in terms of the reduced pressure $p^* = p\, r_0^3/\varepsilon_0$, and a unique equation of state is obtained for all fluids whose intermolecular potential is described by the same function f and expressed with reduced variables (Problem 4.2): this is the law of corresponding states.

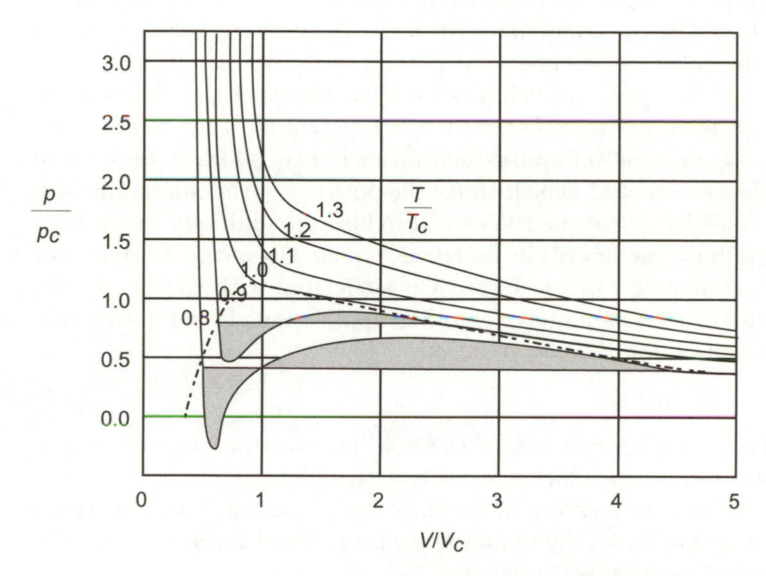

Fig. 4.2. Law of corresponding states. The system of isotherms for fluids obeying the van der Waals equation is shown. Reduced coordinates $p/p_c, T/T_c, V/V_c$ were used to write the equation of state. The isotherms are identical for all fluids when their behavior is described by the van der Waals equation.

The law of corresponding states has been generalized in a variety of ways by introducing a third parameter. One was introduced by Lunbeck and de Boer and it takes into account the corrections introduced by quantum mechanics utilizing a perturbation method. The other takes into account the difficulty of describing the properties of many molecules with spherical potentials.

If the law of corresponding states is rigorously applied to a family of fluids, we should find that their critical coordinates, expressed in reduced variables:

$$p_c^* = \frac{r_0^3}{\varepsilon_0} p_c; \quad T_c^* = \frac{kT_c}{\varepsilon_0} \quad \text{and} \quad V_c^* = \frac{V_c}{N_0 r_0^3}$$

where V_c is the volume of one mole and N_0 is Avogadro's number, should be strictly identical for all fluids. However, this has not really been found, even for a series of relatively simple fluids whose atoms or molecules are approximately spherical in shape; thus: T_c^* is equal to 0.90 for H_2, 1.25 for Ne, 1.29 for CH_4, 1.31 for Xe, and 1.33 for N_2; p_c^* is equal to 0.064 for H_2, 0.111 for Ne, 0.126 for CH_4, 0.132 for Xe, and 0.131 for N_2; V_c^* is equal to 4.30 for H_2, 3.33 for Ne, 2.96 for CH_4, 2.90 for Xe, and 2.96 for N_2. Relatively important dispersion of the values of these critical coordinates is observed.

These deviations from the law of corresponding states are even more important for nonspherical and polar molecules like H_2O and C_2H_5OH and for molecules in the shape of chains like hydrocarbons. In these cases, as could be expected, the spherical symmetrical potentials do not satisfactorily describe reality, and likewise, the behavior of molecules such as CCl_4 cannot be satisfactorily described by a Lennard–Jones potential, a function of two parameters ε_0 and r_0. A better approximation consists of adding a term $a\,r$ to the Lennard–Jones potential, which shifts the position of the minimum with respect to r_0. This led Pitzer in 1955 to introduce an additional parameter for describing a fluid, the **acentric factor** ω, which was meant to represent the nonspherical character of the force field associated with each molecule. Later it was used indiscriminately for other application. Pitzer postulated a practical estimation for ω by the relation:

$$\omega = -\log_{10}(p^s/p_c) - 1 \tag{4.7}$$

where p^s is the saturation pressure of the fluid at temperature T such that $T/T_c = 0.7$ (selected values which are easily accessible).

Pitzer noted that the pressure of the vapor phase at saturation p_r^s could be expressed as a function of the temperature (in reduced variables) with the following empirical relation utilizing ω:

$$-\log p_r^s = A - \frac{B}{T_r} + \frac{C}{T_r^2} + \frac{D}{T_r^3} + \omega\Delta(T_r) \tag{4.8}$$

where A, B, C, D are constants and Δ is a correction factor which is a function of $1/T_r$.

For monoatomic fluids, ω is always very small ($\omega = 0.001$ for argon), likewise for fluids whose molecules are spherical ($\omega = 0.011$ for CH_4), and it is relatively large for polar fluids ($\omega = 0.344$ for H_2O and 0.644 for C_2H_5OH) and hydrocarbons ($\omega = 0.3978$ for n-octane).

The acentric factor is empirically determined from measurements of the saturation pressures of fluids. As we will show, it is used as a correction factor in the equations of state.

4.2.3 Behavior Near the Critical Point

The description of the liquid–vapor phase transition would be incomplete if we did not specifically consider the behavior of the fluid in the vicinity of the critical point C. In general, new reduced variables are introduced:

$$\tilde{p} = (p - p_c)/p_c, \quad \tilde{V} = (V - V_c)/V_c \quad \text{and} \quad \tilde{T} = (T - T_c)/T_c$$

which allow expressing the van der Waals equation of state in the form:

$$\tilde{p}(2 + 3\tilde{V}) = 8\tilde{T} - \frac{3\tilde{V}^3}{(1 + \tilde{V})^2} \tag{4.9}$$

used for studying the behavior of the fluid in the vicinity of the critical point.

The studies of liquid–gas transitions near the critical point led to better comprehension of phase transitions. Experimentally, it was found that a liquid/gas system becomes "milky" near the critical point: this is the phenomenon of **critical opalescence** caused by scattering of light by dielectric constant inhomogeneities in the fluid induced by density fluctuations. This phenomenon was observed for the first time in CO_2 by Andrews in 1869. These density fluctuations can be calculated with statistical physics (Problem1.4). The experimental study of the behavior of a fluid near a critical point has greatly evolved since the 1960s, particularly due to light scattering techniques using lasers. It allowed showing that the similar isotherms and curves determined in the immediate vicinity of a critical point did not obey a van der Waals equation of state, for example, written like (4.9). They can only be represented by series expansions, but with fractional exponents. This nonanalytical behavior of the thermodynamic quantities is represented by what are called **scaling laws**. They led to a large number of theoretical developments that allowed better understanding of the behavior of a phase transition in the vicinity of a critical point.

For example, the critical isotherm ($T = T_c$) is much flatter than the result obtained from a classic model such as (4.9) would indicate, and it is written as:

$$\tilde{p} = D|\tilde{\rho}|^{\delta}, \quad \text{with} \quad \tilde{\rho} = (\rho - \rho_c)/\rho_c \quad \text{where} \quad \rho \quad \text{is the density} \tag{4.10}$$

The exponent δ, called the **critical exponent**, has the value of 4.80 ± 0.02, while the van der Waals equation gives a value of 3, as (4.9) indicates. This difference between the theoretical and experimental values was demonstrated for the first time in 1900 by Verschaffelt. These power laws are justified by theories called **scaling laws**.

The fundamental principle of these scaling laws is to replace the free energy for a system with N particles by an equivalent expression for $N/2$ particles by changing the coupling constants between them and the field in which they are placed. This transformation is called **renormalization**. It can be repeated an infinite number of times. In each stage, the constants or coupling parameters between the particles of the system are renormalized so that the free energy retains the same value. The value of the critical exponents, such as δ, near the critical point can be determined from these operations. The values found by this method are very close to those obtained experimentally. These methods, called the **renormalization group**, were initially introduced by Kadanoff, Widom, Domb, and Wilson in the 1960s. We will discuss the principle in Chap. 7.

To conclude, we emphasize that the study of the laws of critical behavior with determination of the critical exponents is closely related to the study of the "cooperative" behavior of the molecules in the fluid near the critical point. In this region, the effective indirect interaction between the molecules becomes very long-range, despite the fact that the direct interaction potentials such as the Lennard–Jones potential are short-range. This is manifested in the behavior of the spatial correlation function $S(\boldsymbol{k})$ which approaches infinity at the critical point if $\boldsymbol{k} \to 0$, as the critical laws show.

4.3 Thermodynamic Conditions of Equilibrium

4.3.1 Liquid–Gas Equilibrium

We note that there are three classic formal conditions for equilibrium between the different phases of a system: the temperature, pressure, and chemical potentials must be equal. These three conditions determine the coexistence curve, for example, in (p, T)-plane, which is the projection of the (p, V, T)-surface. Each phase has a different specific volume (except at the critical point): the transition is first order (Sect. 1.2.3).

If a gas is contained in a cylinder and the volume is reduced by compression with a piston at constant temperature, droplets will appear in the cylinder at a certain volume: this is the **dew point**. This phenomenon only occurs if the temperature is low enough, that is, if it is below the critical point. With a greater reduction in the volume, the amount of gas will decrease and the amount of liquid will increase. This process evolves at constant pressure and movement of the piston requires no work. When the space below the piston is completly filled with liquid, it will be necessary to apply very high pressure to reduce the volume of liquid. The volume at which the vapor disappears is called the **bubble point** because the last bubbles of vapor appear at this point. This description of a condensation experiment is based on the fact that the isotherm in the (p, V)-plane has a horizontal part between point V_1 (bubble point) and point V_2 (dew point), as indicated in Fig. 4.3. As condensation takes place at constant temperature and pressure, the thermodynamic potential $G(p, T, N)$ should remain constant. If we designate the number of molecules in the liquid state by N_l and the number of molecules in the vapor phase by N_v, we then have:

$$G = N\mu = N_l\mu_l + N_v\mu_v \tag{4.11}$$

This potential remains constant during condensation if:

$$\mu_l = \mu_v \tag{4.12}$$

This condition expresses the fact that if one molecule is removed from the gas phase and inserted in the liquid phase (or vice versa), no Gibbs free energy is gained or lost. Although the energy in the liquid is lower the entropy is higher. This condition can be easily verified using the relation $G = F + pV$:

$$F_1 + pV_1 = F_2 + pV_2 \tag{4.13}$$

that is,

$$\frac{F_1 - F_2}{V_1 - V_2} = -p = \frac{\partial F}{\partial V} \tag{4.14}$$

The thermodynamic conditions for equilibrium between liquid and vapor phases in a system with one constituent are as follows: the pressures, temperatures, and chemical potentials must be equal.

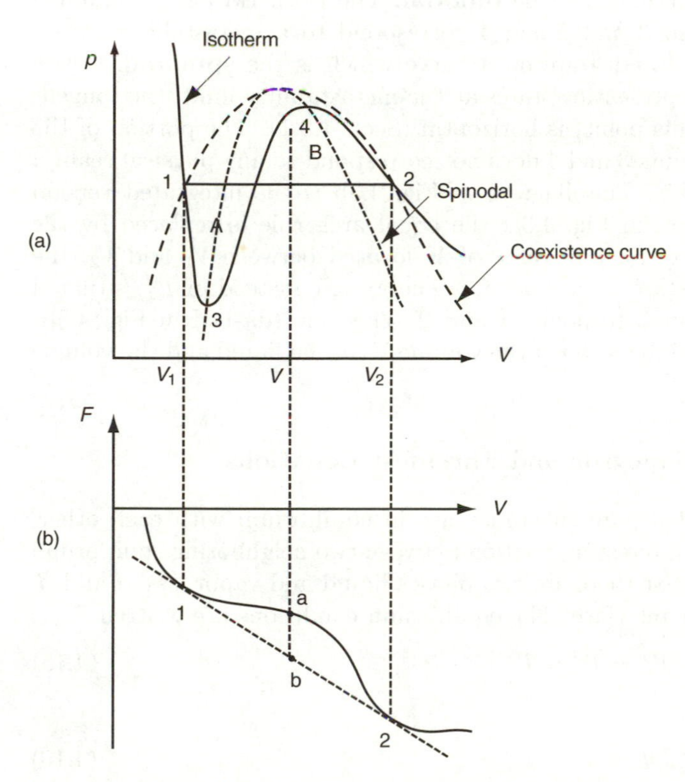

Fig. 4.3. Maxwell's rule and construction of the double tangent. (**a**) Points 1 and 2 are on the coexistence curve, with the isobar bordering two areas A and B which have equal areas; points 3 and 4 correspond to the metastability limit of liquid and vapor and are on a curve which is the spinodal. (**b**) The free energy F is obtained by integration $F = -\int p dv$, where points 1 and 2 correspond to an isobar, since they have a common tangent.

4.3.2 Maxwell's Rule

These equilibrium conditions can be used with van der Waals (4.3), which allows determining the value of the volume on an isotherm at a given pressure. Since (4.3) is a cubic equation in V, it allows three roots for certain pressure values and for $T < T_c = 8a/(27bk)$ as shown in Fig. 4.3. On the isotherm, the pressure p at which the liquid–vapor transition occurs is determined by the horizontal line positioned such that the areas of the surfaces between the isotherm and the isobar at p are equal (Fig. 4.2, areas in gray): this is **Maxwell's rule** (Problem 4.5).

Points 1 and 2 on the isotherm shown in Fig. 4.3 are on the liquid and gas phase coexistence curve, called the **binodal**. The isotherm parts contained between points 1 and 3 and 2 and 4 correspond to a metastable thermodynamic state, and 3 and 4 are on a curve which is the **spinodal**: this is the locus of points representing states at the metastability limit; the tangent to the isotherm at this points is horizontal (Sect. 1.2.1). The portion of the isotherm between points 3 and 4 does not correspond to any physical reality: the system is unstable. The diagram in Fig. 4.3b is the integrated version $F_T(V)$ of the diagram in Fig. 4.3a: the equal area rule is replaced by the common tangent rule: for any value of V located between V_1 and V_2, the system can be in a state with lower free energy (F_b instead of F_a) situated on the tangent common to points 1 and 2. This line (dashed in Fig. 4.3b) represents a linear combination of the volume V_1 of the liquid and the volume V_2 of the vapor.

4.3.3 Clausius–Clapeyron and Ehrenfest Equations

When two phases of a pure substance are in equilibrium with each other, there is a useful and interesting relation between two neighboring equilibrium situations. Assume that there are two phases liquid and vapor, say α and β, separated by a plane interface. The equilibrium conditions are written:

$$p^\alpha = p^\beta = p, \quad T^\alpha = T^\beta = T \tag{4.15}$$

and

$$\mu^\alpha(p,T) = \mu^\beta(p,T) \tag{4.16}$$

If the pressure and temperature change by δp and δT, the chemical potential can be developed in a Taylor expansion in the vicinity of the equilibrium point (p_0, T_0):

$$\begin{aligned}\mu(p,T) = \mu(p_0,T_0) &+ \frac{\partial\mu}{\partial p}\delta p + \frac{\partial\mu}{\partial T}\delta T \\ &+ \frac{1}{2}\frac{\partial^2\mu}{\partial p^2}\delta p^2 + \frac{\partial^2\mu}{\partial p\partial T}\delta p\delta T + \frac{1}{2}\frac{\partial^2\mu}{\partial T^2}\delta T^2 + \dots\end{aligned} \tag{4.17}$$

where $\delta p = p - p_0, \delta T = T - T_0$. Equation (4.17) should be applied to μ^α and μ^β in (4.16) and to all orders.

To the first order, we thus obtain:

$$\frac{\partial \mu^\alpha}{\partial p}\delta p + \frac{\partial \mu^\alpha}{\partial T}\delta T = \frac{\partial \mu^\beta}{\partial p}\delta p + \frac{\partial \mu^\beta}{\partial T}\delta T \tag{4.18}$$

which can be rewritten as:

$$\frac{\delta p}{\delta T} = \frac{s^\alpha - s^\beta}{v^\alpha - v^\beta} \tag{4.19}$$

where $(\partial \mu / \partial p)_T = v$ is the molar volume and $(\partial \mu / \partial T)_p = -s$ is the molar entropy. Introducing the molar heat transition from one phase to the other $L = T(s_\alpha - s_\beta)$, we derive from (4.19), the **Clausius–Clapeyron** equation that determines the slope of the coexistence phase curve or vapor pressure curve in the (p, T)-plane:

$$\frac{\delta p}{\delta T} = \frac{L}{T(v_\alpha - v_\beta)} \tag{4.20}$$

Near the critical point, the differences between the first order derivatives are cancelled: $s_\alpha \to s_\beta$ and $v_\alpha \to v_\beta$. However, (4.20) holds and the slope $\delta p / \delta T$ remains finite. This relation applies also to metastable situations beyond the triple point until the system becomes unstable.

In binary liquid–gas systems and in superfluid helium, we have a line of critical points rather than an isolated critical point because there is an additional degree of freedom. In this case, we use the second term in the Taylor series given by equation (4.17) which leads to the additional conditions along the critical line:

$$\frac{\partial \mu^\alpha}{\partial p} = \frac{\partial \mu^\beta}{\partial p} \tag{4.21}$$

$$\frac{\partial \mu^\alpha}{\partial T} = \frac{\partial \mu^\beta}{\partial T} \tag{4.22}$$

Replacing the second order derivatives in (4.17) by the notation $\mu_{pp} = (\partial^2 \mu / \partial p^2)$, etc., and inserting (4.21) and (4.22), the equilibrium condition are then written as:

$$\Delta \mu_{pp} \delta p + \Delta \mu_{pT} \delta T = 0 \tag{4.23}$$

$$\Delta \mu_{pT} \delta p + \Delta \mu_{TT} \delta T = 0 \tag{4.24}$$

where $\Delta \mu_{pp} = \mu^\alpha_{pp} - \mu^\beta_{pp}$, etc. Introducing the expansion coefficient $\alpha = V^{-1}(\partial V / \partial T)_p$, and the compressibility $\kappa_T = -V^{-1}(\partial V / \partial p)_T$, we then obtain:

$$\frac{\partial p}{\partial T} = \frac{\alpha^\alpha - \alpha^\beta}{\kappa^\alpha - \kappa^\beta} \tag{4.25}$$

$$\frac{\partial p}{\partial T} = (V_m T)^{-1} \frac{C_p^\alpha - C_p^\beta}{\alpha^\alpha - \alpha^\beta} \tag{4.26}$$

These are the **Ehrenfest equations** for a second order phase transition. Since equations (4.25) and (4.26) must be identical, we obtain the Prigogine–Defay equation (Problem 5.3).

4.4 Main Classes of Equations of State for Fluids

4.4.1 General Principles

Equations of state for fluids are mathematical relationships for the pressure of fluids as function of volume, temperature and composition; they are important for predicting some of the properties of fluids and in particular the thermodynamic conditions in which phase transitions occur. These equations are widely used to predict the behavior of fluids in chemical process industry, in the recovery of oil and natural gas and in nuclear engineering.

4.4.2 One-Component Fluids

One may consider the van der Waals equation of state as half-way between the desire of "purist" physicists, specialists in statistical mechanics, who would like to obtain an equation of state starting from the calculation of the partition function, and supporters of a heuristic treatment yielding higher accuracy at the expense of more adjustable parameters (approximately 20 in the case of fluids such as ethylene).

With Kamerlingh Onnes, we can write the **compressibility factor** for a fluid $Z = pV/RT$, as a series expansion:

$$Z = 1 + \frac{B}{V} + \frac{C}{V^2} + \ldots \tag{4.27}$$

where $B, C, \ldots$ are only, for a pure substance (only one component), functions of the temperature and are called second, third, ... **virial coefficients**.

These virial coefficients can be determined empirically and for spherical molecules, they obey the law of corresponding states, although there are exceptions. They represent interactions of pairs, triplets, etc. of molecules. Systematic calculation of the virials become very complicated beyond third.

Strictly speaking, equations of type (4.27) cannot be considered as equations of state since they comprise an infinite number of terms, most of which cannot be calculated; since the series does not simply converge, Z cannot be written in a close form. If we only use a limited number of coefficients, the truncated series then corresponds to a **virial equation of state**. It is necessary to emphasize that an equation of this type, limited to the second term, only allows describing the vapor phase and thus does not account for the vapor/liquid transition. As for the equation limited to the third term, although it allows describing both the liquid and vapor phases, however, it does not satisfactorily account for the liquid to vapor transition: this is an approximation similar to the van der Waals equation.

If we use the van der Waals equation as reference model, we find that it contains two fundamental ingredients that allow predicting condensation (or vaporization): the short-range repulsive term and the long-range attractive term. The repulsive term, of the form $RT/(V-b)$ was introduced with the simple argument that part of the volume was not accessible to every molecule. To obtain this result with expansion of the virial, an infinite series is required. The attractive term, a/V^2, introduced by van der Waals to account for the existence of uncompensated forces in the vicinity of the surface (it is also equivalent to the internal pressure), leads with the preceding to a cubic equation (4.3). Solving problem (4.3), we will find the statistical basis of the calculation of the van der Waals equation. In this equation, parameter a, representing the attraction between molecules, as well as the co-volume b, are assumed to be independent of the temperature. The co-volume b corresponds to a hard core interaction potential at $V = b$, the pressure is infinite, and the system is totally jammed.

The free energy F and pressure p hence are in the form of the contribution of two terms: one describing short-range repulsion, the other describing long-range attraction.

4.4.3 Variants of the van der Waals Equation

There are several variants for treating the contribution of the repulsive term (the hard-sphere part of the interaction potential). Scott used a computer to test the different hard-core equations of state which have been proposed by comparing them to the expansions of the virial calculated for fluids consisting of hard-spheres only. He found that the second virial coefficient is always the same, but the third virial coefficient obtained from the **Carnahan–Starling equation** is much better than the one obtained from the van der Waals equation.

Based on a hard-sphere model for the interaction potential and using the Percus–Yevick equation (Appendix B) for the statistical mechanics calculation, Carnahan and Starling proposed the following expression for the contribution of the hard-sphere part of the potential to the compressibility factor in 1969:

$$Z = 1 + \frac{4\eta - 2\eta^2}{(1-\eta)^3} \quad \text{with} \quad \eta = \frac{b}{4v} = \frac{v^0}{v} \tag{4.28}$$

where v^0 is the volume of one mole of hard-spheres and v is the molar volume of the fluid. If we add to (4.28) the equation proposed for the attractive term, for example, a/V^2 (van der Waals), we then obtain a complete form for Z that can be compared with the form obtained with the van der Waals equation.

It was possible to show that the results obtained were equivalent if the parameters in the two types of equations were determined on the same empirical bases. Still, the critical coordinates calculated with these equation are not satisfactory.

The equation given by **Redlich–Kwong** introduces a different volume dependence for the attractive term:

$$p = \frac{RT}{V-b} - \frac{a/T^{0.5}}{V(V+b)} \tag{4.29}$$

This equation was studied in detail by Deiters to test its overall behavior, that is, the different forms of phase diagrams it gives when parameters a and b are modified. Berthelot proposed another relatively similar form of equation:

$$p = \frac{RT}{V-b} - \frac{a}{TV^2} \tag{4.30}$$

Since the results obtained from equations like (4.29) and (4.30) were found to lead to important deviations when they were used to predict the behavior of light hydrocarbons, important fluids in industry, Soave replaced factor $a/T^{0.5}$ in (4.29) by factor $a\,\alpha(T)$ with:

$$\alpha = [1 + (0.48 + 1.57\omega - 0.176\omega^2)(1 - T_r^{0.5})]^2 \tag{4.31}$$

where $T_r = T/T_c$ is the reduced temperature and ω is the acentric factor introduced previously; it accounts for the nonspherical character of the molecules. Parameter a is a constant dependent on the critical coordinates. Values of α and ω are found in R. C. Reid, J. M. Prausnitz, and B. E. Poling, *The Properties of Gases and Liquids*, McGraw-Hill (1987). This is the **Soave–Redlich–Kwong** equation.

In the industry, the equation of state most widely used, together with the preceding, one is the **Peng–Robinson** equation:

$$p = \frac{RT}{V-b} - \frac{a(T)}{V(V+b) + b(V-b)} \tag{4.32}$$

Among other things it gives a better estimation of the second virial coefficient.

In addition to these cubic equations, there are also equations which have a totally different form, for example, the Dieterici equation:

$$p\,e^{[a/RT]} = \frac{RT}{V-b} \tag{4.33}$$

The critical compressibility factor associated with this equation has the value $Z_c = 0.2706$, which is closer to the factor observed for most hydrocarbons, i.e., 0.29 (for the van der Waals equation, $Z_c = 8/3 = 0.375$, as can be found from (4.4)), this is a huge improvement.

Chemical engineers still prefer to use cubic equations of state which are in a way generalizations of the van der Waals equation. We again emphasize that these equations do not satisfactorily account for the behavior of the fluid in the vicinity of a critical point, but this situation is generally not important for chemical engineering since one operates relatively far from the critical point. A practical question nevertheless still remains unsolved: is there a simple rule for finding a correlation between the real critical coordinates and those incorrectly predicted by an analytical equation? We know that the use of (p, V, T) data for liquid and vapor and critical point data give different results for calculations of the parameters contained in the equations of state.

There are other types of approaches for obtaining an equation of state. For some of them, the contributions of the repulsive (hard-sphere part of the potential) and attractive (long-range part) terms in the free energy and pressure are no longer separated. These are called **nonseparable** models. One of these methods is derived from the **Bethe** lattice model in which the particles are assumed to occupy lattice sites; this method is an improvement in the molecular field models because the correlations existing between particles are taken into consideration. Sandler used a potential of the square well type ($u(r) \neq 0$ for $r_0 < r < r_1$) to describe the interactions between molecules. We will return to this type of description in treating the case of mixtures. Another type of equation was obtained by J. P. Hansen, who used the results of a computer simulation method for molecules with a Lennard–Jones potential as given by (3.4):

$$p/kT\rho = 1 + B_1 x + B_2 x^2 + B_3 x^3 + B_4 x^4 + B_{10} x^{10} \tag{4.34}$$

with $x = \rho^*/(T^*)^{1/4}$, $\rho^* = \rho r_0^3$, and $T^* = kT/\varepsilon_0$, where ε_0 and r_0 are the usual parameters of the Lennard–Jones potential. The tenth-order term in x was introduced to account for the very large increase in the pressure at very high density.

4.5 Metastable States: Undercooling and Overheating

4.5.1 Returning to Metastability

If a system is cooled from a temperature just above a phase transition, the phase which is stable at high temperature may subsist below the transition temperature despite the fact that the new phase has the lowest free energy and is thus stable: this is the general phenomenon of **metastability**. We say that the system is **undercooled** (for the liquid/solid transition this is often referred as supercooling).

Inversely, it is often possible to maintain the low-temperature stable phase above the transition temperature: it is **overheated**. To describe this phenomenon in detail, using the case of vapor, for example, we assume that above the condensation point, small droplets have already formed in the homogeneous phase: their size fluctuates and they appear and disappear because they are not stable. Below the condensation temperature, they have a chance of existing and they can coalesce, thus inducing condensation.

The stable and unstable states are called local properties of the free energy or $F(V,T)$ surface because they are represented by the values of the (V,T)-coordinates corresponding to a minimum, respectively a maximum, in F. Moreover, there is a metastable state when two states located at two different points of this surface correspond to minima, one corresponding to a value smaller than F than the other (Sect. 1.2.1): the lowest minimum is stable, the other is metastable. There is a barrier between these two states whose height is responsible for the difficulty encountered in passing from a metastable state to a stable state. We are indebted to Ostwald for distinguishing between these three types of states.

4.5.2 Formation of Drops and Bubbles

When there are two phases present, the system is inhomogeneous and the theory of phase transitions leads us to take into account the free energy of the interface between phases; this is the nucleation principle (Sect. 2.2.3).

Assume that we are investigating condensation of a vapor and that the drops formed are spheres of radius r. As shown previously (2.15) and (2.16), the drops formed are only stable if the vapor is metastable (undercooled, $\Delta g_v < 0$). Then introducing grand potential Ω (sometimes also called J) and designating the pressure of the metastable vapor phase by p, and the pressure inside a drop of area S and volume $V_1 = 4\pi r^3/3$ by p_1, Ω is written:

$$\Omega = -p_1 V_1 - pV_2 + \gamma S \tag{4.35}$$

V_2 is the volume of the vapor phase and γ is the surface tension. As the total volume $V_1 + V_2$ remains constant, the equilibrium condition is written:

$$\mathrm{d}\Omega = -(p_1 - p)\mathrm{d}V_1 + \gamma \mathrm{d}S = 0 \tag{4.36}$$

It is convenient to assume that the droplets is a sphere of radius r ($V_l = 4a\pi r^3/3$ and $S = 4\pi r^2$) so that the pressure difference between the inside and the ouside of the droplet is given by:

$$r^* = 2\gamma/(p_1 - p) \tag{4.37}$$

The radius r^* (solution of (4.36)) corresponds to the critical radius defined in (2.24). Droplets of radius $r < r^*$ will evaporate and those of radius $r > r^*$ will become larger. Equation (4.37) gives the value of the excess pressure existing in the drop.

The probability w that a fluctuation will give rise to a drop of size r^* can be calculated using nucleation theory (Sect. 2.2.3):

$$w \propto \exp\left\{-\frac{16\pi\gamma^3 T_0}{3L_v k^2(\Delta T)^2}\right\} \tag{4.38}$$

T_0 is the condensation temperature, $\Delta T = T_0 - T$ is the difference with respect to this temperature, and L_v is the latent heat of vaporization per unit of volume.

This description only holds for droplets suspended in the vapor, and no undercooling will be observed if the supersaturated vapor is in contact with a solid surface totally wetted by the liquid. A situation of heterogeneous nucleation is then observed, equations (2.29) and (2.30).

4.6 Simulation of Phase Transitions

4.6.1 Principles

Homogeneous nucleation in a vapor forming a liquid or in a liquid forming a vapor or a solid is a phenomenon that occurs in the absence of contaminants or any contact with a wall. For this reason, this phase transition phenomenon is very suitable for a computer simulation calculation.

These calculations consist of simulating the behavior of a macroscopic system by studying the evolution of a large number of particles of the material (of the order of 10^6). The complete study of simulation methods is beyond the scope of this book. However, as they contributed to improving understanding of the behavior of liquids, we will schematically describe the principle of these methods. There are in fact two very different categories of techniques; **molecular dynamics** and the **Monte Carlo** method.

4.6.2 Molecular Dynamics

Molecular dynamics is one of the computer simulation methods used in the physics of phase transitions, primarily in fluids. It is based on the solution of equations of motion for a finite number of particles. They are generally placed in a box using periodic boundary conditions in order to artificially increase the number of particles.

It is assumed that the particles are exposed to a force field $f(\boldsymbol{r}_{ij})$ which represents the forces of interaction between neighboring particles (i) and (j) and which is derived from potential $u(\boldsymbol{r}_{ij})$.

The most widely used method is the Gear "predictor-corrector" method.

For a continuous trajectory, the positions $\boldsymbol{r}$, velocities $\boldsymbol{v}$, etc., near a time $t + \delta t$ are obtained for each particle using a Taylor expansion from time t:

$$\begin{aligned}
\boldsymbol{r}^p(t+\delta t) &= \boldsymbol{r}(t) + \delta t\boldsymbol{v}(t) + \frac{1}{2}\delta t^2\boldsymbol{a}(t) + \frac{1}{6}\delta t^3\boldsymbol{b}(t) + \dots \\
\boldsymbol{v}^p(t+\delta t) &= \boldsymbol{v}(t) + \delta t\boldsymbol{a}(t) + \frac{1}{2}\delta t^2\boldsymbol{b}(t) + \dots \\
\boldsymbol{a}^p(t+\delta t) &= \boldsymbol{a}(t) + \delta t\boldsymbol{b}(t) + \dots \\
\boldsymbol{b}^p(t+\delta t) &= \boldsymbol{b}(t) + \dots
\end{aligned} \tag{4.39}$$

where $\boldsymbol{a}$ is the acceleration and $\boldsymbol{b}$ is the third-order derivative of $\boldsymbol{r}^p$ with respect to time. The superscript p means that this involves predicted values which must be corrected. The equation of motion is introduced in the corrector stage, that is, the corrected acceleration $\boldsymbol{a}^c$ which determines the "error" from the forces at time $t + \delta t$:

$$\Delta\boldsymbol{a}(t+\delta t) = \boldsymbol{a}^c(t+\delta t) - \boldsymbol{a}^p(t+\delta t) \tag{4.40}$$

This "error" is used to determine the corrected positions and velocities by means of the following relations:

$$\begin{aligned}
\boldsymbol{r}^c(t+\delta t) &= \boldsymbol{r}^p(t+\delta t) + c_0\Delta\boldsymbol{a}(t+\delta t)\\
\boldsymbol{v}^c(t+\delta t) &= \boldsymbol{v}^p(t+\delta t) + c_1\Delta\boldsymbol{a}(t+\delta t)\\
\boldsymbol{a}^c(t+\delta t) &= \boldsymbol{a}^p(t+\delta t) + c_2\Delta\boldsymbol{a}(t+\delta t)\\
\boldsymbol{b}^c(t+\delta t) &= \boldsymbol{b}^p(t+\delta t) + c_3\Delta\boldsymbol{a}(t+\delta t)
\end{aligned} \tag{4.41}$$

Coefficients c_0, c_1, c_2, c_3 are tabulated in the specialized literature, where algorithms based on this method are found for performing the calculations.

There is another method of integrating the equations of motion, called the Verlet method. The position of each particle is calculated with the equation:

$$\boldsymbol{r}^p(t+\delta t) = 2\boldsymbol{r}(t) - \boldsymbol{r}(t-\delta t) + \delta t^2\boldsymbol{a}(t) \tag{4.42}$$

With this algorithm, the velocity does not have to be calculated. This method also has the advantage of permitting reversal of the direction of the time.

It should be noted that if the precision in determination of the trajectory of the particles is to be improved, it is then necessary to take the terms of order greater than 3 in (4.39), which allows taking larger δt intervals but this has the drawback of increasing the calculation time.

The real difficulty of the molecular dynamics method is that the trajectory of particles in phase space (in the sense of statistical mechanics, here in $6N$ dimensions) are very sensitive to small changes in the initial conditions (velocities and coordinates), which makes the process of integrating the equations of motion with a finite difference methode essentially unstable. Changing the calculation steps implies modifying the initial conditions. The smaller δt, the better the accuracy of the calculation will be, but it is necessary that the total time interval $n\,\delta t$ (if n is the number of steps) on which it is executed be greater than the relaxation time of the characteristic physical quantities of the material.

For example, if we want to calculate the equation of state for a fluid, we can use the virial theorem (Problem 4.1), which allows writing:

$$pV = NkT - 1/6 < \sum_{i \neq j} \boldsymbol{f}_{ij}.\boldsymbol{r}_{ij} > \tag{4.43}$$

where $< >$ is an overall average and where it is assumed that the kinetic energy was purely translational.

The more particles in the sample used, the greater the chance the simulation has of representing reality, but the longer it takes by the calculation. We are thus guided by considerations of saving calculation time. In general, a molecular dynamics simulation is executed with several thousand particles.

The choice of potential $u(\boldsymbol{r}_{ij})$ is obviously essential for performing the calculations, a bad potential will give bad results no matter whether it intervenes in $\boldsymbol{a}$. We will return to this problem repeatedly.

The first molecular dynamics calculation was performed on a system of some one hundred hard-spheres by Alder and Wainwright in 1957. The classic potential in this case is relatively simple: it consists of taking a potential $u(r)$ and representing an infinite force for $r < r_0$ and being zero for $r > r_0$. It is thus a law for forces between rigid particles without attraction: when they are separated by a distance greater than r_0, they do not interact. For $r = r_0$, the force becomes infinite and they repel each other.

This hard-sphere potential is not very realistic. The form that has been most widely used corresponds to the **Lennard–Jones** potential.

It comprises an attractive part which varies in r^{-6} and a part describing the repulsive forces which are assumed to vary in r^{-12}. The Lennard–Jones potential is thus written as:

$$u(r) = 4\varepsilon_0[(r_0/r)^{12} - (r_0/r)^6] \tag{4.44}$$

For argon atoms, for example, we have $\varepsilon_0/k = 120$ K and $r_0 = 0.34$ nm. We show the general shape of the curve representing the Lennard–Jones potential in Fig. 3.8b.

Hansen and Verlet are among the first to have developed equation of state calculations with the molecular dynamics method using a Lennard–Jones potential.

This method is not limited to the study of the behavior of fluids. Molecular dynamics calculations were recently performed on ice in conditions of high pressure. While ice has been studied up to pressures of the order of 1 Mbar, these calculations predict the existence of a new crystalline form of ice, ice XI, which is stable up to 4 Mbar and a temperature of 2000 K.

4.6.3 Monte Carlo Method

The molecular dynamics method by definition allows simulating the dynamic behavior of a system in short time intervals. In general, all dynamic considerations can be abandoned (particular when the macroscopic aspect of a phase transition is of interest), and only the equilibrium behavior is thus of interest.

The Monte Carlo methods of simulation completly avoid describing the dynamic behavior of a system. They were introduced in 1953, when computer science was in its infancy, and they consist of producing states $i, j, k, \ldots$ by a stochastic process so that the probability of obtaining one of these states is the appropriate statistical distribution function.

One of the goals of statistical mechanics is to calculate the mean value $< X >$ of the thermodynamic quantities characteristic of a system (energy, pressure, magnetization, etc.). $< X >$ is written as a summation over all of the states in the phase space:

$$< X >= \sum_i X_i \exp(-\beta \mathcal{H}_i) \,/\, \sum_j \exp(-\beta \mathcal{H}_j) \tag{4.45}$$

where $\mathcal{H}_i$ is the Hamiltonian of the system in configuration i. For example, if we take a magnetic material represented by the Ising model (two values for the spins) and for a lattice with N sites, in principle, we must perform a summation in (4.45) on 2^N states. This number of states, or configurations, is a rapidly increasing function of N. For $N = 9$, we have $2^9 = 512$ possible states and for $N = 100$, the number of possible configurations is $2^{1000} \cong 10^{30}$, etc. The numerical calculation of the mean values very quickly exceeds the capabilities of the most powerful modern computers and is practically unfeasible.

In fact, this difficulty can be turned around if we consider that in statistical mechanics, we know that the system will pass most of its time in states with which values of the thermodynamic quantities close to the equilibrium values can be associated. We can thus imagine a calculation strategy that would allow limiting the number of states used to perform the averaging operations (4.45).

The Metropolis algorithm is generally used for Monte Carlo simulation calculations. The principle of this calculation can be described on an Ising spin model with the following stages:

1. Begin with an initial state for the N spins representing a certain state (i) by causing one or more spins to randomly flip, and we obtain a new state (f). If E_i and E_f designate the initial and final energies of the states, then $\Delta E = E_f - E_i$ is the change in energy associated with this spin flip, and we calculate $w = \exp(-\Delta E/kT)$.
2. Randomly select a number r such that $0 < r < 1$.
3. Compare r and w.
4. • If $w < r$, the final state (f) obtained by spin inversion is rejected and the operation of selecting a new state is begun again.
   • If $w > r$, the final state (f) is retained and is accepted with a probability equal to 1 if $\Delta E < 0$ and a probability equal to $\exp(-\Delta E/kT)$ if $\Delta E > 0$, then the operation is begun again by flipping other spins.
5. The average values of the thermodynamic quantities are calculated on states (f) obtained in the process.

This method is stochastic, hence its name, and it generates states in a **Markov** chain.

We note that a Markov process is a chain of randomly occurring events so that a new configuration of the system is only a function of the previous state and not of former configurations.

The Monte Carlo method and the Metropolis algorithm can be used to simulate the behavior of a system using the Gibbs distribution function. The probabilities of a transition from one state to another should satisfy the microreversibility principle, that is, detailed balancing.

If $P(i \to f)$ is the probability of a transition from one state to another and w_i and w_f are the probabilities of being in states (i) and (f), we should have:

$$w_i\, P(i \to f) = w_f\, P(f \to i) \tag{4.46}$$

The first states obtained with the stochastic process from the initial state of the system should not be used for calculating the mean values (step 5) in a Monte Carlo calculation; they retain a certain memory of this state in effect.

The principal difference between the Monte Carlo method and a molecular dynamics calculation is that the molecular motions in a Monte Carlo simulation do not correspond to the "natural" evolution of the system in time. Dynamic information cannot be drawn from such a simulation, which only produces static quantities at equilibrium (the pressure, for example) and permits calculating the equation of state.

The Monte Carlo method was used and perfected by Panagiotopoulos for simulating liquid/vapor equilibrium by introducing two boxes in which particles with identical pressures and chemical potentials are confined. The calculation is begun with different temperatures and densities and ends at equilibrium at a given temperature with different densities.

Monte Carlo methods have also been used to determine the phase diagrams of metal alloys. In these cases, a crystalline system with 10^4 to 10^5 sites is generally used, taking only interactions between closer neighbors into consideration.

4.7 Binary Mixture of Two Components

Gas/liquid systems with several components of the type often found in the petroleum industry can be observed in the form of several coexisting phases. The study of the phase changes for such systems thus has undeniable practical applications. To simplify matters, we will first treat the case of binary systems that can coexist in the form of three phases.

4.7.1 Conditions of Phase Equilibrium in a Binary Mixture

The general rules of phase equilibrium in mixtures were given by Gibbs. At a fixed temperature, the pressure and the chemical potentials of the components should be equal in the coexisting phases:

$$T^\alpha = T^\beta \quad \text{and} \quad \mu_i^\alpha = \mu_i^\beta \tag{4.47}$$

where α and β are the two phases with constituents i ($i = 1, 2$ for a binary mixture). Additional conditions must be satisfied to ensure the stability of the equilibrium. As the Gibbs function is the thermodynamic potential that is a function of all variables in the system, We have also:

$$G(p^\alpha, T^\alpha, \mu_1^\alpha, \mu_2^\alpha) = G(p^\beta, T^\beta, \mu_1^\beta, \mu_2^\beta) \tag{4.48}$$

If N is the total number of moles, N_1 and N_2 are the number of moles of constituents (1) and (2). Designating the concentration of (2) by $x \equiv x_2 = N_2/N$, the concentration of (1), can be written as $x_1 = N_1/N = 1 - x$.

Introducing the Gibbs molar function $G_m = G/N$, we then obtain (Appendix A):

$$\begin{aligned} \mu_1 &= \frac{\partial G}{\partial N_1} = G_m - x\frac{\partial G_m}{\partial x} \\ \mu_2 &= \frac{\partial G}{\partial N_2} = G_m + (1-x)\frac{\partial G_m}{\partial x} \end{aligned} \tag{4.49}$$

That is

$$\mu_2 - \mu_1 = \frac{\partial G_m}{\partial x} = \frac{\partial F}{\partial x} \tag{4.50}$$

where F is the Helmoltz free energy per mole. Using the definition of G and the Gibbs–Duhem relation (1.1) one obtains (using $p = -N(\partial F)/(\partial V)$):

$$\begin{aligned} G &= NF + pV = N_1\mu_1 + N_2\mu_2 \quad \text{or} \\ G_m &= F - V\frac{\partial F}{\partial V} = (1-x)\mu_1 + x\mu_2 \end{aligned} \tag{4.51}$$

resulting in

$$\mu_1 = F - V\frac{\partial F}{\partial V} - x\frac{\partial F}{\partial x} \tag{4.52}$$

The condition that this quantity is the same for each phase is equivalent to Maxwell's rule for a one-constituent system. This is expressed by the condition that the plane tangential to $F_T(v, x)$-surface, as expressed in equation (4.52), must contain the line connecting the two contact points (we have the equivalent of Fig. 4.3b for a one-component system). The conditions of equality of pressures and differences in chemical potential $\mu_2 - \mu_1$ require that the planes tangential to the surface at these two contact points coincide.

The general condition of stability of phase equilibrium according to Gibbs–Duhem imposes a simple stability condition:

$$\left(\frac{\partial^2 G_m}{\partial x^2}\right)_{p,T} > 0 \tag{4.53}$$

4.7.2 Systems in the Vicinity of a Critical Point

The Gibbs phase rule of variance (1.3) applied to a binary system ($c = 2$) which can be in equilibrium in three different phases ($\varphi = 3$) corresponds to variance $v = 1$. In this case, we no longer find an isolated critical point but a line of critical points in (p, T)-plane.

A series of phase diagrams in the (p, T)-plane of the mixture N_2–CH_4 for different concentrations x (these curves, Fig. 4.4, with $x = C^{te}$ are called isopleths). The curves for pure systems ($x = 0$ and $x = 1$) correspond to the left and right parts of the diagram. These curves are not liquid–gas coexistence curves since vapor and liquid coexist at different concentrations. One branch of a curve corresponding to the dew points with concentration x_V and a curve corresponding to the bubble points and concentration x_L intersect at each point of the (p, T)-plane. The curves with their two branches (dew and bubble) are tangential to the critical line C_1C_2. The condition of phase equilibrium assumes that the chemical potentials of each constituent are equal in each phase.

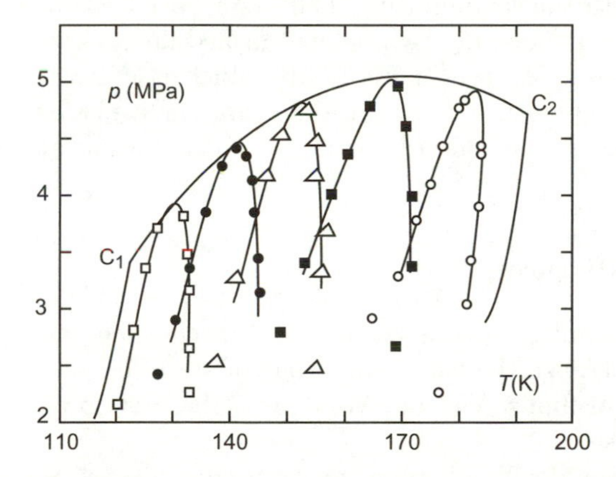

Fig. 4.4. Critical line for a binary mixture. There are five liquid/gas lines for a N_2–CH_4 mixture at the concentrations (from left to right) 0.1002–0.2897–0.5088–0.697–0.8422 (isopleths). There is a line of critical points which is upper curve C_1C_2. The isopleth lines were calculated with a model.

In a binary mixture, it is possible for three phases to coexist along a line of triple-points. Near these points one may have three phases: a vapor phase and two liquid phases. If we attempt to mix two liquids A and B, we can either obtain a totally homogeneous mixture or an incomplete mixture with two liquid phases separated by a meniscus. In the last case, the fluid consists of a mixture with a phase strongly concentrated at A and weakly concentrated at B on one side, and on the other side, the opposite situation prevails. In general, we obtain a homogeneous mixture at high pressures and temperatures and inhomogeneous mixtures at low temperatures.

An example of a system that can be in the one, two, or three phases as a function of the external conditions is shown in Fig. 4.5. The diagrams are projections in the (p, x), (T, x), and (p, T)-planes of the (p, T, x)-surface representing the system at equilibrium. This can be in the form of a gas, a homogeneous mixture of two liquids L_1 and L_2, or two separate liquid phases L_1 and L_2. We thus have two separate critical lines. Line PQ (Fig. 4.5a) goes from point P, the critical point of pure fluid A ($x = 0$), to point Q, the critical point of pure fluid B ($x = 1$). The critical points along this line correspond to all values of x ($0 < x < 1$) and are of the liquid/gas type. The second critical line CWD is the locus with a series of critical points of the liquid–liquid type. Points C and D are respectively called lower critical end point(LCEP) and upper critical end point (UCEP); at those points unmixing starts and stops. The projection of this line on the (T, x)-plane is the line NFWR, the dashed line below NFWR corresponds to metastable states. H is the liquid L_1 with low concentration and K the liquid L_2 with high concentration. The critical point W corresponds to **unmixing**, where the homogeneous mixture L_1 and L_2 is separated into two liquid phases L_1 and L_2, and the concentration dependence varies somewhat with the temperature. The lower part CED of the diagram is a line of triple points (gas, L_1, L_2); below (dashed line), there is a line of unstable critical points. We see in Fig. 4.5b, which is a cross-section of (p, V, T)-surface at temperature T_E, that there are three phases with concentrations x_G, x_H, and x_K at this temperature which coexist at pressure p_{GHK}.

4.7.3 Equation of State of Mixtures

The most important applications of equations of state are in chemical engineering. It is necessary to generalize the usual equations of state for pure fluids to describe the mixtures of fluids. Van der Waals was the first to do this (1890). Following him we use:

An equation of state of the van der Waals type can be used:

$$\left(p + \frac{N^2 a(x)}{V^2}\right)(V - Nb(x)) = NkT \tag{4.54}$$

with

$$a(x) = (1 - x)^2 a_{11} + x(1 - x)a_{12} + x^2 a_{22} \tag{4.55}$$

where x is the concentration of the second constituent of the mixture in the first, which is generally the less volatile constituent. A linear or bilinear expression in x is frequently selected for $b(x)$.

A term corresponding to the entropy of mixing must be introduced in the free energy F.

$$F = F_0 + NkT\left[(1 - x)\ln(1 - x) + x\ln x\right] \tag{4.56}$$

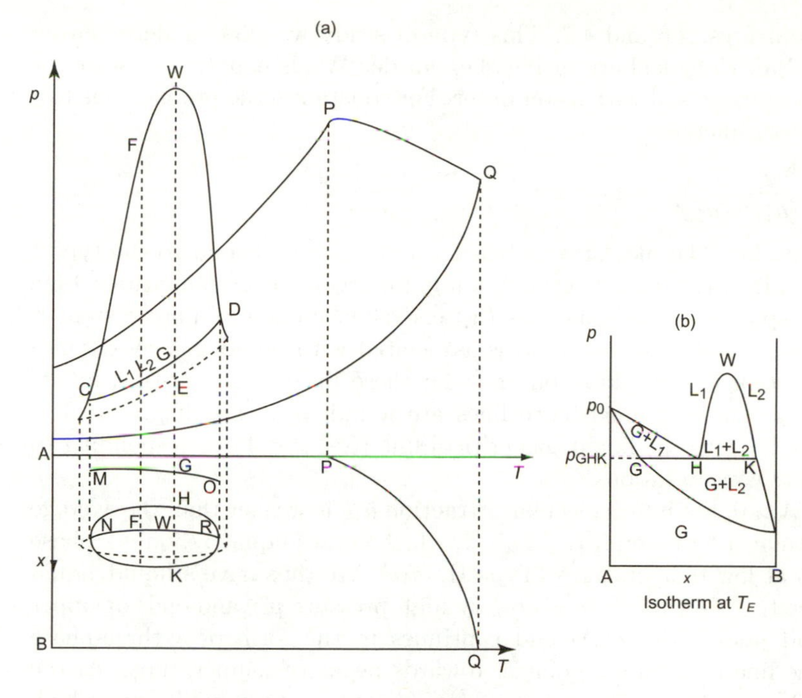

Fig. 4.5. Binary mixture diagram. (**a**) Binary (p, T) and (x, T) diagrams, the line PQ is a line of critical points, CD is a coexistence line of three phases L_1 (liquid 1), L_2 (liquid 2), G (gas), and line CWD corresponds to liquid/liquid critical points. (**b**) Phase diagram $p(x)$ which is a vertical section of diagram (**a**) at the temperature T_E corresponding to points W, E, G, H, K. The dashed line below CED corresponds to metastable states. NWR is the projection of CED.

At the beginning of the 20th century, a_{11} and a_{22} were determined experimentally from the critical temperatures and densities of the pure solvent (component 1) and the pure solute (component 2) and the following expression was taken for a_{12}:

$$a_{12} = \sqrt{a_{11}a_{22}}$$

(Lorentz–Berthelot rule). This choice was later justified by the fact that this quantity comes from the dipole moments mutually induced in the molecules (they correspond to the **van der Waals forces**). The predictions of the phase diagrams that can be made with this method are very sensitive to the precise value of a_{12}. The chemical engineers use the following expression for a_{12}:

$$a_{12} = (1 - k)\sqrt{a_{11}a_{22}} \tag{4.57}$$

where factor k is an empirical coefficient.

One might consider the global behavior of an equation of state, that is, its behavior for different values of interaction parameters a. The different situations that can be obtained for molecules of equivalent size ($b(x)$ = constant) are shown in Figs. 4.6 and 4.7. This type of study was first undertaken by Scott and Van Konynenburg, using the van der Waals equation of state for binary mixtures, solved with a computer. For constant b the problem has two adjustable parameters:

$$\zeta = \frac{(a_{22} - a_{11})}{(a_{11} + a_{22})} \quad \text{and} \quad \Lambda = \frac{(a_{11} + a_{22} - 2a_{12})}{(a_{11} + a_{22})}$$

used in Fig. 4.6. The plot areas labeled I and I-A correspond to the type I diagram in Fig. 4.7. Such a diagram is found for argon/krypton mixtures. Line lg(A) corresponds to the liquid/gas (lg) coexistence curve of pure system A (pure argon, $x = 0$) and line lg(B) is associated with liquid/gas coexistence for pure system B (pure krypton, $x = 1$); these lines end at critical points C_A and C_B. These lg-coexistence lines are found in all six diagrams. The dashed line l $=$ g is the liquid/gas critical line ($0 < x < 1$) corresponding to intermediate concentrations.

When $\Lambda > 0$, the intermolecular attraction a_{12} is weaker than the average intermolecular interaction $(a_{11} + a_{22})/2$, which causes liquid 1/liquid 2 phase separation at low temperatures (Type II, etc.). We thus have a liquid/liquid critical line $l_1 = l_2$ which begins at very high pressure p^{∞} and ends at upper critical end point $T_{UCEP}(\Delta)$ and continues in the form of a three-phase coexistence line to terminal point U towards negative temperatures. At the $T_{UCEP}(\Delta)$ point, the system separates into three phases: liquid l_1 with a high concentration of argon, gas, and liquid l_2 with a low concentration of argon. There is a coexistence line l_1 l_2 g. There are different concentrations x at each point of these lines. Type I diagrams can be considered as a special case of type II where the l_1 l_2 g branch is located in regions with $T < 0$ which have no physical meaning.

The lines $l_1 = l_2$ and l_1 l_2 g have an upper critical end point, represented by triangle $\triangle$, and a lower one, represented by $\bigtriangledown$.

Consider the upper right part of Fig. 4.6 (corresponding to $a_{22} > a_{11}$); this is in a type III region. The corresponding diagrams differ from type II diagrams since terminal points C_A, C_B, p^{∞}, and U are connected differently. C_A is connected to U and C_B is connected to p^{∞}. This last branch may veer to the right and these critical lines represent sequences of gas–gas critical points. Near these points the gases are unmixing. This was discovered by Kamerlingh Onnes and predicted by van der Waals. Actually, there is no sharp distinction between a gas and a liquid, as it is already clear from the situation in a one-component system above the critical point. One also may have the situation that a gas bubble is sinking in the liquid. The curve labeled "g.m" in Fig. 4.6 corresponds to the geometric mean in (4.57) of coefficient a_{12}, that is, to $k = 0$, while the curve labeled "a.m" corresponds to the arithmetic mean $a_{12} = (a_{11} + a_{22})/2$.

For type IV diagrams, critical line $l_2 = g$ originating from C_B dips down into line l_1 l_2 g as found in type III and hence the point U (the end point for $T < 0$) is now connected to point p^{∞}. The other critical line goes from

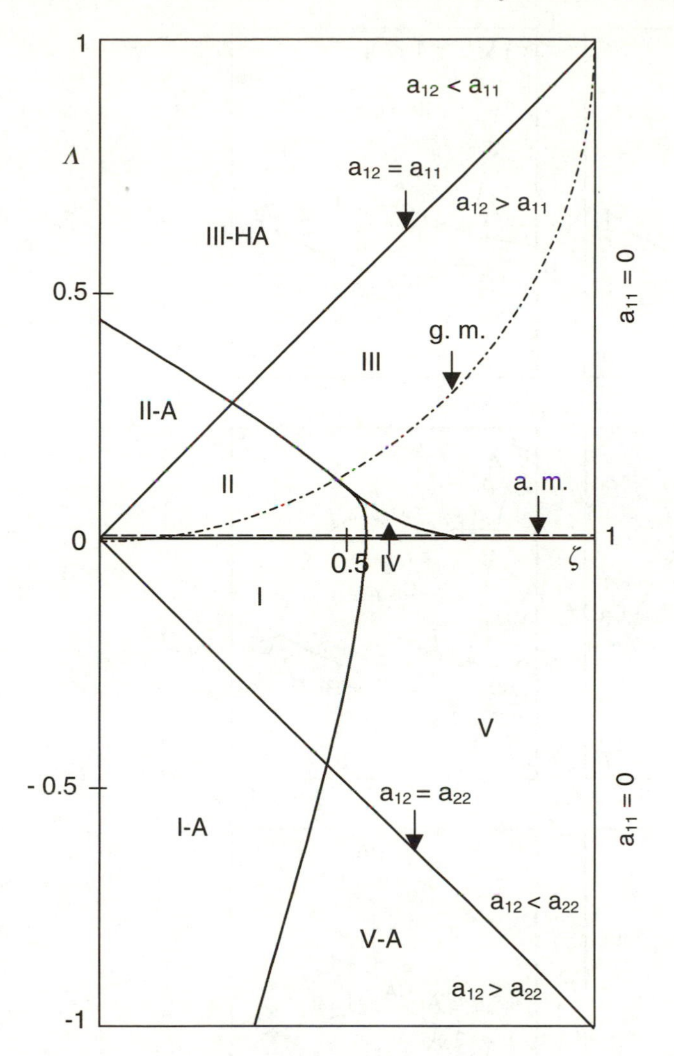

Fig. 4.6. Binary mixtures and van der Waals equation.
These diagrams (van Konynenburg and Scott) describe the behavior of a binary fluid of molecules of the same size as a function of the parameters:

$$\zeta = \frac{(a_{22} - a_{11})}{(a_{11} + a_{22})} \quad \text{and} \quad \Lambda = \frac{(a_{11} + a_{22} - 2a_{12})}{(a_{11} + a_{22})}$$

There are nine regions corresponding to different forms of (p, T) diagrams separated by solid lines.

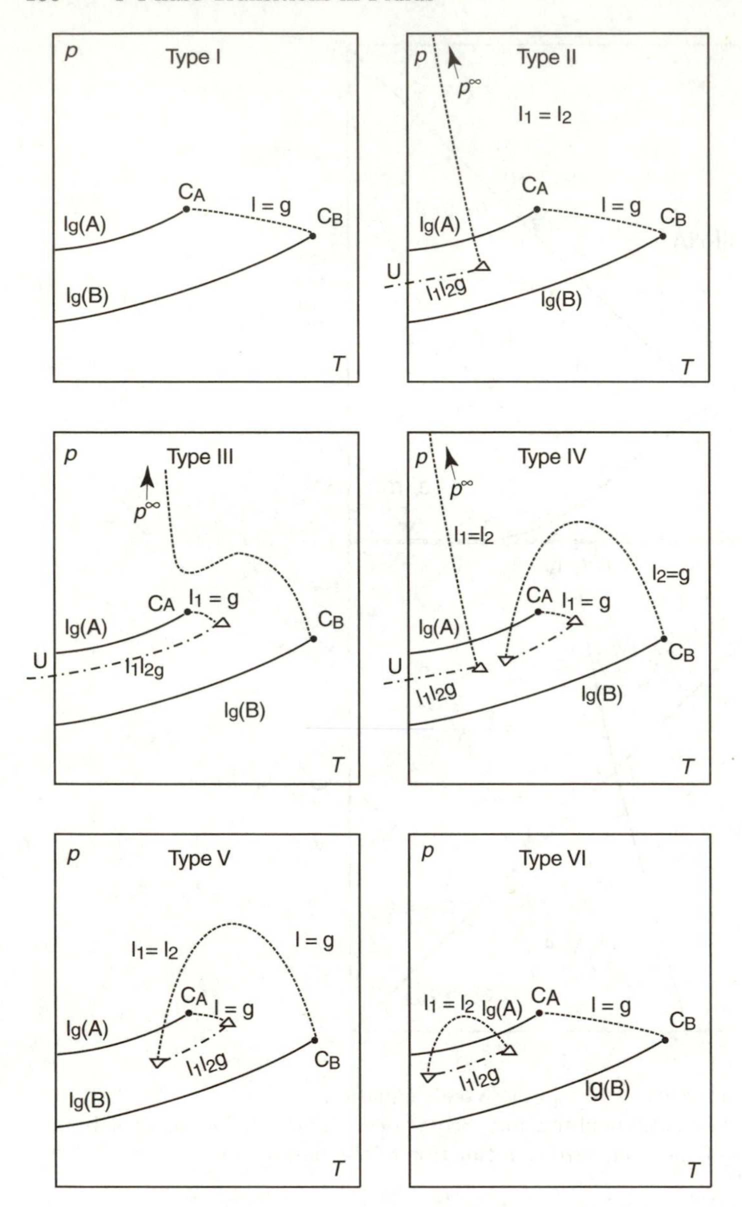

Fig. 4.7. Phase diagram topology. The types of diagrams correspond to the regions in Fig. 4.6. Lines lg(A) and lg(B) refer to liquid/gas coexistence for pure compounds A and B. They end in critical points C_A and C_B. The dashed lines correspond to specific critical lines $l = g$, $l_1 = l_2$, etc. The triangles are the critical end points of these critical lines . The dash-dotted lines are associated with the coexistence of three phases. Type VI corresponds to Fig. 4.5 [P. H. van Konynenburg and R. L. Scott, *Philos. Trans. R. Soc.*, **298**, 495–594 (1980)].

C_A to C_B via a three-phase line l_1 l_2 g which begins at uper critical end point (triangle pointing up) and ends at the lower critical end point (triangle pointing down). A type II diagram is retrieved when the segment connecting these two critical end points disappears.

Fig. 4.8 represents a p–T plot with the critical point indicated. This plot represents experimental work in a schematic way. A mixture at constant concentration is placed in a closed Toricelli tube with mercury on the bottom (indicated as black in the figure), such that the volume can be changed and the pressure measured (this volume change is not indicated in the figure). The tube is placed in a thermostat. The inside of the loop represents the coexistence of vapor (V) and liquid (L). The black dot is the critical point. The curve to the right of the critical point is the dew-point curve; the curve to the left the bubble-point curve. The two figures on the left show the most common behavior: increasing the pressure leads to formation of a droplet and finally at the top nearly all vapor is gone and a bubble remains. The inserts on the right of the critical point show the so called "retrograde condensation". If the pressure is increased a droplet forms, and subsequently some liquid, but the liquid disappears again at higher pressures. Retrograde means reversal of the previous direction. The word was coined by Kuenen in 1892, the first to observe this sequence. Retrograde condensation occurs in case were the critical line starting from the point C_B has initially a positive slope before veering to the left instead of taking off with a negative slope as shown in Fig. 4.7. Consequently there will be two critical points at a given

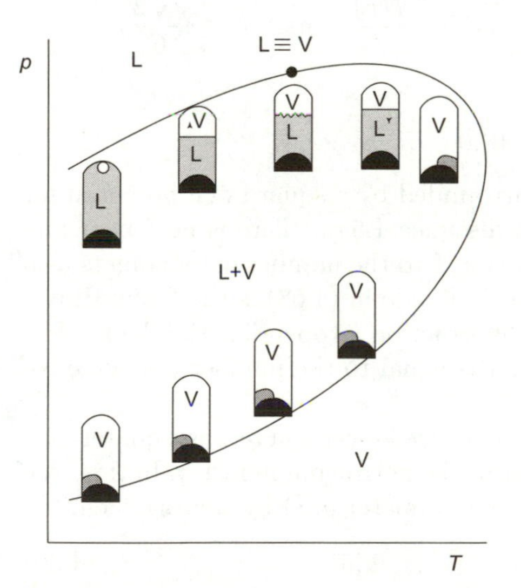

Fig. 4.8. Behavior of a binary mixture. The diagram shows the different phases of the mixture coexisting at fixed x. The black dot is the liquid/gas critical point. The liquid (L) and vapor (V) phases are mixtures of two constituents with different concentrations. The position of the meniscus has not been adjusted in the figure.

temperature slightly above T_{C_B}. In Fig. 4.8, which is at fixed x, one will notice on the right side that, upon increased pressure, the liquid will come and go.

For type V diagrams, line $l_1 l_2$ of type IV is shifted to negative temperatures. As for type VI diagrams, they can only be obtained with a more complicated model than given by (4.54) and (4.55); they correspond to those in Fig. 4.5.

4.7.4 Mixtures of Polymers or Linear Molecules

Alkanes are typical of the approximately linear molecules encountered in the chemical industry and in petroleum engineering. It is necessary to utilize specific equations of state to predict their behavior. Polymers also can be considered as n chains of monomers, but the chain is generally not linear.

The classic theory of polymer is Flory's theory; it is based on a lattice model transposition for liquids: each molecule occupies several cells within the liquid.

The most widely used equation of state for chain molecules is derived from the so-called perturbed rigid chain theory. It is in the following form:

$$p = \frac{RT}{V} + \frac{cRT}{V}\frac{4\eta - 2\eta^2}{(1-\eta)^3} - \frac{RT}{V}\frac{z_m c V^* Y}{V + V^* Y} \tag{4.58}$$

with

$$\eta \equiv \frac{\pi r_0^3 N}{6V} = \frac{\pi}{6}\frac{\sqrt{2}V^*}{V} = \frac{\tau V^*}{V}; \quad (V^* \equiv \frac{N r_0^3}{\sqrt{2}} \quad \text{and} \quad \tau \equiv \frac{\pi\sqrt{2}}{6})$$

and

$$Y = \exp\left(\epsilon q / 2ckT\right) - 1; \quad z_m = 18.0$$

The hard core is of size r_0; it is surrounded by a square well potential for $r_0 < r < 1.5\, r_0$ and depth ε. Beyond distance $1.5\, r_0$, there is no longer any interaction. The parameter q is proportional to the number of monomers n in the molecule. The second term on the right side in (4.58) is the **Carnahan–Starling** expression for the replusive interaction (Appendix B). The parameter c was introduced by Prigogine and is equal to the number of degrees of freedom.

The preceding expansions can be generalized for use in ternary, quaternary mixtures, etc. They are very important in the petroleum industry. In this case one will generalize (4.54) and (4.55) for parameter $a(x)$ by writing them:

$$a = \sum_{i,j} a_{i,j} x_i x_j \quad \text{with} \quad a_{i,j} = (1 - k_{i,j})\sqrt{a_i a_j} \tag{4.59}$$

and we will also write:

$$b(x) = \sum_i b_i x_i$$

where x_i and x_j are the concentrations of constituents (i) and (j) in the mixture.

4.7.5 Binary Mixtures far from the Critical Point

Mixtures which have separated (demixing) are in the form of two coexisting states. The associated equation of state $p = p(T, V, x)$ is a surface in four-dimensional space which can only be described by projection in two- or three-dimensional subspaces. In general, representations in the (p, T)-plane and the (V, x)-plane are used. Figure 4.8 is thus a (p, T) diagram in which the liquid/gas critical point has been shown at the vertex of the curve. A phase transition experiment was also schematically shown. A binary mixture of fixed concentration was captured in a Toricelli tube; mercury was placed at the bottom of the tube (it was turned upside down in a cuvette containing the mercury) and the volume could thus be varied and the pressure measured. The tube was thermostated. There is liquid (L)/vapor (V) equilibrium inside the coexistence curve. The curve to the right of the critical point (L $\equiv$ V) corresponds to the dew curve and the left part corresponds to the bubble curve. The two tubes shown to the extreme left of the figure (top and bottom) show the most frequent behavior: an increase in pressure at constant temperature induces the formation of a drop and in the final phase, all of the vapor has almost disappeared and a bubble remains. Similar experiments were conducted on board the space shuttle (D. Beysens).

The tubes to the right of the critical point exhibit so-called retrograde condensation: if the pressure is increased, a droplet forms, then a little liquid, but at higher pressure, it disappears. This phenomenon was observed by Kuenen in 1892.

The location of the critical point on the curve is a function of the nature of the constituents of the mixture. The critical point can also be located on the lower part of the curve. This is the case in natural gas. In this case, retrograde condensation means that vapor is formed first when the pressure is reduced, liquid reappears. This phenomenon is observed in natural gas lines.

The dew and bubble curves are often plotted in the (p, x)-plane (Figs. 4.9 and 4.10). Figure 4.9a shows the curves for a temperature below the critical temperatures of the constituents. If a mixture of composition x is decompressed from point a, a bubble forms at point l of composition y: the vapor is richer in constituent 2. At point a', the liquid has composition x' and the vapor has composition y' (points l$'$ and v$'$ in the diagram), and the concentration of the vapor decreases. The number of molecules in each phase and the volume of the liquid and vapor phases can be determined from these concentrations and using the **lever rule**. Finally, when the pressure corresponds to point v$''$, a small drop of liquid with concentration corresponding to l$''$ remains. These sequences, repeated for different values of x, correspond to the principle of fractional distillation; the mixture is gradually enriched with the least volatile constituent.

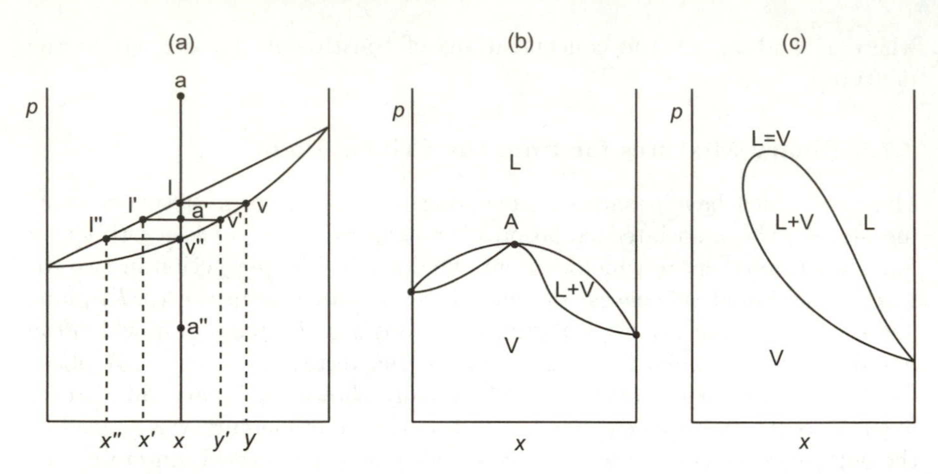

Fig. 4.9. Dew and bubble curves. $p(x)$ diagrams at different temperatures: (**a**) $T < T_{C_1} < T_{C_2}$. (**b**) $T < T_{C_1} < T_{C_2}$, with an azeotropic point. (**c**) $T_{C_1} < T < T_{C_2}$.

Figure 4.9b shows a diagram in which two branches touch at point A, called the azeotropic point (the bubble and dew curves have identical maxima); in its vicinity, distillation is impossible because there are no differences in concentration between liquid and vapor. Diagram of Fig. 4.9c is still different, it corresponds to a temperature T between the two critical temperatures T_{C_1} and T_{C_2} of the components of the mixture ($T_{C_1} < T < T_{C_2}$). There is no longer a contact with the axis $x = 0$ and it has a liquid/gas critical point.

A three-dimensional diagram in (p, T, x)-space describing a binary fluid corresponding to type V in Fig. 4.7 is shown in Fig. 4.10a.

At T_1 the situation is similar to Fig. 4.9a. At T_2 (this cross section is also shown in Fig. 4.10c) the upper part of the "banana" shaped curve has deformed and developed an inflection point with horizontal tangent, labelled (d) both in the main figure as well as in Fig. 4.10b. This is the onset of the formation of the second "hump" and hence this is, per definition, the lower critical end point (LCEP) of the three phase line.

On the far left side of this cross section the two phase region still is attached to the $x = 0$-axis by a curve in the form of a "wedge". At slightly higher temperature one reaches T_{c1} and the curve will detach from the axis (as illustrated by Fig. 4.9c). This is the start of the V–L_1-critical line shown as bc in subfigure (**b**) going through the peaks of the left hump in Fig. 4.10d. At this temperature (T_3) one has three phases as indicated by the horizontal line. The projection of this line on the $x = 1$-plane forms a locus indicated by a dash-dotted line, which starts at d_p (the projection of d) and ends at c_p (the projection of (c)), the upper critical end point (UCEP). At this temperature T_4 (see cross section Fig. 4.10e) the left V–L_1-phase separation has shrunk into a point, again with an inflection point with horizontal tangent. It is near

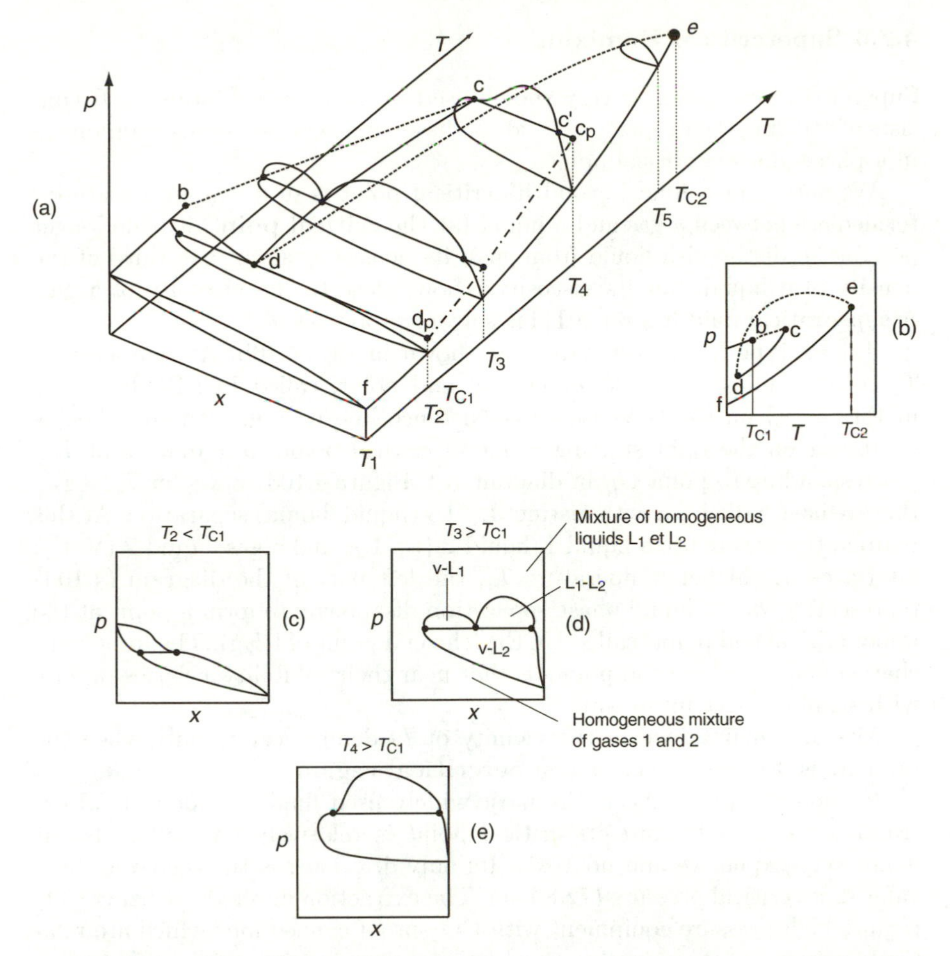

Fig. 4.10. The p, T, x-diagram of a binary mixture. (**a**) Diagrams in p, T, x space with cross-sections corresponding to different special temperatures. (**b**), (**c**), [**d**), [**e**) Cross-sections for these temperatures. The diagrams are type V, $T_2 = T_{LCEP}, T_4 = T_{UCEP}$ [V. J. Krukonis and M. A. McHugh, *Supercritical Fluid Extraction*, Fig. 3.5, Butterworth, Boston (1986)].

such a point that the solubility is extremely sensitive to small changes in pressure and hence is very useful for supercritical extraction.

At T_5 the L_1–L_2-critical line, which originated at the point (d), still continues and collapses into the $x = 1$ plane at the point (e) (At temperature T_{c2}).

The curve f–e represents the vapor liquid equilibria for pure $x = 1$ and is used for reference in Fig. 4.10b. This curve is similar to the one in Fig. 4.7 type V where it is labelled as: lg(B).

4.7.6 Supercritical Unmixing

Supercritical unmixing, is very widely used in the chemical industry. It consists of utilizing the property of fluids of dissolving very numerous compounds in a phase above a critical point.

We note that a fluid beyond its critical point is in a way in a state intermediate between a gas and a liquid (at the **critical point**, it is no longer possible to distinguish liquid from gas): its density is about one third of the density of a liquid, but its viscosity is low, close to the viscosity of a gas. A supercritical fluid has remarkable solvent properties.

Take a type V phase diagram as shown in Fig. 4.10b. At temperature T_2 corresponding to the lower critical end point, called LCEP, (line l_1 l_2 in Fig. 4.7V) on the three-phase coexistence line l_1 l_2 g, a "hump" begins to appear on the right starting a line of critical points terminating at T_{C_2} (corresponding to point C_B in diagram V). Figure 4.10d shows for $T_3 > T_{C_1}$ that we have a diagram with distinct L_1–L_2 (liquid/liquid) separation. At this temperature, there are a liquid 1/liquid 2 (L_1–L_2) and a gas/liquid 2 (V–L_2) interfaces. At higher temperature T_4, the left part of the diagram (4.10d) representing vapor/liquid phase separation disappears to form a point at the upper critical end point, called UCEP (the end point of $l_1 l_2$g). The important characteristic of these end points is that near their solubility x varies rapidly with small changes in pressure.

Mixtures maintained in the vicinity of T_4 demix very rapidly when the pressure is decreased. This is a **supercritical region**.

Carbon dioxide, CO_2, is the most widely used fluid in supercritical extraction methods because its critical point is relatively low (301.4 K), it is not very expensive and nontoxic. Its only drawback is the relatively high value of its critical pressure (72.8 bar). The extraction methods utilizing CO_2 require high-pressure equipment with CO_2-proof connections which are relatively expensive; they are thus used for extracting various high added value substances. The principle of the supercritical extraction method is simple: it is sufficient to pass the supercritical fluid (CO_2, for example) through a solution from which desirable solutes in a complex mixture will be extracted. The various substances are recovered by reducing the pressure stepwise.

This process, initially developed in Germany at the end of the 1960s, is used particularly to extract natural aromas (such as caffeine in coffee, attar of roses), pectin (a polysaccharide extracted from fruit used as gelling agent in jams), certain vitamins, and flavonoids. It is also used to remove the bitterness from hops. Water in the supercritical state also has particularly important solvent properties; for example, water at high pressure and high temperature notably dissolves the silica in glass. However, the rather high values of the critical coordinates of water ($p_c = 218$ bar, $T_c = 374$°C) are a handicap for its use in extraction processes. However, supercritical utilization of water as solvent for manufacturing ultrafine powders by rapid pressure reduction of supercritical solutions and for high-pressure and high-temperature treatment of organic wastes (450° and 220 bar, conditions in which the density of water is only 100 g liter^{-1} which is the tenth of the density at ambient conditions) has been contemplated. Supercritical water is also an important fluid in geological phenomena. The water from hydrothermal springs, due to the great depth at the bottom of oceans at very high temperature ($T > 300$°C), is in a state close to near-critical conditions and deposits mineral compounds (sulfides in particular) dissolved in the geological strata near these springs.

Finally, we note that chromatographic analytical techniques utilize supercritical CO_2 as eluent (the mobile phase) in chromatography columns; the fluid dissolves the compounds to be analyzed which are deposited on the solid support packing the column (generally silica). This method is faster than classic chromatography.

The efficiency of supercritical CO_2 extraction processes can be improved by using CO_2-soluble surfactants which will disperse the mixture in the supercritical fluid. These are fluorated compounds or copolymers.

4.7.7 Tricritical Points

A critical point in the thermodynamics of phase changes is the terminal point of the two-phase coexistence line for which these two phases (a liquid and a gas, for example) become identical. By analogy, a **tricritical point** can be defined as the terminal point of a three-phase coexistence line (a triple line) where three phases become identical (they can no longer be distinguished); this situation imposes one condition: the chemical potentials of the three phases must be the same ($\mu_1 = \mu_2 = \mu_3$, i.e., two relations). If we apply the Gibbs phase rule, it is written $v = c+2-\varphi-2$ (with $\varphi = 3$) in these particular conditions. One needs $c \geq 3$ in order that $v \geq 0$. With three constituents ($c = 3$), $v = 0$, the system is invariant, and there can be a tricritical point.

Such a situation is encountered in mixtures of helium (^{3}He and ^{4}He) and metamagnetic systems. A $T(x)$ phase diagram can then be plotted for helium, where x is the concentration of ^{3}He in ^{4}He (Fig. 4.11). There are three regions: the region for a normal fluid, the region corresponding to a superfluid phase separated from the first region by line $T_\lambda(x)$ (The λ-points correspond to the critical points of the transition from normal liquid to superfluid liquid), and a third region where the system separates into two phases with high and low concentrations of ^{3}He. Line T_λ has a end point P corresponding to the top of the coexistence zone. This point P can be considered as the junction of three critical lines, two corresponding to a fictive parameter related to the characteristic parameter of the superfluid phase which is the complex wave function of the superfluid phase, as Griffiths proposed in 1970.

The situation is similar in metamagnetic solids, which are systems with two sublattices that undergo a first order phase transition at low temperature on application of a strong magnetic field $\boldsymbol{H}$, passing from a weakly magnetized state to a strongly magnetized state. In these materials, there is

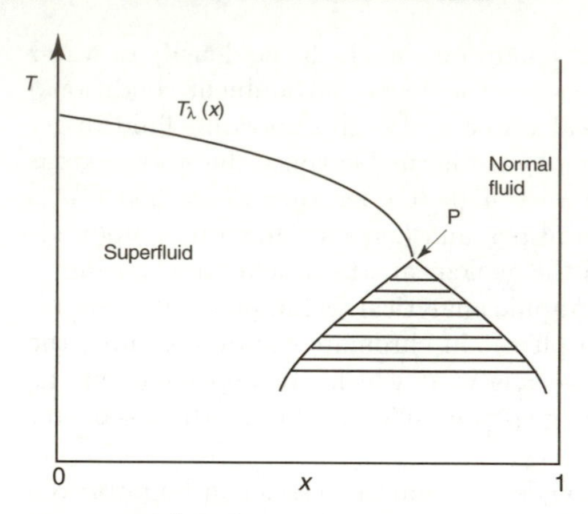

Fig. 4.11. ^{3}He–^{4}He binary mixture. The $T(x)$ phase diagram was plotted, where x is the concentration of ^{3}He in ^{4}He. The λ-points correspond to the normal liquid/superfluid liquid critical points. $T_\lambda(X)$ is a line of critical points. P is the end point ($x = 0.67$) and is also the vertex of the ^{3}He–^{4}He separation curve.

magnetic coupling between spins within a plane, on one hand, and antiferromagnetic coupling between spins in neighboring planes on the other hand; these two interactions are thus in competition. A $H(T)$ phase diagram has two regions: in one, the system behaves like a paramagnetic system, but it has antiferromagnetic behavior in the other; they are separated by a region of coexistence. The separating line is partially a line of critical points and partially a line of triple points which join at the tricritical point. This behavior can be comprehended by introducing a fictive field $\boldsymbol{H}_s$ (or "staggered" field) which acts in one direction on one sublattice and in the opposite direction on another sublattice. We then have a three-dimensional $H(H_s, T)$ diagram with three critical lines that meet at the tricritical point. Situations of this type are experimentally observed in a crystal exhibiting a mixed state where paramagnetic and antiferromagnetic domains coexist (DyAl garnets).

Kohnstamm suggested in 1926 the possibility of observing the tricritical points when he was studying liquid/gas binary mixtures. It is necessary to satisfy a series of conditions for such an observation to be possible; they will only be satisfied for binary mixtures of normal liquids (this is the case with superfluid helium on the other hand). A tricritical point can also be observed when the upper and lower critical end points of the diagrams for three phases coincide; this situation occurs at the transition of diagram IV to type II in Fig. 4.7. At this point, the two meniscuses separating three phases α, β, γ disappear simultaneously at a certain pressure and temperature.

Fig. 4.12 shows the sequence of events that occur when the system evolves from the upper critical end point T_{UCEP} from which phases β and γ ($l_1 + l_2 = g$) begin to separate to the lower critical end point T_{LCEP} where phases α and β ($g + l_1 = l_2$) become identical. In a $T(V, x)$-diagram, the state

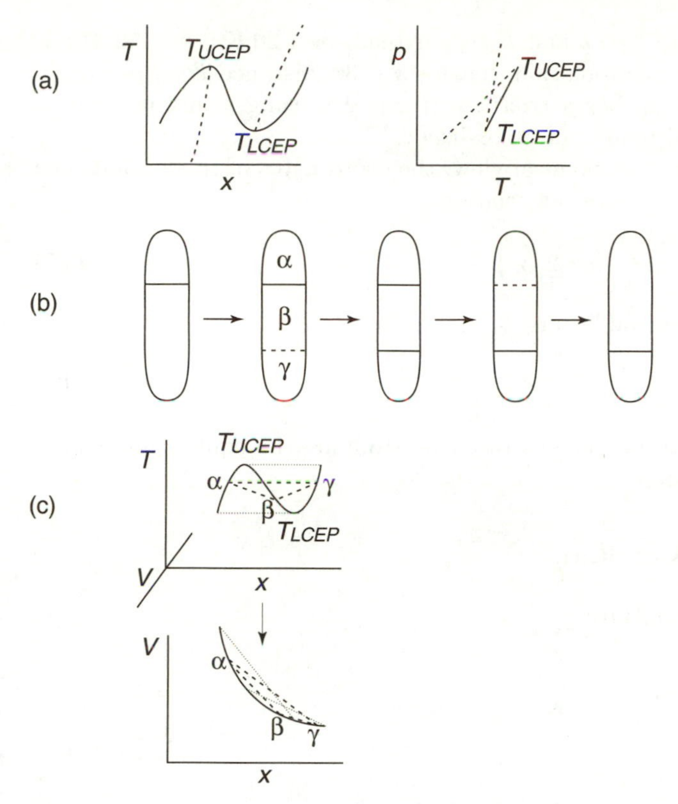

Fig. 4.12. Tricritical point approach. (**a**) Points T_{UCEP} and T_{LCEP} designate the high- and low-temperature critical end points on the three-phase coexistence lines (dashed). (**b**) When T_{UCEP} is attained, phases β and γ separate (appearance of a meniscus at T_{LCEP}), and phases α and β merge. (**c**) At fixed T, the three phases are represented by three points in space (T, V, x). A tricritical point is observed when T_{UCEP} and T_{LCEP} coincide: the two meniscuses disappear simultaneously.

with three phases α, β, γ is represented by a triangle (Fig. 4.12c). When the temperature is raised, the $\alpha\beta$ side of the triangle disappears, and distance $\beta\gamma$ vanishes when the temperature is lowered.

If we assume that the interaction between molecules can be controlled, we will want to see what happens in passing from type IV to type II diagrams (or from type V to I) by contracting the three-phase coexistence line, which leads to coincidence of the two terminal critical points T_{UCEP} and T_{LCEP}. At this point, the two meniscuses disappear simultaneously and the three phases become identical ($l_1 = l_2 = g$).

The phase rule, as we saw, requires the presence of a minimum of three components to be able to observe this phenomenon. One way of adding a degree of freedom to a binary system is to use molecules of different size. For a mixture of propane and n-alkanes, the region in which there can be three phases between points T_{UCEP} and T_{LCEP} extends over 20 K if $n = 50$. If n is decreased, we find by extrapolation that for $n < 30$, distance T_{UCEP}–T_{LCEP} vanishes and the system has a tricritical point. By using a mixture of C_{29} and C_{30} the tricritical point can be realized.

From a thermodynamic point of view, the coordinates of these points can be determined with the following relations:

$$\left(\frac{\partial^i G}{\partial x^i}\right)_{p,T} = 0 \quad \text{for} \quad i = 2, 3, 4, 5 \tag{4.60}$$

with the local stability condition:

$$\left(\frac{\partial^6 G}{\partial x^6}\right)_{p,T} > 0 \tag{4.61}$$

These relations can be demonstrated by studying the equilibrium conditions of a ternary system.

Table of Equations of State

1. Van der Waals (mixture)

$$\left(p + \frac{N^2 a(x)}{V^2}\right)(V - Nb(x)) = NkT$$

with

$$a(x) = (1-x)^2 a_{11} + x(1-x)a_{12} + x^2 a_{22}$$

2. Soave–Redlich–Kwong

$$p = \frac{RT}{V-b} - \frac{a/T^{0.5}}{V(V+b)}$$

3. Peng–Robinson

$$p = \frac{RT}{V-b} - \frac{a(T)}{V(V+b) + b(V-b)}$$

4. Dieterici

$$p\,\mathrm{e}^{[a/RT]} = \frac{RT}{V-b}$$

5. Equation for molecular chains (or *Perturbed Hard Chain Theory* PHCT)

$$p = \frac{RT}{V} + \frac{cRT}{V}\frac{4\eta - 2\eta^2}{(1-\eta)^3} - \frac{RT}{V}\frac{z_m c V^* Y}{V + V^* Y}$$

with

$$\eta \equiv \frac{\pi r_0^3 N}{6V} = \frac{\pi}{6}\frac{\sqrt{2}V^*}{V} = \tau V^*/V;\ \left(V^* \equiv N r_0^3/\sqrt{2} \text{ and } \tau \equiv \frac{\pi\sqrt{2}}{6}\right)$$

and

$$Y = \exp(\epsilon q/2ckT) - 1; \quad z_m = 18.0$$

Problems

4.1. Virial Theorem

Write a general equation of state for a gas of N particles of volume V assuming energy of the form $E = \sum_i \frac{\mathbf{p}_i^2}{2m} + \frac{1}{2}\sum_{i,j} u(\boldsymbol{r}_{i,j})$ where $u(\boldsymbol{r}_{ij})$ is a binary interaction potential between particles. The quantity $\sum_i \boldsymbol{p}_i.\boldsymbol{r}_i$ is by definition the virial. With the total time conserved, the average value of the virial at equilibrium is invariant in time. Each particle is assumed to be exposed to two forces: an external force due to the pressure exercised by the walls; an internal force $\boldsymbol{f}_i$ due to interaction with other particles.

1. Write the relation between T and the average kinetic energy.
2. Using conservation of the virial, write the relation between the forces at equilibrium.
3. Introducing force $\boldsymbol{f}_{ij}$ exercised by particle j on particle i, derive the equation of state.
4. If $g(r)$ is the distribution function for a pair and $n(r) = \frac{N}{V}^2 g(r)$ is the probability density for a pair, derive a new form of the equation of state expressed with u.

4.2. Law of Corresponding States

Assume that in a fluid, the intermolecular potential is $u(r) = \varepsilon_0 f(\frac{r}{r_0})$. Assume that the kinetic energy of N molecules is purely translational.

1. Calculate the partition function for the system and show that it can be expressed with an integral function of f.
2. Introducing the reduced coordinates of the fluid, show that the pressure is expressed by a function which is only dependent on these variables and the form of f. Determine the reduced pressure.
3. What conclusions can be drawn from this for the equations of state?

4.3. Van der Waals Equation of State

Assume that the total energy of a fluid with N particles is in the form $E = E_c + U_N^0 + U_N^1$, where E_c is the kinetic energy, U_N^0 is the unperturbed potential energy, and U_N^1 is a perturbation. U_N^0 will be represented by a hard-sphere potential and U_N^1 will be a perturbation corresponding to attractive interaction. V is the total volume of the fluid.

1. Calculate the total partition function Z_N and show that it is expressed with a mean $\langle \exp(-\beta U_N^1)\rangle_0$ of function $\exp(-\beta U_N^1)$ in the unperturbed system.
2. Assuming $U_N^1 \ll kT$, write $\langle \exp(-\beta U_N^1)\rangle$ simply and give its expression as a function of the energy of a pair interaction $u^1(\boldsymbol{r})$ and the probability density of the presence of pairs $n_2(|\boldsymbol{r}_1 - \boldsymbol{r}_2|) = \frac{N^2}{V^2} g(|\boldsymbol{r}_1 - \boldsymbol{r}_2|)$, where g is the distribution function.
3. Calculate $\langle \exp(-\beta U_N^1)\rangle_0$ with the hard-sphere potential. Assume $a = -2\pi \int_{r_0}^{\infty} u^1(r) r^2 \mathrm{d}r$, where r_0 is the exclusion distance for this potential.
4. Write the free energy of the fluid and determine its equation of state.

4.4. Spinodal for a fluid

Using the van der Waals equation of state, write the spinodal equation in the form $p(V)$.

4.5. Maxwell's Rules and the Common Tangent

The thermodynamic state of a fluid is represented by an isotherm in (p, V)-plane (Fig. 4.3a).

1. Show that the liquid/gas transition at pressure p corresponds to equality of the areas between the isotherm and the isobar.
2. Show that on the $F(T)$ curve (Fig. 4.3b), points 1 and 2 corresponding to volumes V_1 and V_2 have a common tangent.

5 The Glass Transition

5.1 Glass Formation

Most solid mineral compounds and elements form liquids of low viscosity (several centipoises) when they melt, and when the temperature is reduced, they solidify again to form a crystalline solid. Alternatively, there are materials which become liquids with a very high viscosity (10^5–10^7 P) when melted. When they are cooled below their melting point, these liquids do not solidify instantaneously but remain in a supercooled state, the viscosity of the liquid increases significantly when the temperature is reduced, and they then "freeze" in the form of a **glass**, which is a noncrystalline solid state. We say that the liquid has undergone a **glass transition** and that a **glassy** or **vitreous** state has formed.

For thousands of years it has been known that many liquids are transformed into glasses when cooled. Glasses can be formed in this way from liquids whose structures correspond to strong covalent bonds (typically oxides such as SiO_2, GeO_2, B_2O_3; window glass is a mixture, SiO_2–CaO–Na_2O), but also to hydrogen bonds, such as glycerol, and ionic bonds such as certain molten salts like KNO_3–$Ca(NO_3)_2$.

We still do not understand very well why some molecules can form glasses. This question was first broached in depth by Zachariasen (1932). We can first say that certain liquid structures correspond to topologies that favor the formation of glasses. We also know that addition of another element, even in traces, can prevent crystallization: this is the case with As and Cl, for example, which are added to stabilize glassy (or amorphous) films, and Se used in xerography and which thus leads to formation of a glass. Crystallization can also be prevented by deposition of a substance in the vapor state on a solid substrate or by rapid quenching followed by a shock.

The topology of a glass is designated by the term "continuous random network" and is shown in Fig. 5.1a for a two-dimensional system formed from one element. Contrary to the situation in a crystal, there is significant scattering in the angles formed by the bonds. Most simple glasses are of type A_nO_m, where A is an element like Si, Ge, B, As, or P. Figure 5.1c shows a network corresponding to A_2O_3 where the empty circles are oxygen atoms which are divalent and form bridges in the center of a triangle. Such a structure allows introducing disorder without bringing cations A too close

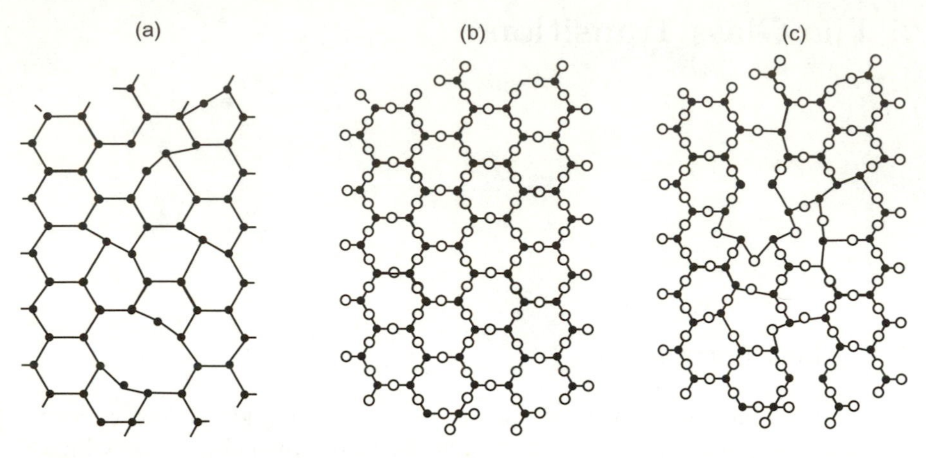

Fig. 5.1. Random two-dimensional network. (**a**) Glass constituted of one element with coordination equal to three. (**b**) Crystalline compound of type A_2B_3. (**c**) Glass of type A_2B_3. Atom A, with coordination of two, is bound by more flexible bonds to neighboring atoms B.

together. In a three-dimensional structure, A must be tetravalent to form tetrahedral structures, and it can only be formed if we assume that the angles are slightly different from those corresponding to the ideal crystal structure. Zachariasen emphasized that this requires internal energy slightly higher than the crystal energy.

Schematically, we can say that the **glass structure and crystal have two essential differences**:

- there is scattering of bond angles in glass, which is prohibited in the crystal;
- long-range order does not exist in glass.

This last property is manifested by the diffuse character of the X-ray diffraction spectra in glasses.

In melting that ends in vitrification, the initial SiO_2 network is broken up, "depolymerized" in a way, and a more disordered network favorable to the appearance of a glass will be formed.

Despite its ubiquity and the very large number of observations and experimental data, the nature of the glass transition is far from understood. It can be studied from several points of view. The time scales should first be examined. Schematically, there are two different time constants that control the transition. The first, designated by τ_1, characterizes the time necessary for crystallization of a given volume of liquid phase. This will first decrease when the liquid is supercooled then increase as a function of the temperature. This is observed on the *TTT* (Time–Temperature–Transformation, Fig. 2.8) curves. The shape of these curves results from two competing phenomena (Fig. 5.2). The upper part (high temperature) shows that the "thermodynamic force" including crystallization tends to increase, while the lower part, corresponding to a decrease in the temperature, shows that the kinetics of transport through interfaces slows crystallization. This curve can be used to determine the critical rate of formation of a glass, that is, the cooling or quenching rate necessary to prevent a given fraction of the liquid from crystallizing. If this rate is high enough (fast quenching), we pass below the crystallization curve and a glass is formed. This critical quenching rate is generally low for oxides of the SiO_2 and GeO_2 type (10^{-4}–10^{-1}Ksec^{-1}), it is slightly higher for organic glasses, H_2O and metals (10^2–10^{10}Ksec^{-1}).

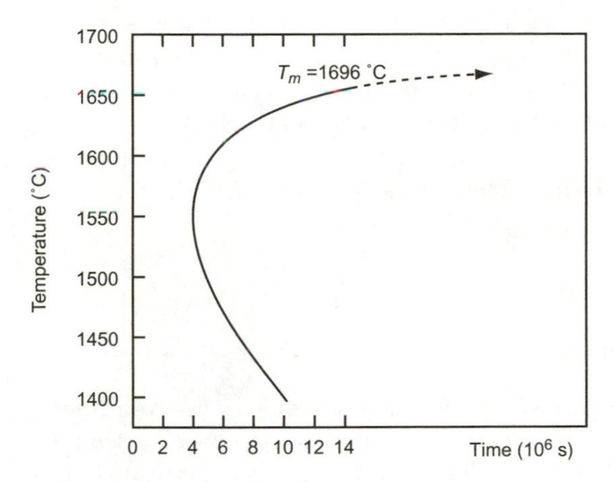

Fig. 5.2. Time–Temperature–Transformation curve. This curve is plotted for SiO_2; it corresponds here to a given fraction of 10^{-6}. This curve represents the time needed to cristallize a given fraction as a function of the temperature [D. R. Uhlmann, *J. Non. Cryst. Solids*, **7**, 337 (1972)].

We emphasize in passing that fast quenching is not the only method used for obtaining a glass. A glass transition can be induced by deposition of an undercooled vapor on a substrate, cathode sputtering, or by ion implantation (method also used for fabricating electrets). Rapid compression at fixed temperature can also be conducted instead of fast quenching: this is the method of cold compression of a solid.

The second time constant τ_2 is the internal relaxation time, also called structural relaxation time. It characterizes the time necessary for rearrangements of molecules in the material to attempt to find equilibrium.

If we compare a liquid to a set of "cages" occupied by molecules, the volume of these cages will decrease when the temperature decreases and increased cooperation between neighboring molecules will be necessary for one of them to be able to escape from its initial cage and diffuse between its neighbors, which takes a longer time, the internal relaxation time. In supercooling, τ_2 increases very significantly and even attains astronomical values. This time constant is more or less proportional to the viscosity, but is not rigorously defined, as it implies that the phenomenon will vary exponentially with time, which is only observed in the initial stage: one has actually a linear combination of exponential terms. It is necessary to note that a glass is often distinguished from an **amorphous** phase by the fact that an amorphous phase (generally a thin, uncrystallized film) crystallizes rapidly when heated, while a glass will progressively pass into a liquid state when its temperature is raised. Amorphous phases are often obtained from deposition of thin layers of metal alloys or metal–metalloid mixtures from a vapor phase.

Glasses have been known since prehistoric times when obsidian was used for manufacturing cutting tools such as knives. The first glasses were undoubtedly made in Mesopotamia in approximately 4500 B.C., then in Egypt. The enamels used for coating pottery also should have been produced at the beginning of metallurgy (10,000–12,000 B.C.). The methods of fabricating glasses, to which we will return, have undergone many changes and improvements since then, especially since the end of the 19th century.

5.2 The Glass Transition

5.2.1 Thermodynamic Characteristics

A large number of liquids become syrupy when supercooled and even freeze; their viscosity increases very strongly in the supercooled phase so that no motion at all can be detected in them: they became a glass. Viscosity η can attain $10^{12} - 10^{13}$ P. The liquid can remain in this frozen stage for decades or even thousands of years; we cannot properly speak of a solid state, and classic thermodynamics does not strictly apply because the system is not in a state of thermodynamic equilibrium with respect to changes in configuration; we say that it is a system **out of equilibrium**. We can nevertheless assume that glass is partially in a state of **metastable** thermodynamic equilibrium because the vibrational energy in the glass, whose temperature is uniform, is equilibrated and thermal conduction ensures local redistribution of the temperature. Glass can be assigned an entropy even if it is not identical to the entropy of a liquid under the same (p, v, T)-conditions. The entropy of a glass can be calculated by integrating the specific heat at constant pressure and comparing the result with the entropy of the crystal at the same temperature. As expected, glass does not have the lowest entropy, it has excess entropy ΔS with respect to the crystalline state, but this difference between the two entropies decreases when the temperature is lowered.

$$\Delta S = S_{(\text{liq or glass})} - S_{\text{cryst}}$$

$$\Delta S = L_f/T_m - \int_T^{T_m} (C_{pl} - C_{ps}) \mathrm{d}T/T \tag{5.1}$$

L_f is the latent heat of fusion of the solid at T_m and C_{pl} and C_{ps} are the specific heats of the liquid and the solid. We find (Fig. 5.3) that at the glass

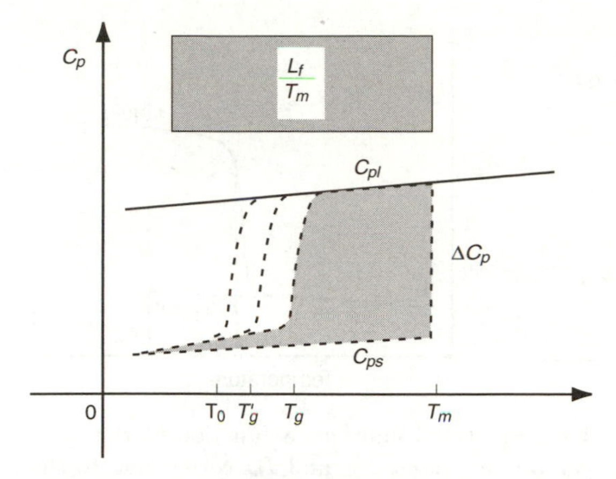

Fig. 5.3. Calculation of the excess entropy of a glass. C_{pl} is the specific heat of the liquid for different cooling rates, C_{ps} is the specific heat of the crystal. The cross-hatched area of the rectangle represents the entropy of melting of the crystal. The temperature is plotted on a logarithmic scale.

transition at T_g, there is an entropy corresponding to the configurational disorder persisting in the congealed liquid during formation of the glass.

The glass transition can be characterized by observing the behavior of several thermodynamic quantities. If a liquid is cooled at constant pressure, the volume decreases and this decrease will continue into the supercooled state unless crystallization is induced (for example, by seeding with crystal nuclei). Either the supercooled liquid will then crystallize spontaneously or the reduction in volume will suddenly change tempo: at temperature T_m, the slope of the $V(T)$ curve changes abruptly (Fig. 5.4a), which corresponds to a discontinuity in the thermal expansion coefficient α_p, and T_g is the **glass transition** temperature. Note that at very low temperatures, coefficients α_p are the same for glass and crystal (Fig. 5.4b). We also find that the value of T_g is a function of the cooling rate of the liquid. The glass transition occurs earlier if the quenching rate is high, and if it is very low, it is possible to pass directly to the crystalline state, avoiding the formation of glass, which is in agreement with the principles of thermodynamics. It is clear that the glass transition is not abrupt, contrary to crystallization and melting. It is not accompanied by any latent heat.

The behavior of the enthalpy parallels the behavior of the volume at constant pressure: a sudden change in the course of the variation of the function $H(T)$ with a discontinuity in the slope T_g (that is, the specific heat). As previously, this variation occurs in a narrow region and the value of T_g is a function of the quenching rate (Fig. 5.5). The change in C_p is associated with the decrease in the number of degrees of freedom of the system, as some become frozen. The difference between C_p in the liquid and glass is always positive. The situation is different for the thermal expansion coefficients

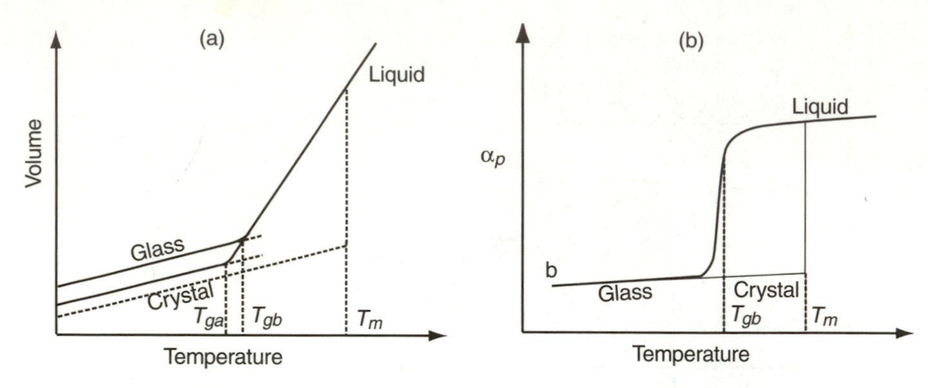

Fig. 5.4. Glass transition. (**a**) Change in the volume as a function of the temperature in the liquid, glassy and crystalline states. T_{ga} and T_{gb} correspond to the glass transition temperatures obtained for high and low cooling rates. (**b**) Thermal expansion coefficient in different phases. It is discontinuous at T_g. T_m is the melting point.

where the differences between the liquid and the glassy states may have any sign. It is not possible to say which degrees of freedom are frozen (that is, blocked for a very long time).

The sudden decrease in C_p when the temperature is decreased and inversely the sudden increase when the temperature is raised are in general considered the signature of a glass transition. The glass transition temperature T_g is often defined as the temperature at which a specific heat peak appears during heating (a heating rate of 10 $\mathrm{K sec}^{-1}$ is selected).

As emphasized above, the difference ΔS between the entropies of the supercooled liquid (or glass) and the crystal continuously decreases, which raises the following question, called the **Kauzmann paradox**: what hap-

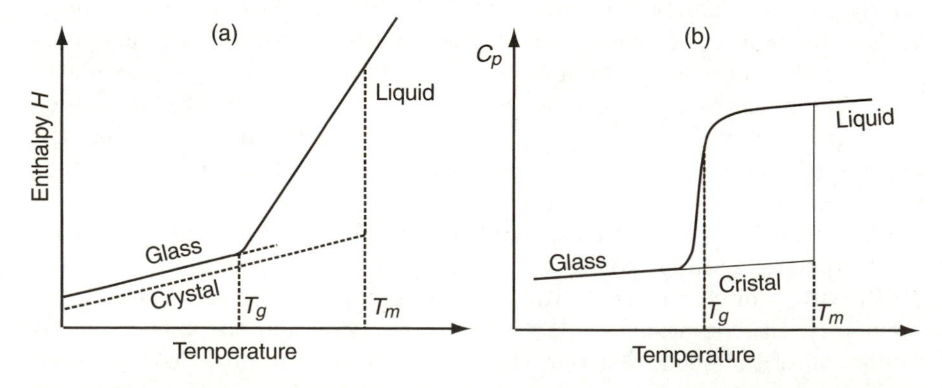

Fig. 5.5. Thermodynamic quantities in vitrification. (**a**) Isobaric enthalpy. (**b**) Isobaric specific heat. It exhibits a discontinuity at the glass transition.

pens when these entropies become equal? Temperature T_0 at which this phenomenon occurs is sometimes called the ideal glass temperature. Beyond T_0, ΔS is negative: the liquid (or glass) would have a lower entropy than the crystal, the most ordered phase, which is physically impossible. Temperature T_0 could be considered as the thermodynamic limit of the glass transition, impossible to attain experimentally. We note that, at this point, since the enthalpy of the liquid is still higher than the enthalpy of the crystal, the free enthalpy $H - T_0S$ will be also higher, which excludes the coexistence of supercooled liquid and crystal in equilibrium, and a transformation, for example, into a glassy state, must thus necessarily occur between T_m and T_0.

We can summarize the situation by stating that the glass transition is poorly defined. The most common criterion (which has been used in particular by C. A. Angell) for defining the glass temperature or glass transition temperature is the observation of a rapid increase in the viscosity ($\eta \approx 10^{13}$P). This temperature is slightly dependent on the cooling rate. Almost simultaneously, but not rigorously at the same temperature, the thermal expansion coefficient and the enthalpy undergo a rapid change, as well as the dielectric polarizability and other quantities characteristic of relaxation.

In passing, we mention the following empirical rule: molecular liquids with a melting point T_m less than two times their boiling point T_v ($T_m/T_v < 2$) are generally very viscous at the solidification temperature T_m so that the nucleation rate during low-temperature quenching is insufficient to induce crystallization. These liquids form glasses more easily.

The glass (or glass transition) temperatures T_g are generally high for oxides or mixtures of oxides (500–1500 K). They are lower for glasses made from elements (244 K for S, 302–308 K for Se) and for organic substances (180–190 K for glycerol, 90–96 K for ethanol).

The quantity most directly related to internal relaxation is the viscosity, which expresses the duration of shear resistance. It seems infinitely long in a solid, but rapidly approaches zero in a normal liquid. The low-frequency shear viscosity is usually characterized by a relaxation time defined by:

$$\eta = G_\infty \tau \qquad (5.2)$$

In estimating G_∞, the instantaneous shear modulus, from solids, we can derive τ. τ is infinitely long in solids and very short in normal liquid ($\tau \approx 10^{-13}$sec); in glasses, τ is estimated at the order of 10^{32} years at ordinary temperature and is thus almost equivalent to the τ of a solid.

5.2.2 Behavior of the Viscosity

The viscosity plays a central role in the behavior of a glass. It is correlated with the internal relaxation time, expressing the duration of shear resistance. The viscosity η of a viscous liquid follows a classic **Arrhenius** law of the type $\log \eta = A + B/T$, but we find that the viscosity of glasses only follows this type of behavior in a limited temperature range. Most of the results can be described with the **Vogel–Tammann–Fulcher** empirical equation:

$$\eta^{-1} = A \ \exp[-a/(T - T_{0\eta})] \tag{5.3}$$

This describes the fluidity, the inverse of the viscosity, in the vicinity of the glass transition temperature. $T_{0\eta}$ is the temperature at which fluidity vanishes, which can be near the temperature at which the excess entropy ΔS disappears. This expression can be found by establishing a correlation between the decrease in the heat capacity C_p near the glass transition with the semi-empirical theory of Adams and Gibbs.

We note the elementary definition of the viscosity with Newton's macroscopic law for viscous flow:

$$\boldsymbol{F} = -\eta \text{grad } \boldsymbol{u} \tag{5.4}$$

where $\boldsymbol{F}$ is the force and $\boldsymbol{u}$ is the flow rate.

The general relation between the deformation rate $(\partial u_l/\partial x_k)$ and the pressure tensor is:

$$p_{i,j} = \sum_{k,l} \alpha_{ij}^{kl} \frac{\partial u_l}{\partial x_k} \tag{5.5}$$

The tensor α has 81 coefficients. Assuming the medium is isotropic and subtracting the pure rotation and translation, there are now only two coefficients κ and η, respectively the bulk and shear viscosities. These quantities introduced in the Navier–Stokes fluid mechanics equation give:

$$m\rho \frac{\partial \boldsymbol{u}}{\partial t} = -\nabla \boldsymbol{p} + \nabla^2 \boldsymbol{u} + \frac{1}{3}\eta + \kappa \nabla(\nabla . \boldsymbol{u}) \tag{5.6}$$

By Fourier transformation of (5.6), we obtain the following equation of the motion for the components of the current $\boldsymbol{J}_\perp$ perpendicular to $\boldsymbol{k}$: $\boldsymbol{J}_\perp$ obeys the following equation of motion:

$$m\rho \left(\frac{\partial \boldsymbol{J}_\perp}{\partial t} \right) = -k^2 \eta \boldsymbol{J}_\perp(\boldsymbol{k}, t) \tag{5.7}$$

This equation can be rewritten by introducing a time-dependent correlation function and taking the ensemble average. This results in a simple equation that defines viscosity η

$$\eta = \frac{1}{VkT} \int_0^\infty < J(t)J(0) > \mathrm{d}t \tag{5.8}$$

where

$$J = \sum_{j=1}^{N} (\frac{p_{xj} p_{zj}}{m} + z_j F_{jx}) \tag{5.9}$$

It is possible to calculate η from (5.9) using the so-called mode-coupling theory which will be discussed later. This concludes the description of the viscosity characterized by two relaxation times. However, it is difficult to find divergent behavior for the viscosity with this calculation (of the type of [5.3], for example), but a self-consistent method that reveals this divergency can be found.

5.2.3 Relaxation and Other Time Behaviors

The Debye theory leads to the hypothesis of the existence of a relaxation time: it postulates that the viscosity is a friction term in the equation of motion. Many experiments nevertheless indicate that the behavior of a liquid in time is much more complicated, particularly in the vicinity of T_g. In fact, instead of exhibiting behavior with simple, purely exponential relaxation (or with two exponentials), functions of type (5.8) can be represented with characteristic exponents of the Kohlrausch–Williams–Watts relaxation function:

$$f(t) = \exp[-(t/\tau)^{\beta}] \tag{5.10}$$

where $\beta < 1$ is an adjustable constant which is a function of the nature of the liquid phases but which also decreases when the temperature decreases.

There are some theoretical models that can be used to reproduce temporal behavior of this type.

Another type of time-dependent behavior is given by a power law found with mode coupling theory:

$$\begin{aligned} f(t) &= A_1(t/\tau)^{-a}, \qquad (t/\tau) \ll 1 \\ f(t) &= -B_1(t/\tau)^{b}, \qquad (t/\tau) \gg 1 \end{aligned} \tag{5.11}$$

where $B > 0$. Such laws could also be obtained by computer simulation on Lennard–Jones liquids.

5.3 The Structure of Glasses

The fact that glasses are supercooled liquids implies that they do not have a periodic structure: this is a characteristic of the glassy state. However, totally random distribution of the positions of particles in the medium is not observed, as in the case of an ideal gas, for two reasons. The atoms or molecules have a size of the hard-sphere type, which implies that they are more or less the same distance apart, and the second reason is that they have a fixed coordination number (or valence). In fact, this is not always strictly the case, because dangling bonds can also be present. Using the two preceding restrictions, a continuous random network (with a coordination number of 3) similar to the one in Fig. 5.1c can be generated. A statistical count in such a network shows that there are many rings with six bonds but few with three or nine bonds. The bond angles are close to 120°. If the structure would have been periodic, the lattice would have been of the honeycomb type consisting of rings with six bonds, having angles between the bonds that are exactly 120°. The local structure in the glass is thus not strictly different from the structure of a microcrystalline solid. This is also revealed in spatial correlation function $g(\boldsymbol{r})$ and structure function $S(\boldsymbol{k})$ (Appendix 4.B). It is difficult to distinguish the exact form of these functions for glass and crystalline solid even if they differ in details. On the other hand, Bragg reflections only appear in crystalline solids because they result from strict periodicity of the crystal lattice. The existence of medium-range local order (< 1 nm) can be demonstrated by experimental techniques such as high-resolution electron microscopy and small-angle X-ray scattering. The basic ideas for interpreting the structure glasses were elaborated by Zachariasen (1932). Consider the case of silica glass, SiO_2, which is formed from tetrahedrons composed of a silicon atom occupying the center of (SiO_4) tetrahedrons and four oxygen atoms located on the vertices. The corner oxygen atoms are bound so that they form bridges between the silicon atoms in neighboring tetrahedrons. While the O–Si–O angle corresponding to tetrahedral bonds ($\approx 109°$) remain identical, the Si–O–Si bond corresponding to a bond between neighboring tetrahedrons is relatively variable, between 120° and 180°, with little cost in energy.

Small microcrystalline regions are observed in silica glass by X-ray and neutron diffraction; they correspond to microcrystals of tridymite or cristobalite with interfaces comprising oxygen atoms doubly bonded with silicons (Fig. 5.6). The continuous random network model with intertetrahedral Si–O–Si bond angles satisfactory accounts in a first approximation for the structure of silica glasses. When a non-glass-forming oxide, Na_2O, for example, is added to silica SiO_2, the additional oxygen atoms introduced are intercalated in the network, causing rupture of certain Si–O–Si bonds.

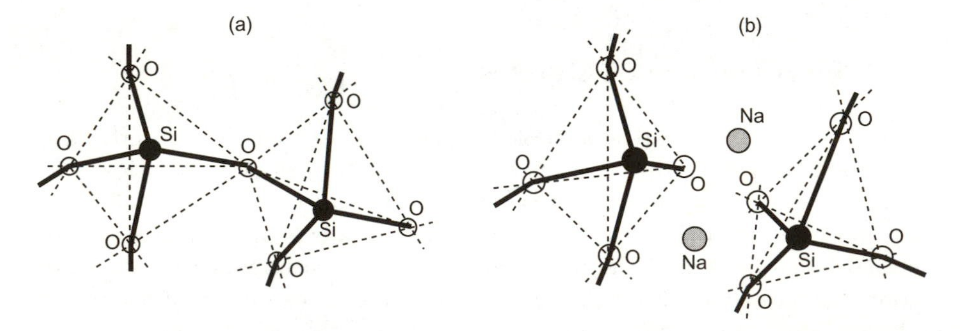

Fig. 5.6. Silica network. (**a**) intact silica network, the SiO_4 tetrahedrons have a common vertex, but there is disorder in the Si–O–Si bonds. (**b**) Introduction of an oxide, Na_2O, a pair of nonbridging oxygen atoms formed in the glass.

An example of an interpretation of a structure function $S(\boldsymbol{k})$ obtained by a scattering experiment was given in the case of a glass transition in a metal/metalloid mixture, $Ni_{76}P_{24}$. This structure factor is particularly simple (Fig. 5.7a); it has a very pronounced peak for $k \approx 0.23\ \mathrm{nm}^{-1}$ which can be represented by a gaussian (dotted curve). The inverse transform of $S(\boldsymbol{k})$ corresponding to $G(\boldsymbol{r})$ is shown in Fig. 5.7b and it can be obtained only if a relatively wide range of values of k is known; any termination in the spectrum of $S(k)$ will be manifested by parasitic fluctuations of function $G(\boldsymbol{r})$. The contribution of the gaussian is indicated by the dashed lines. In this case, the length of the bond between nearest neighbors can be determined, which is something of an exception since the spectra of $S(\boldsymbol{k})$ are the superposition of several different bond lengths corresponding to the nearest neighbors and next nearest neighbors, which makes their interpretation difficult.

Although a glass is in a certain way the equivalent of a snapshot of a liquid whose motion is frozen, there is still motion within it which is very difficult to describe accurately. The most common image would be that of atoms which move around while confined in a cage. This is given by the **free volume** theory, which has enjoyed some success. A simple equation can be obtained for the viscosity (Doolittle calculation). The Cohen and Grest cage theory was used to explain the glass transition as follows:

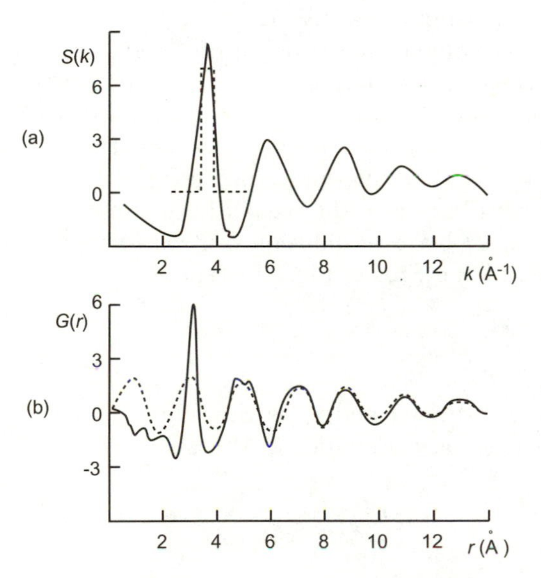

Fig. 5.7. Structure function for metallic glass. (**a**) Structure function $S(k)$ was shown for $Ni_{76}P_{24}$ alloy, where the peak corresponding to the first maximum was replaced by a gaussian (dotted curve). (**b**) Function $G(r)$, inverse Fourier transform of the preceding function.

- the liquid is compared to an ensemble of cells in which atoms can move freely;
- the existence of cages where atoms corresponding to the solid are trapped is assumed.

If we calculate the free energy as a function of the concentration of these two types of structures and if we have a second minimum that decreases below the first minimum, the substance will undergo a first order glass transition associated with discontinuities of volume, entropy and enthalpy. Although this description has some merit, it unfortunately does not completely describe the transition (it takes place without latent heat or volume discontinuity).

5.3.1 Mode-Coupling Theory

Vitrification can be considered a transition from a state in which the dynamics of relaxation of density fluctuations exhibits **ergodic** behavior to another state with **nonergodic** behavior. In the first case, all microscopic configurations (in the sense of statistical mechanics) are accessible, while in the second case entire regions of phase space become inaccessible because structure relaxation is blocked. This is the statistical translation of passing from a liquid system in which particle movements are totally free to a system where they are "frozen." This is the principle of mode-coupling theory, initially applied to supercooled liquids (Leutheusser, Bengtzelius, 1984).

Geszti described the physical phenomena at the origin of the glass transition more precisely with a self-consistent equation for the viscosity. He assumes that relaxation is simultaneously of vibrational and structural origin. The second term contributes to relaxation over a long time and the first contributes to transient effects with a short lifetime which are not a function of the structural part. The contribution of the second term is inversely proportional to diffusion coefficient D because diffusion controls structural relaxation. Introducing a relation of the Stokes–Einstein type for diffusion (2.6):

$$\eta = \eta_0(T) + B(T)\eta \tag{5.12}$$

where $B(T)$ is a function to be determined and η_0 is the "pure viscosity" corresponding to the purely vibrational part. Equation (5.12) can be rewritten as:

$$\eta = \frac{\eta_0(T)}{1 - B(T)} \tag{5.13}$$

which expresses the fact that η_0 is augmented by the feedback mechanism. We can estimate that $B(T) \propto b/T$. η diverges for $T = T_0$ so that $B(T_0) = 1$. Expanding $B(T)$, (5.13) gives the Batchinski–Hildebrand law for the viscosity:

$$\eta \simeq A/(T - T_0) \tag{5.14}$$

where $A = \eta_0(T_0)/|B'(T_0)|$; (5.14) describes the increase in the viscosity in the previtrification region. This result differs from the Vogel–Tammann–Fulcher empirical equation (5.3). This semi-phenomenological description was completed by more sophisticated approaches based on the notion of feedback mechanisms: the viscosity controls the shear-related relaxation time which in turn acts on the viscosity. Equation (5.14) predicts an increase in the viscosity when the temperature decreases, and the difference with T_0 is interpreted as the manifestation of the glass transition.

In glass theory, the van Hove density–density autocorrelation function plays an important role. This function $G(\boldsymbol{r}, t)$ is defined as:

$$G(\boldsymbol{r}, t) = N^{-1} < \sum_i \sum_j \int \delta[\boldsymbol{r}' + \boldsymbol{r} - \boldsymbol{r}_i(t)]\delta[\boldsymbol{r}' - \boldsymbol{r}_j(0)]\mathrm{d}\boldsymbol{r}' > \tag{5.15}$$

$< >$ is the ensemble average. Equation (5.15) is rewritten as:

$$G(\boldsymbol{r}, t) = N^{-1} < \sum_i \sum_j \delta[\boldsymbol{r} + \boldsymbol{r}_j(0) - \boldsymbol{r}_i(t)] > \tag{5.16}$$

The van Hove function $G(\boldsymbol{r}, t)\mathrm{d}\boldsymbol{r}$ is proportional to the probability of observing a particle i with $\boldsymbol{r} \pm \mathrm{d}\boldsymbol{r}$ at time t, knowing that there was a particle j at time $t = 0$ at the origin.

The correlation function has two representations. If the summation is over i and j simultaneously, we obtain the "total function"; if the summation is restricted to i, the function is "incoherent." X-ray scattering experiments can be used to measure the first function, and the second is measured with neutron scattering measurements (if the substance contains hydrogen atoms). The incoherent function can also be obtained by simulation calculations with methods of molecular dynamics, for example, which are also faster.

Density $\rho(\boldsymbol{r})$ at $\boldsymbol{r}$ (the number of molecules per unit of volume) can also be introduced to obtain a new definition of $G(\boldsymbol{r}, t)$:

$$G(\boldsymbol{r}, t) = \rho^{-1} < \rho(\boldsymbol{r}, t)\rho(0, t) > \tag{5.17}$$

Its Fourier transform is designated by $F_{\boldsymbol{k}}(t) = N^{-1} < \rho_{\boldsymbol{k}}(t)\rho_{-\boldsymbol{k}}(0) >$.

This function expresses the coupling of the two density fluctuation modes. This function $F_{\boldsymbol{k}}$ divided by its value at $t = 0$

$$\Phi_{\boldsymbol{k}} = F_{\boldsymbol{k}}/S_{\boldsymbol{k}} = \frac{\langle \rho_{\boldsymbol{k}}(t)\rho_{-\boldsymbol{k}}(0)\rangle}{\langle \rho_{\boldsymbol{k}}(0)\rho_{\boldsymbol{k}}(0)\rangle} \tag{5.18}$$

plays a crucial role in glass theory. Its development in time, particularly its limiting value when $t \to \infty$, allows determining whether a glass or a liquid is involved.

A liquid is an ergodic system in the sense of statistical mechanics. By definition, an average of any physical quantity $g(t)$ associated with the system over a very long time is equal to its ensemble average:

$$< g(t) >= \frac{1}{T} \int_0^T g(t) \mathrm{d}t \tag{5.19}$$

when $T \to \infty$. This identity is obtained when the phase space has the appropriate metric. Using the metric of fluctuations (that is, the metric associated with density fluctuations characterized by function $\Phi_{\boldsymbol{k}}$), the behavior of function $\Phi_{\boldsymbol{k}}$ for long times is inversely proportional to the time and will thus eventually approach zero. In glasses, density fluctuations are blocked because structure relaxations no longer occur and function $\Phi_{\boldsymbol{k}}$ retains a nonzero value.

The situation can be schematically represented as follows:

$$\lim_{t \to \infty} \Phi_{\boldsymbol{k}}(t) = 0, \quad \text{liquid}$$

$$\lim_{t \to \infty} \Phi_{\boldsymbol{k}}(t) \neq 0, \quad \text{glass}$$

This behavior has been observed in simulation calculations.

We can also show that function $\Phi_{\boldsymbol{k}}$ obeys an equation of motion which has an integro-differential form:

$$\ddot{\Phi}_{\boldsymbol{k}}(t) + \nu_0 \dot{\Phi}_{\boldsymbol{k}}(t) + \Omega_{\boldsymbol{k}}^2 \Phi_{\boldsymbol{k}}(t) + \Omega_{\boldsymbol{k}}^2 \int_0^t \Gamma_{\boldsymbol{k}}(t-t') \dot{\Phi}_{\boldsymbol{k}}(t') \mathrm{d}t' = 0 \tag{5.20}$$

with the following initial conditions:

$$\dot{\Phi}_{\boldsymbol{k}}(0) = 0 \quad \text{and} \quad \Phi_{\boldsymbol{k}}(0) = 1 \tag{5.21}$$

$$\nu_0 = (kT/m)^{1/2} \quad , \quad \Omega_{\boldsymbol{k}}^2 = (kT/m)^{1/2} \tag{5.22}$$

where $\Omega_{\boldsymbol{k}}$ is a characteristic frequency and m is the mass of the molecule and ν_0 a damping constant. Equation (5.20) describes a damped harmonic oscillator and contains memory term Γ in the last integral corresponding to a feedback mechanism. Γ must be described as a function of $\Omega_{\boldsymbol{k}}$ through coupling of two relaxation modes. The instantaneous fluctuation density relaxation rate is a function of its history through this feedback term.

Equation (5.20) can be solved using the two-mode approximation for the memory term

$$\Gamma_{\boldsymbol{k}}(t) = (2\pi)^{-3} \int V^2(\boldsymbol{k}, \boldsymbol{k'}) \Phi_{\boldsymbol{k}}(t) \Phi_{|\boldsymbol{k}-\boldsymbol{k'}|}(t) \mathrm{d}\boldsymbol{k'} \tag{5.23}$$

where function $V^2(\boldsymbol{k}, \boldsymbol{k'})$ is a function of $S(\boldsymbol{k})$.

Results obtained on a fluid with a Lennard–Jones intermolecular interaction potential using two values of $\boldsymbol{k}$ and different values of the density for function $\Phi_{\boldsymbol{k}}(t)$ as a function of time are shown in Fig. 5.8. The temperature is assumed to be fixed and equal to $kT/\varepsilon = 0.6$, where ε is the energy constant

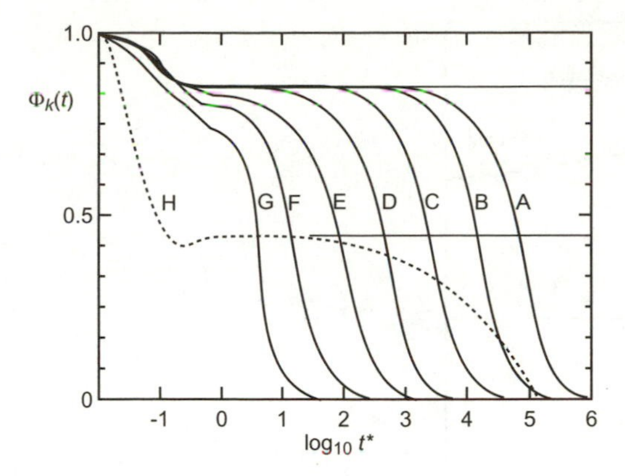

Fig. 5.8. Mode coupling model. The results of calculation of function $\Phi_{\boldsymbol{k}}(t)$ for relaxation of a Lennard–Jones supercooled liquid for temperature $kT/\varepsilon = 0.6$. Time t^* is in units $r_0(m\varepsilon)^{1/2}$. The solid curves are for $kr_0 = 7$ and the dashed curve is for $kr_0 = 6$. Curves A, B, C, D, E, F, G, H are calculated for $\Delta = (\rho_x - \rho)/\rho_x =$ 0.00031 – 0.00073 – 0.0023 – 0.0054 – 0.0106 – 0.021 – 0.052 – 0.00031. ε and r_0 are parameters of the Lennard–Jones potential. (U. Bengtzelius, *Phys. Rev. A*, **34**, 5059 (1986), copyright Am. Inst. of Phys.).

of the potential. The solid curves correspond to $k = 7/r_0$ and the dashed curve corresponds to $k = 6/r_0$, where r_0 is the intermolecular distance at which the intermolecular potential decays. In all cases, the evolution of the curves as a function of time attains a plateau where they remain for a certain time before again decreasing to zero. This behaviour is most pronounced for the curve corresponding to $\Delta = (\rho_x - \rho)/\rho_x = 0.00031$ where ρ_x is the density at which structural blockage occurs i.e. at which the self-diffusion stops.

Simulation calculations on supercooled Lennard–Jones mixtures give the same profile for the curves: relaxation occurs in two stages during which function $\Phi_{\boldsymbol{k}}$ seems to attain a stationary plateau before decreasing again. It remains to be proven that this is indeed the glass transition.

Mode coupling theory also allows predicting the behavior of self-diffusion coefficient D and thus the viscosity. We then find:

$$D \text{ (or } \eta^{-1}) \propto \Delta^{\gamma} \tag{5.24}$$

where $\gamma > 0$ and $\Delta = (\rho_x - \rho)/\rho_x$ or $(T_x - T)/T_x$, where T_x is the temperature at which structural blockage occurs.

One of the most important results of mode coupling theory is the existence of a change in the dynamics for a temperature T_C above the glass transition temperature T_g. This change appears if the variation of the structural relaxation time τ is investigated as a function of the temperature, and a very clear change in the variation of $\mathrm{d}\log\tau/\mathrm{d}T$ in salol (Fig. 5.9) at temperature $T_C = 1.15T_g$. This change in rate occurs at higher temperatures for glasses of covalent structure such as SiO_2.

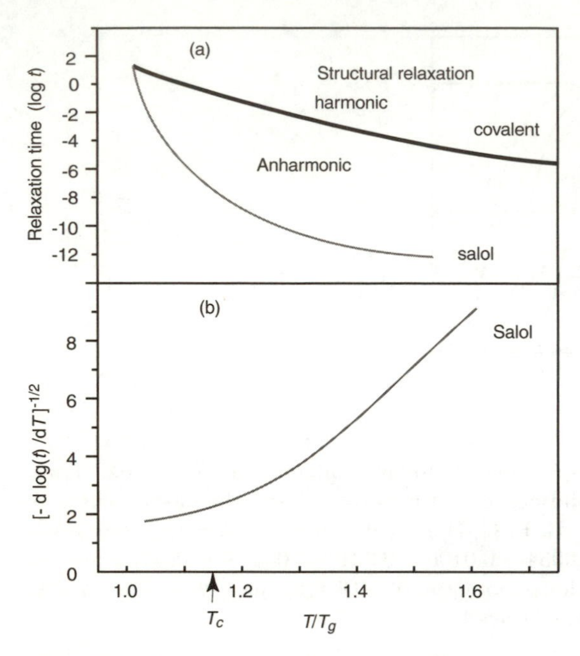

Fig. 5.9. Dynamics of glasses. The figures represent the behavior of relaxation rates. (**a**) Variation of the structural relaxation time with the temperature for a van der Waals liquid such as salol on one hand and a covalent liquid on the other. Relaxation of the harmonic type corresponds to quasi-local vibrations, while anharmonic relaxation corresponds to motions in a two-well potential at higher frequency. (**b**) This curve seems to indicate a change in the relaxation rate at a temperature $T_C \cong 1.15 T_g$ for salol ($T_g = 220K$) (A. P. Sokolov, *Science*, **273**, 1675 (1996), copyright Am. As. Ad. Sci.).

It would thus seem that above temperature T_C, glass-forming systems again tend to behave like normal liquids and at T_C undergo a transition toward a state of the "solid" type with spatial and dynamic heterogeneities that begin to appear.

Even leaving aside the fact that T_g is a function of the cooling rate, it is still necessary to know which transport coefficient should be used for defining the glass transition temperature. Most engineering studies utilize the viscosity, which is what we did at the beginning of this chapter, while recent theoretical studies are based on the density–density correlation function. Unfortunately, there is no theoretical method of linking the singularities of one transport coefficient with those of other transport coefficients. This evidently contrasts with the situation of equilibrium in statistical mechanics, where it is easy to understand that if a transition occurs in one physical property it will also show up in all other properties since they all reflect the behavior of the underlying free energy.

As we just saw, mode coupling theory has the advantage of showing that an important phenomenon occurs at a temperature just above T_g. The function Φ_k seems to approach a singularity when decreasing. It was also later shown that this behavior could be interpreted as follows; the diffusion mechanism changes nature, going from a long-range mode to a local mode. This behavior is understandable if the dynamics of supercooled liquids is described with two processes (Stillinger, 1995): slow collective motions associated with very deep configuration energy minima (potential wells); more rapid motions corresponding to individual movements of particles associated with local energy minima in the vicinity of a deep free energy minimum that the particles "recognize" in a way (Fig. 5.10). The relaxation due to collective motions is called mode α, and the much more rapid individual relaxations correspond to mode β (or Johari–Goldstein process). Although this model is very attractive, it is not certain that it will lead to a theory of the glass state. Simulation calculations on supercooled liquids indeed lead to predicting an increase in the viscosity and for even low viscosity values, they indeed represent a state resembling a glass; this would only have a limited lifetime since it would end in " flow" at the end of several hours. Beyond this point and the corresponding temperature, simulations become very difficult and take a great deal of time so that the glass transition temperature cannot really be determined.

Recent simulations performed with Lennard–Jones potentials on mixtures of liquids cooled with different quenching rates reveal three different regimes when the temperature is decreased. First, the system will exhibit behavior with nonexponential relaxation of the KWW type (5.10). Then when the temperature continues to decrease, the potential energy of the system exhibits increasingly pronounced minima, and finally, it is blocked in one of these minima; this corresponds to the behavior described above.

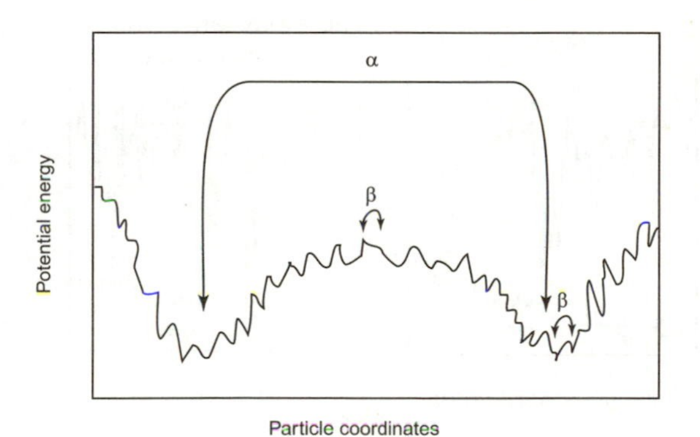

Fig. 5.10. Relaxation process in glass. The potential energy is shown as a function of the particle coordinates for a weak glass. The long-range collective transitions correspond to relaxation mode α and the local transitions by jumping between two neighboring minima correspond to mode β and are more rapid (F. H. Stillinger, 1995).

These considerations led Angell to propose a classification of glasses (1985) also based on the study of the behavior of the viscosity of a very large number of supercooled glass-forming liquids. Angell called liquids whose viscosity follows an Arrhenius law over the entire temperature range **strong liquids**; liquids with different behavior correspond to **so-called weak or "fragile" liquids**. SiO_2 and GeO_2 are typically strong liquids with four-coordinated network structures. They are generally resistant to thermal degradation. Ionic systems or systems with hydrogen bonds (glycerol, ethanol, etc.) are weak or fragile liquids; they are much more thermally fragile: heated above T_g, all frozen microstructures in the medium tend to disappear.

Strong liquids correspond to systems in which the potential energy has a limited number of minima (Fig. 5.11a) which are separated by relatively high potential barriers. Fragile liquids are characterized by a very large number of minima coexisting with "local" minima separated by relatively small "peaks" (Fig. 5.11b). Liquids with hydrogen bonds such as alcohols exhibit thermodynamic fragility associated with the existence of numerous minima separated by high potential barriers (Fig. 5.11c). When the potential barrier separating the maxima of the potential surface shown in Fig. 5.11 is greater than kT, the relaxation modes are activated by infrequent jumps from one minimum to another. Mode-coupling theory thus applies to weak liquids for $T > T_g$ corresponding to type α and β relaxations defined above, which are thus characteristic of this type of liquid. This classification of glasses in two broad categories is based on an interesting concept and it remains to be seen whether it will end in a unified theory of supercooled liquids and glasses.

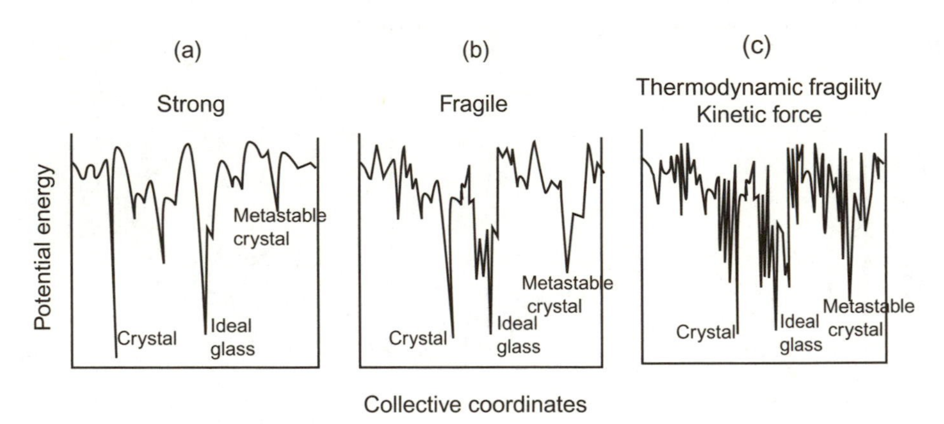

Fig. 5.11. Potential energy of supercooled liquids. (**a**) Strong liquids with a restricted number of potential minima separated by high potential barriers. (**b**) Fragile liquids with a large number of minima. (**c**) Hydrogen-bond liquids have numerous minima (thermodynamic fragility) separated by high potential barriers (kinetic force) (C. A. Angell, NATO ASIE, **188**, 133 (1990).

5.3.2 Industrial Applications

It is their optical properties that give classic glasses, particularly those obtained from oxides, their technical and industrial importance. As materials that are transparent to light, they can be used for manufacturing containers for different applications or devices such as optical fibers for transmitting a light signal over a long path with no losses. Both a fiber 100 μ in diameter and a telescope mirror several meters in size can be made with glass.

The quality of the optical properties of glasses is a function of the homogeneity of their refractive index (and thus the absence of impurities or bubbles in the material) and their method of manufacture. If defect-inducing stresses are to be avoided in a mirror, it is necessary to use ultraslow cooling into the glassy state with a rate of the order of $10^{-5}\text{K} \cdot \text{sec}^{-1}$.

Oxide glasses are particularly important in practical applications. Industrial glasses are silicate glasses with the approximate composition 70% SiO_2, 20% Na_2O, 10% CaO. Pyrex glass contains B_2O_3. Optical fibers, which have evolved considerably stimulated by telecommunications and data processing technologies (they allow transmitting signals in the infrared), are SiO_2–B_2O_3 mixtures. The core of the fiber is generally constituted by a SiO_2–GeO_2–B_2O_3 mixture. Fluorinated glasses based on ZrF_4 and ThF_4 that have the property of good transmission in the infrared up to approximately 9 μ, of interest for optical fiber telecommunications, have also been developed. The oxide-based industrial glasses are obtained at high temperatures (1500–2000 K for SiO_2, 800–820 K for window glass SiO_2–Na_2O–CaO).

Although classic industrial glasses can be manufactured by simply cooling a liquid by using air in installations producing several hundred tons, alternatively, it is necessary to utilize more draconian cooling methods for manufacturing certain special glasses, that is, very fast quenching.

The most brutal method, called "splat cooling" (or gun method Sect. 3.6.1), consists of projecting the cooled liquid with a shock wave onto a copper support. Powerful lasers can be used for conducting by melting a surface-vitrification (or glazing) operation corresponding to very fast quenching with a rate of $\approx 10^{14}$ K sec^{-1} The domain of glasses is not limited to substances of the oxide type, as we noted in the introduction to this chapter. Water, H_2O, is a special case of an oxide that can be obtained in the form of a glass either from the vapor phase or from ice at very high pressure (Sect. 3.5.3). A very large number of polymers in the solid state are amorphous structures (Plexiglas is an organic polymer used for its optical properties), and there are also semiconductors in the amorphous state (silicon, for example) and metal alloys. We note that amorphous systems, contrary to classic glasses, pass into the crystalline state when their temperature is raised.

Amorphous metallic materials are particularly interesting for their magnetic, electrical, and mechanical properties. There are two categories here:

- metal/metalloid alloys;
- metal/metal alloys.

The first category includes transition metals (Au, Pd, Pt, Ni, Mn) and a metalloid (Si, Ge, P, C, B), generally in a ratio of 80/20. The second category involves compounds of the $Mg_{65}Cu_{35}$, $Zr_{72}Co_{28}$, $Ni_{60}Nb_{40}$ type. These glasses, or amorphous metallic substances, have a higher mechanical strength than similar crystallized phases. They are generally more resistant to corrosion due to the low density of grain boundaries and dislocations. Many of them are ferromagnetic or ferrimagnetic. They are used as films and bands in transformers and electric motors. These materials are sensitive to devitrification, which modifies their properties. As for amorphous silicon, it has semiconducting properties and is used in photocells.

In the case of polymers in the glassy or amorphous state, aging is particularly important due to its consequences. A first type of aging is induced by crystallization, which causes nucleation of microcrystals in the polymer, particularly when it is heated above T_g. A second type corresponds to physical aging, which originates in structural relaxation of the glass to a state of metastable equilibrium over a very long period of time (this type of relaxation takes an almost infinite time in inorganic glasses).

The mechanical and optical properties of glassy or amorphous polymers are modified by physical aging, although they are widely used in industry (composites for automobiles, plastic films, packaging materials, etc.).

Unmixing must also be prevented in glasses which are mixtures, particularly mixtures of oxides like SiO_2–B_2O_3. Phase separation due to unmixing modifies the texture of a glass on a scale of 1–100 nm and thus its optical properties due to the appearance of heterogeneities. For this reason, these phenomena, spinodal decomposition in particular, must be prevented (Sect. 2.3). Alternatively, unmixing is used in manufacturing enamels, which are glassy emulsions with disperse metallic compound phases.

In conclusion, it is necessary to note the use of vitrification for storage of radioactive wastes. The interest in storing radioactive materials from reprocessing nuclear fuels in vitrified form comes from the stability of the vitreous substances made from liquid solutions of these wastes, as they are not sensitive to the thermal fluctuations generated by decomposition of the wastes.

5.3.3 Models for Biological Systems

Glass transition phenomena are far from being totally understood, although important progress has been made in the past twenty years in the physics of glasses. Nevertheless, glass theory and the models used for explaining the glass transition can be transposed into related domains such as biophysics.

Macromolecules of biological interest such as DNA, proteins, and polypeptides have individual behaviors approaching those of glasses. Proteins, for example, can have different very close energy conformations (for example, corresponding to helix or coil forms). As we will see (Sect. 6.4.1), the helix/coil transition in collagen plays an important role in gelation (a gel is a different material from a glass) which is collective in nature since it involves polymers in solution.

The individual biological molecule comprises many units (amino acids, for example), whose conformations can be rearranged with respect to each other (going from a coil form to a helical form, for example), and the dynamics of these rearrangements and thus the type of relaxation will be a function of the temperature. This is the stage where the analogy with the behavior of glasses and the glass transition intervenes. At a given time, an individual protein, for example, is in a state with a specific conformation; it generally does not remain in this state, but "jumps" into other very close energy states, thus scanning its energy spectrum.

These transitions are more important and frequent the higher the temperature is. By analogy, this potential energy spectrum can be described by a hypersurface of conformations in space which is similar to the surface representing the potential energy for glasses (Fig. 5.11). In a relaxation process, the system will undergo transitions from a nonequilibrium state to a state of equilibrium. For a protein, there will thus be type α relaxation involving motions of large segments of the polymer and type β relaxations corresponding to local motions (local reorientation of units) with very different characteristic times, as in a glass. Passing from mode α (at high temperature) to mode β (at lower temperature) is equivalent to the glass transition and it is observed at a temperature equivalent to T_g. We can also say that the glass transition in a protein corresponds to blocking of collective chain motions associated with changes in conformation.

This type of behavior has been demonstrated in proteins in the native (they have a helical conformation, for example) or denatured state (they have lost their helical conformation and are in the coil form) by specific heat measurements in the presence or absence of water. The transition is much more pronounced in denatured protein. These experiments have been conducted on vegetable casein, gluten, and poly-L-asparagine.

The interest of these models is not purely academic because there is a close correlation between the conformation of biological molecules such as enzymes, for example, and their function.

Problems

5.1. Unmixing in a Glass

Consider a glass composed of a binary mixture (for example, SiO_2–Na_2O) in which highly connected heterogeneous microstructures have been revealed by electron micrographs.

1. What could the origin of these heterogeneities be?
2. Give a simple model to explain their presence.

5.2. Kauzmann Paradox
Consider the boundary conditions of existence of a supercooled liquid phase corresponding to ideal vitrification temperature T_0 (Sect. 5.2.1).

1. Calculate the difference in entropy between liquid and crystal phases at the temperature T_g.
2. Calculate the specific entropy of the glass at $T = 0$.

5.3. Prigogine–Defay Relation
Consider a second order transition between phase α and phase β and designate the variations of compressibility, specific heat, and expansion coefficient between the two phases at the transition by $\Delta\kappa_T$, Δc_p, and $\Delta\alpha_p$.

1. Using the Ehrenfest equations (Sect. 4.3.3), find the correlation between these quantities.
2. Assume that phase α is a supercooled liquid and β is a glass and find that
$$\frac{\mathrm{d}T_g}{\mathrm{d}p} < \frac{\Delta\kappa_T}{\Delta\alpha_p}$$
3. What does the preceding Prigogine–Defay equation become? T_g is the glass transition temperature. What can we conclude from this for the glass transition?

5.4. Supercooling and Glass Transition
Assume that a liquid can be described by an ensemble of cells: some of volume v_0 per molecule are occupied, others v_f are empty. The viscosity η can then be written:

$$\eta^{-1} = A\exp(-\frac{bv_0}{v_f})$$

1. Write a relation between $\eta(T)$ and $\eta(T_g)$, the viscosities at temperature T and the glass transition temperature T_g from this equation, which will be expressed with
$$f = \frac{v_f}{v_0}$$
2. $\Delta\alpha_p$ being the difference in the thermal expansion coefficients of liquid and glass, calculate $\eta(T)/\eta(T_g)$, assuming that:
$$f = f_g + \Delta\alpha_p\ (T - T_g).$$

6 Gelation and Transitions in Biopolymers

6.1 The Gel State and Gelation

The appearance and the very particular consistency of solid gelatin are well known since it is very widely used in the food industry. Gelatin is a solid phase with specific properties; it is easily deformed under weak pressure, but the deformation ceases when the pressure stops, and it is homogeneous: it is a **gel**. We can say that it is a disordered phase or a **soft substance**. There are a great number of examples of gel phases more or less similar to gelatin. One finds gelled milks, in the food industry, and paints and cosmetics which are gels. There are numerous applications of gels. For example, gelatin is used to lay down the light-sensitive emulsion on the photographic films.

6.1.1 Characterization of a Gel

To obtain a gel, it is necessary to begin with at least two distinct constituents: a solvent and a solute. The initial phase consists of a liquid (water, for example), in which a molecular compound, usually a polymer (a protein in the case of gelatin), is dissolved. The initial fluid state is the **sol**. Under certain conditions of temperature, pH, and concentration, the sol undergoes a phase transition: it passes into the **gel** state. This is the **sol/gel transition** or **gelation**. We also call the transition **gel formation**.

The solvent and solute form a three-dimensional network in the gel state. The constitution of this network is characteristic of gelation. The solvent and solute cannot be distinguished in the gel phase, which is homogeneous.

Gels can also be formed from mineral particles (hydrated metal oxides, clays, etc.) in the form of spheres or rods.

The "consistency" of a gel, as characterized by a mechanical quantity such as the modulus of elasticity E, varies significantly as a function of its chemical composition and factors such as the temperature and pH of the initial solution: there is an entire gamut of gels, ranging from very viscous fluids to elastic solids.

In general, the term gel is reserved for systems in which the solvent is the dominant component.

The gel state was described for the first time in 1861 by T. Graham in a study of colloidal phases, but since then numerous studies have been published on gels and gelation. A gel can be characterized as follows (characterization proposed by P. H. Hermans in 1949):

- A gel constitutes a homogeneous phase with two components: a solid solute dissolved in a liquid phase.
- The dissolved (or dispersed) constituent and the solvent occupy the entire volume of the gel phase: they are interconnected in the form of a three-dimensional network (but there is no longer any liquid phase within the gel per se).
- The gel behaves like a solid under the effect of microscopic forces.

The research of J. Flory in the fifties also contributed to clarifying the properties of gels and the concepts that allow explaining them.

6.1.2 The Different Types of Gels

The molecules of the phase dispersed in the solvent can be mineral or organic. Mineral gels are obtained in this way from solutions of colloidal (crystalline or amorphous) metal hydroxides whose size is of the order of 5–10 nm. Organic gels are obtained from solutions of polymers such as polyacrylamide, polystyrene, or biopolymers such as collagen (which forms gelatin).

Several types of possible structures for gels are shown in Fig. 6.1. The gels formed from colloidal particles (milk, silica, metal hydroxides) correspond to Fig. 6.1a. These spherical particles are generally crystalline or amorphous aggregates of small molecules and their size can vary from 5 to 10 nm. Figure 6.1b corresponds to particles in the form of fibers or rods. Figures 6.1c and 6.1d respectively represent polymer gels either with weak physical bonds or with strong covalent bonds. The properties of gels are a function of the nature

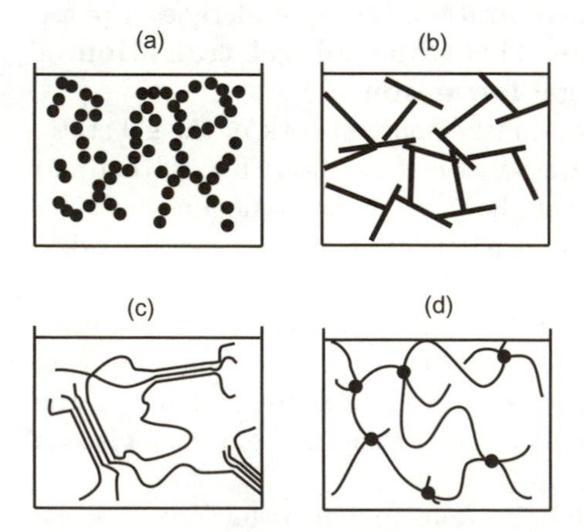

Fig. 6.1. Different types of gel structures. (**a**) Aggregates of spherical colloidal particles; (**b**) rod-shaped particles; (**c**) physical polymer gel with crystalline junctions; (**d**) chemical polymer gel with covalent bonds.

of the bonds between the particles or polymer chains. There are two broad categories of gels: **physical gels** on one hand and **chemical gels** on the other.

In the case of physical gels, the interactions or bonds between the chains that make up the material are physical in nature (van der Waals interactions, hydrogen bonds, $E \approx 2$ kcal/mole). These bonds are weak, involving energy of the order of kT, and for the system to be stable, they must not be localized in one point, but extended in space. Since the bond energy is weak, gelation is thermoreversible. This category includes gelatin, casein (a protein in milk that forms spherical aggregates 20–300 nm in diameter), agarose and pectin gels, and carrageenans (polysaccharides extracted from algae).

In **chemical gels**, the three-dimensional network is formed by a chemical reaction: covalent bonds are established between polymer chains, for example. The bond zone can be considered localized in this case. Chemical gels are formed irreversibly, since destruction of covalent bonds causes degradation of the polymer chains constituting the gel, no longer permitting its reconstitution. Polyacrylamide and polystyrene gels, as well as rubber (polyisoprene) vulcanized in solution with sulfur, are examples of chemical gels. In the vulcanization procedure, bridges are created between the polymer chains by sulfur atoms (Fig. 6.2).

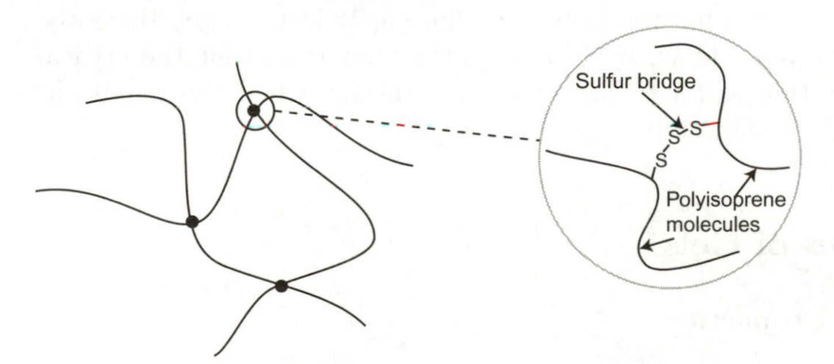

Fig. 6.2. Structure of vulcanized rubber. Sulfur atoms create bridges between polymer chains.

Silica gels also belong to the category of irreversible gels. When sodium metasilicate is dissolved, monosilicic acid is formed according to the reaction:

$Na_2SiO_3 + 3\ H_2O \quad \Leftrightarrow \quad H_4SiO_4 + 2\ NaOH$

Monosilicic acid can be polymerized, liberating water:

$$\mathrm{HO{-}\underset{\displaystyle OH}{\overset{\displaystyle OH}{Si}}{-}OH + HO{-}\underset{\displaystyle OH}{\overset{\displaystyle OH}{Si}}{-}OH \longrightarrow HO{-}\underset{\displaystyle OH}{\overset{\displaystyle OH}{Si}}{-}O{-}\underset{\displaystyle OH}{\overset{\displaystyle OH}{Si}}{-}OH + H_2O}$$

This reaction can continue and a three-dimensional network of Si–O bonds, a gel, is formed. The Si–O chains can be connected amongst themselves by hydrogen bonds which reinforce the structure of the gel, called a **silica hydrogel**. As the polymerization process progresses haphazardly, water accumulates on the surface of the gel: this is syneresis. The sol/gel transition in silica is very sensitive to the pH. Gelation is a gradual process that becomes irreversible when Si–O bonds are formed.

Although the situation is relatively simple (with the solvent in the sol phase or in the gel phase) in certain systems such as gelatin (a polymer of biological origin), which we will treat in more detail, the same is not true of numerous gels constituted of polymers. In effect, numerous polymers tend to form crystalline phases from their folded chains. They can thus form lamellar crystals or spherulites in a solvent (Chap. 3). Gelation can compete with crystallization in this case.

The tacticity, that is, the way in which the monomers are joined in a polymer chain, determines the microstructures that the polymer can form since it has a direct effect on the chain rigidity.

For example, isotactic polystyrene (all side groups are on the same side of the polymer chain) forms lamellar crystals at high temperature in a solvent, cis-decalin, while it forms a gel phase only at low temperature. In fact, the capacity of a polymer to form a rigid chain is the determining factor in gelation. Agarose, a rigid natural polymer, thus easily forms a gel. Recently, Koga and Tanaka have shown by Monte Carlo simulation that the critical gelation concentration shifts to higher concentrations when the persistent length of linear chain molecules is increasing.

6.2 Properties of Gels

6.2.1 Thermal Properties

Gelation is a complex process (it can compete with crystallization, as we noted) and the conditions in which it occurs are a function of the nature of the solvent, concentration of the solute (a polymer, for example), as well as the heat treatment the system undergoes.

For many gels, there is no fixed and unique gelation temperature for a given system (a fixed concentration of solute in the solvent), and the gelation temperature is a function of the thermal history of the material (the cooling rate, for example). This is the case of gelatin, for example. Moreover, when it is in the gel state and is re-heated, it does not again pass into the sol phase at the temperature at which gelation occurred: we have **thermal hysteresis** or "memory effect" here, as if the system "remembered" its previous structural states. In this case, there is some analogy with glasses: the gel is not in a **state of thermodynamic equilibrium**.

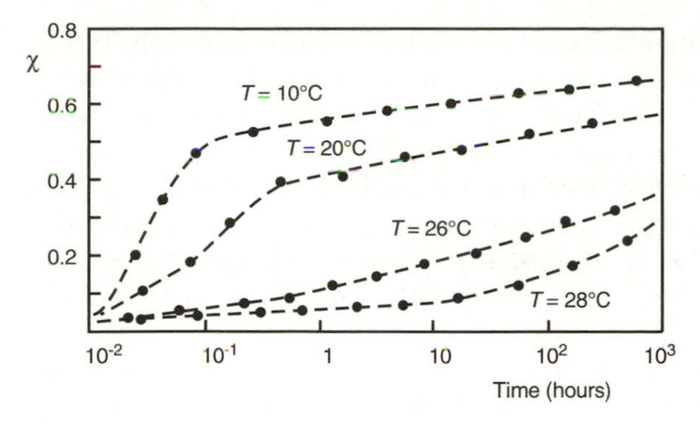

Fig. 6.3. Kinetics of formation of helices in gelatin gels. The fraction of helices χ formed as a function of time is shown. Each curve corresponds to an isotherm. χ is a function of the quenching temperature (After M. Djabourov, *Contemp. Phys.*, **29**, 286 (1988)).

In Fig. 6.3, note that for gelatin, the proportion of helical conformations of the polymer, which strongly influences gelation, depends on the quenching temperature.

Gelation and melting of a gel can be studied by differential thermal analysis in different systems, although this method gives results that are difficult to interpret in the case of gels.

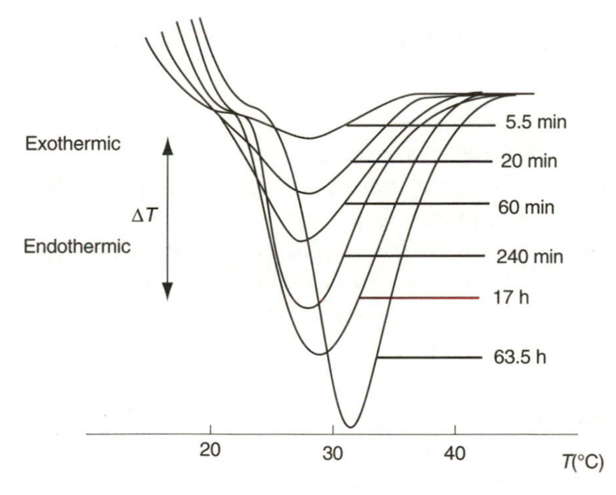

Fig. 6.4. Melting curves of gelatin gel. The melting peaks of gelatin gel were measured by differential thermal analysis for different gelation times. The gelation temperature was 15°C and the scanning rate was $2.5°\mathrm{C/min}^{-1}$. An endothermic peak is observed here (After P. Godard et al., *J. Polym. Sci., Polym. Phys. Ed.*, **16**, 1823 (1978)).

The enthalpy of formation of a gel constituted of isotactic polystyrene and cis-decalin can also be measured (it is of the order of 1 cal/g) and is strongly dependent on the concentration of polymer in the solvent. These measurements show that a polymer-solvent compound is formed.

Melting of a gel has generally been investigated by calorimetry, for gelatin in particular (Fig. 6.4). These measurements show that the melting point and value of the enthalpy of fusion are strongly dependent on the temperature and gelation rate for a given gel and consequently the thermal history of the material.

For some gels, those prepared from PVC (polyvinyl chloride), for example, it is not possible to follow gelation by differential thermal analysis because no exothermic peak is observed at the gel point and thus the enthalpy of gelation.

The studies of the thermal properties of gels tend to show, at least for physical gels, that gels do not form in a state of thermodynamic equilibrium.

6.2.2 Mechanical Properties

The gel state has strong analogies with the solid state and a gel is characterized by its rheological properties: the viscosity and the elasticity. To study the mechanical properties of gels, their elastic response (by measuring the elastic modulus) and viscoelastic behavior are estimated. The measurements are generally performed by two methods: by compressing the gel or by exposing it to periodic mechanical perturbations.

In mechanics, the shear modulus G is generally defined by the relation:

$$G = \frac{\sigma}{e} \tag{6.1}$$

where σ is the stress applied to the material and e is the associated deformation (G is actually the ratio between the stress per unit of area and the shear angle). The regime is assumed to be linear.

For very viscous liquids, setting gels or a supercooled liquid before vitrification, for which the characteristic times of molecular motions are long (long relaxation times), it is necessary to take these relaxation phenomena into account: the behavior is viscoelastic. The shear modulus is a function of time or frequency of the experiment.

For example, a periodic deformation is imposed on the material (Fig. 6.5). The gelling system is placed between a plane surface and a cone which are almost in contact and the lower plate is made to oscillate; the upper cone, held with a torsion bar, oscillates at a lower amplitude. The two movements are recorded.

The movement of the lower plate can be estimated to be proportional to the deformation e, while the motion of the upper cone is proportional to the stress σ. This is the principle of the rheological measurement.

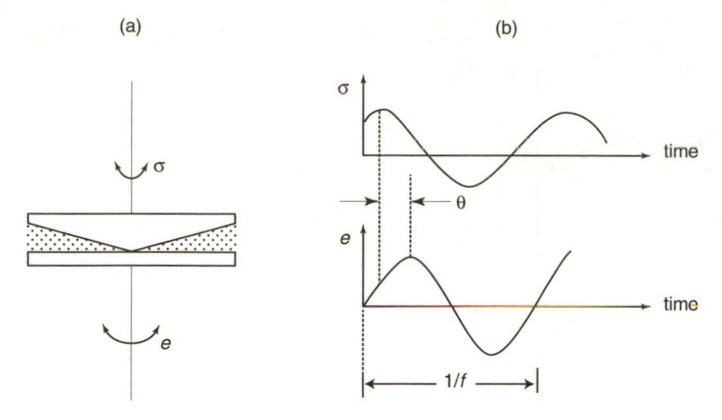

Fig. 6.5. Measurement of the shear modulus of a gel. (**a**) The gel is placed in the space between the plane surface and the cone and the lower plate is made to oscillate. When the upper plate is held in place by a torsion bar, its oscillation is of lower amplitude; (**b**) The two motions are recorded, movement of the lower plate is proportional to the deformation, and movement of the upper plate is proportional to the stress σ. The ratio and phase shift θ of the amplitudes are expressed by $|G|$ and φ given by (6.2) (M. Djabourov, *Contemp. Phys.*, **29**, No. 3, 273-297 (1988)).

Complex shear modulus G^* is then defined as follows, if the deformations are small:

$$G^* = \frac{\sigma}{e} = G' + \mathrm{i}\,G'' = |G|(\sin\phi + \mathrm{i}\,\cos\phi) \tag{6.2}$$

where $G'(\omega)$ is a function of angular frequency ω, and this is the elastic modulus; it corresponds to the elastic energy stored and returned to each cycle. $G''(\omega)$, which is also a function of the frequency, is the viscosity modulus or loss modulus; it corresponds to the viscous energy dissipated in each cycle.

For a Newtonian liquid, $G' = 0$, and it is characterized by its viscosity:

$$\eta = \lim(\omega \to 0)\frac{G''}{\omega} \tag{6.3}$$

Alternatively, a solid is characterized by its elastic shear modulus:

$$E = \lim(\omega \to 0)G' \tag{6.4}$$

where $\tan\varphi = G''(\omega)/G'(\omega)$ is the tangent of the loss angle.

For gelation, it is important to characterize the rheological properties within low frequency limits $(\omega \to 0)$.

For physical gels, as well as for chemical gels, a very important increase in moduli G' and G'' is observed in the vicinity of the sol/gel transition. The change in shear modulus G during gelation as a function of time at different temperatures is shown in Fig. 6.6. This modulus thus varies very rapidly when the gel is formed at temperature $T = 24°\mathrm{C}$, and the rate of increase in

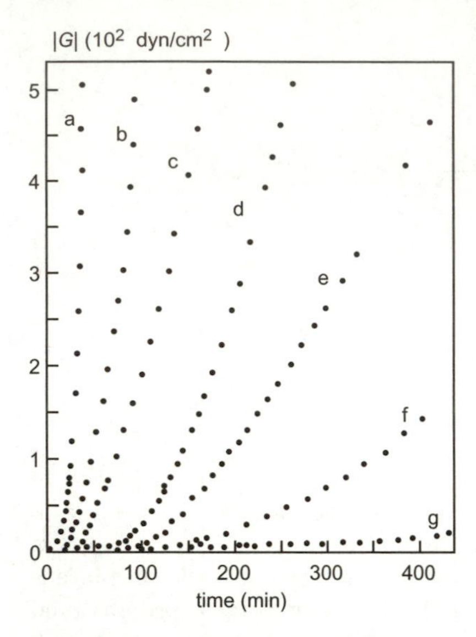

Fig. 6.6. Evolution of the shear modulus during gelation. The shear modulus $|G|$ of gelatin during gelation was measured for different temperatures: (**a**) $T = 24°\text{C}$; (**b**) $T = 25°\text{C}$; (**c**) $T = 26°\text{C}$; (**d**) $T = 26.5°\text{C}$; (**e**) $T = 27°\text{C}$; (**f**) $T = 27.5°\text{C}$; (**g**) $T = 28°\text{C}$. Very rapid variation of $|G|$ is found for gelation at $T < 26°\text{C}$ (M. Djabourov et al., *J. Phys.*, **49**, 333-343 (1988)).

G is divided by a factor of ten when the gelation temperaturc is only raised 4°C. This type of behavior is observed in most gels.

Since this involves the mechanical properties of gels, it is as if quantities such as the viscosity of the sol phase and the elastic shear modulus exhibited **critical behavior** in the vicinity of the sol/gel transition.

6.3 A Model For Gelation: Percolation

In gelation of gelatin, a three-dimensional network is formed by association of polymer chains with each other and with the solvent (water). Rheological measurements (Fig. 6.6) show that the physical, particularly the mechanical, properties of the gel are directly correlated with the existence of this network. This property is not unique to gelatin since it is found in all gels, including chemical gels. It is thus necessary to find a model that accounts for the progressive change in properties when going from the liquid phase (sol) to the gel phase, but also which answers the following question: are there universal laws that account for gelation independently of microscopic processes and the nature of gels? Important progress was made here during the seventies.

6.3.1 The Percolation Model

A new theoretical approach to the gelation problem was proposed by P. G. de Gennes and D. Stauffer in 1976; it consists of applying so-called **percolation** theory to the sol/gel transition (Sect. 2.5.2). This theory, by J. M. Hammersley (1957), initially allowed describing a statistical geometric situation: passage of a fluid through a network of channels randomly distributed in a solid, some of which are clogged. The fluid "percolates" through this more or less filtering network, attempting to find a path through open channels (Fig. 2.21a).

This situation can be described with a scheme (Fig. 2.21b, c) in which the open channels (permitting flow of the fluid) are represented by a bond symbolizing a connection between neighboring sites. A complete percolation path that allows fluid to pass through the system is represented by an infinite series of connected sites (these connected sites are the equivalent of a cluster or nucleus with s branches).

We can also draw a direct analogy between this system of channels in which a fluid percolates and a network of sites connected by electrical resistance: the bonds between sites in Fig. 6.7 are replaced by electrical resistances. If voltage is applied between the two opposite sides of the rectangle, an electric current begins to circulate when fraction p of occupied and connected sites attains a critical threshold p_C. If new electrical connections are established in the network, the current intensity increases rapidly for $p > p_C$.

This percolation model can be transposed for describing the effects related to the increase in the number of connections in a disordered medium (solvent and solute) which gradually forms a three-dimensional network during gela-

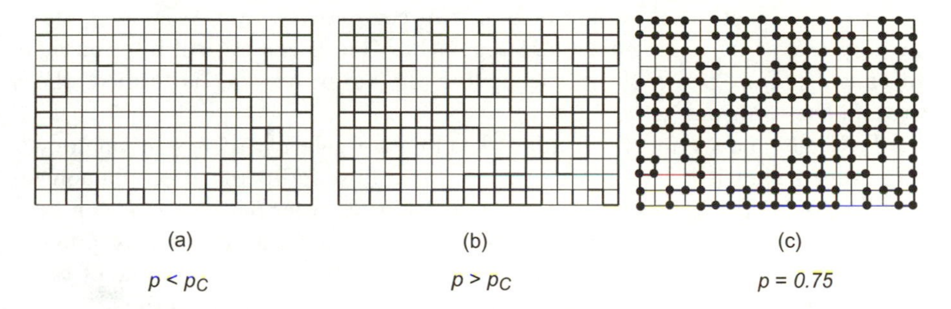

Fig. 6.7. Percolation. Percolation in a two-dimensional rectangular network is simulated: (**a**) All sites are assumed to be occupied (by a particle, for example) in a network and subsequently bonds are introduced between neighbors. If $p < p_C$, only nuclei of finite size appear; (**b**) $p > p_C$, an infinite network appears connected to the four sides of the network; (**c**) here it is assumed that only a fraction p of sites is occupied and bonds are progressively established between them. Here $p = 0.75$ and an infinite cluster is formed. p_C is the percolation threshold.

tion. This model predicts a sudden transition when the number of connections between particles or molecular ensembles crosses a critical threshold.

Percolation can be represented and simulated numerically by imagining that there is a plane regular network constituted of squares or rectangles of lines or bonds joining the nodes of the network constituted by sites. Probability p of occupation of a site is assigned (that is, the proportion of occupied sites). p is the variable of the problem. If a small number of bonds has formed, there will be small clusters constituted by s bonds (small s). If p increases, an increasing number of sites will be connected (we have the equivalent of a polymerization reaction) and beyond a certain **critical threshold** ($p > p_C$), an "infinite" cluster which connects the four edges of the network is formed (Fig. 6.7c).

This percolation situation can be simulated by calculation with a law for the number of nuclei of size s ($s = 1, 2, 3 \ldots$), that is, the number of connected sites, $n_S(p)$, which is a function of the probability of occupation of one site. For example, if we take a square network and select a site, it has 4 occupied neighbors with probability p, thus $n_2(p) = 4p$. The average size of a cluster can be represented with connected sites by the relation (2.90):

$$\langle S_m \rangle = \frac{\sum_{s=1}^{\infty} s^2 n_s(p)}{\sum_{s=1}^{\infty} s n_s(p)} \quad \text{for} \quad p > p_c, \; S_m \to \infty$$

A gel can be considered equivalent to the network of connected sites described previously: the molecules of the solute (the polymer) occupy sites with probability p and establish chemical or physical bonds with their neighbors.

The connected sites belong to clusters of variable but increasing size. When the percolation threshold is reached, a cluster that occupies the entire volume is formed: the gel point is then surpassed.

If we designate the fraction of bonds in the network by p (Fig. 6.7), the average size of the aggregates increases with p. There is a critical fraction p_C after which ($p > p_C$) a cluster occupying the entire network is formed: the system has formed a gel.

P. G. de Gennes and D. Stauffer developed a gelation theory by analogy simultaneously with second order phase transitions and percolation phenomena. For second order transitions, we know that an order parameter P can be defined, which is the magnetization $\boldsymbol{M}$ for a magnetic material, for example. The order parameter is a function of the temperature T and it is one of the state variables that can be acted upon to induce the phase transition.

We have:

$$\begin{aligned} &P \neq 0 \quad \text{for} \quad T < T_C \\ &P = 0 \quad \text{for} \quad T > T_C \\ &P \to 0 \quad \text{if} \quad T \to T_C \quad \text{with} \quad T < T_C \end{aligned} \tag{6.5}$$

where T_C is the transition temperature, often called the Curie temperature. The order parameter varies continuously in the transition.

A critical exponent is defined for the transition so that: $P \propto \varepsilon^{\beta}$ with $\varepsilon = |T - T_C|/T_C$. Similarly, in a magnetic system, the susceptibility $\chi = (\partial M/\partial H)_{T,H=0}$ obeys a critical law with exponent γ: $\chi \propto \varepsilon^{-\gamma}$. As for the coherence length ξ characteristic of fluctuations of the order parameter, it obeys a critical law with exponent ν: $\xi \propto \varepsilon^{-\nu}$.

By analogy, in the case of gelation, an order parameter P which is the probability that an interchain intermolecular bond which has reacted will belong to an infinite cluster is defined. It is also called **the gel fraction**. P is the fraction of particles or molecules attached to the macroscopic aggregate as it is being formed.

Once the macroscopic aggregate has formed, P increases with the fraction of attached molecules or bonds, obeying a law with a critical exponent β:

$$P \propto (p - p_C)^{\beta} \text{ for } p > p_C \text{ and } P = 0 \text{ for } p < p_C \tag{6.6}$$

The parameter p, the fraction of bonds in the network, is the equivalent of thermodynamic variable T here.

In approaching the gel point, the average molecular weight of the gelled clusters M_W increases and it also obeys a law with a critical exponent:

$$M_W \propto (p - p_C)^{-\gamma} \quad p < p_C \tag{6.7}$$

$M_W \to \infty$ if $p \to p_C$ (an infinite cluster obviously never exists since volume V remains finite). Similarly, the average radius $< R >$ of a cluster obeys the following law:

$$\langle R \rangle = (p_C - p)^{-\nu} \quad p < p_C \tag{6.8}$$

where $< R >$ is the equivalent of correlation length ξ of a second-order transition.

Using the percolation model, the mechanical quantities E and η can be correlated with parameter p with the following "critical" laws:

$$\eta \propto (p_C - p)^{-k} \text{ for } p < p_C \text{ and } E \propto (p - p_C)^{t} \text{ for } p > p_C \tag{6.9}$$

The increase in the viscosity near the gel point is due to the very large increase in the size of the clusters.

Using a simulation method for describing a percolation transition, we obtain the following results:

- For a square network: $p_C = 0.5$.
- For a simple three-dimensional network: $p_C = 0.247$ with $\beta = 0.4$, $\gamma = 1.7$, $\nu = 0.85$.

Several models have been proposed for estimating exponents k and t. The different estimations lead to a value of k between 0.75 and 1.35.

By analogy with a percolation model describing a network of electrical resistances, a value of exponent t between 1.8 and 2 is found. A mean field theory predicts a different value: $t = 3$.

Measurements near the gel point are very difficult to perform because gelation is very rapid there. We thus find $t = 1.9$ for the sol/gel transition in the mixture mono- / bisacrylamide (chemical gel), just as in gelatin (physical gel).

6.4 Biopolymers Gels

The gels obtained from polymers of biological origin are an important category because they have the advantage of utilizing water as the solvent and thus have many applications in the food industry. These are physical gels since the gelation process is reversible.

Two types of biopolymer gels are particularly important; **protein** gels and **polysaccharide** gels.

6.4.1 An Important Gel: Gelatin

Gelatin is probably the most well-known physical gel undoubtedly because it has very old and numerous uses in food products. Gelatin is a protein prepared by degradation of collagen by hydrolysis, and collagen itself is the principal protein found in animal tissues such as the skin, but also in tendons and bones. Gelatin is thus also called denatured collagen.

The basic chemical formula of gelatin is complex and depends on its origin (animal species) and method of preparation. However, the polymer chain is constituted of a sequence of three basic monomers which are amino acids: proline, hydroxyproline, and glycine (chemical formula NH_2–CH_2–COOH). The polymer contains approximately 1000 monomers (called residues).

Collagen in the native state is in the form of fibers composed of helices wound in threes, **triple helices**. Each chain in the helix rotates counterclockwise. The triple helix is 300 nm long and the chain constituting it has a molecular weight of approximately 10^5. The triple helices are stabilized by interchain hydrogen bonds (Fig. 6.8); they form parallelly aligned rods attached by covalent bonds located on their ends.

Fig. 6.8. Structure of collagen. Three chains of helical conformation (left-handed helix) are wound together and form a triple helix (right-handed helix), which is collagen in the native state. The pitch is 8.6 nm; it is constant and is stabilized by hydrogen bonds between chains.

Collagen denaturation causes separation of the rods and total or partial separation of the chains by destruction of hydrogen bonds and thus the disappearance of the triple-helix conformation. The polymers are then in the **coil** form, and industrial gelatins are mixtures of different compounds: α gelatin (one polymer chain), β gelatin (two polymer chains), γ gelatin (three polymer chains).

In solution at high temperature, the gelatin chains are in the so-called **coil** conformation. When the temperature is decreased (below 30°C) and the concentration is higher than 0.5 wt.%, the gelatin chains progressively change conformation, forming helices again: this is the **coil ⇔ helix transition**. The solution loses its liquid properties and becomes a soft solid which keeps its shape: this is the gel state. The conformational transformation is associated with the sol/gel transition. The protons of the solvent, water in this case, play an essential role in gelation since they contribute to formation of hydrogen bonds between chains (Fig. 6.9) which form a very entangled network (Fig. 6.10).

The fraction of helices present in the sol or gel, the **helix content**, can be measured by optical polarimetry: the rotatory power (angle of rotation of polarization of an incident light) is directly correlated with the helix content χ, defined by χ = number of residues in helical conformation/total number of residues in solution.

The change of χ can be followed during gelation and for different gelation temperatures (Fig. 6.11).

The equilibrium state is only reached after a very long time; it corresponds in fact to the total re-naturation of collagen triple helices, that is, to the native form. The coil ⇔ helix transition associated with gelation also exhibits thermal hysteresis: the gelatin gel returns to the sol state in heating

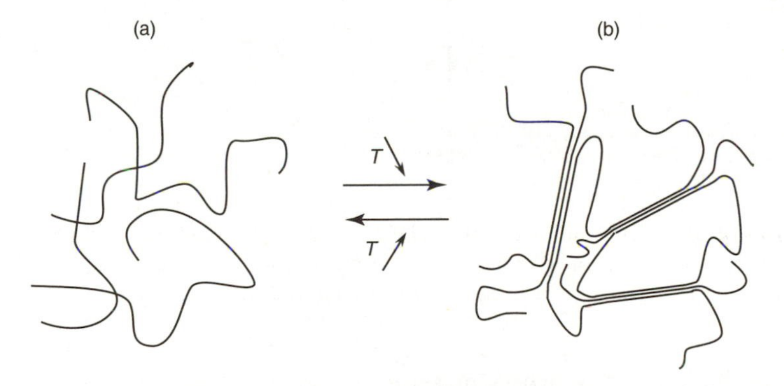

Fig. 6.9. Formation of a gelatin gel. (**a**) Initial polymer chains in a coil; (**b**) If the concentration is high enough and if the temperature is lowered, triple helices form again: a gelatin gel appears. The gel is not in thermodynamic equilibrium since the number of bonds between chains increases with time. Gelation is reversible in this case.

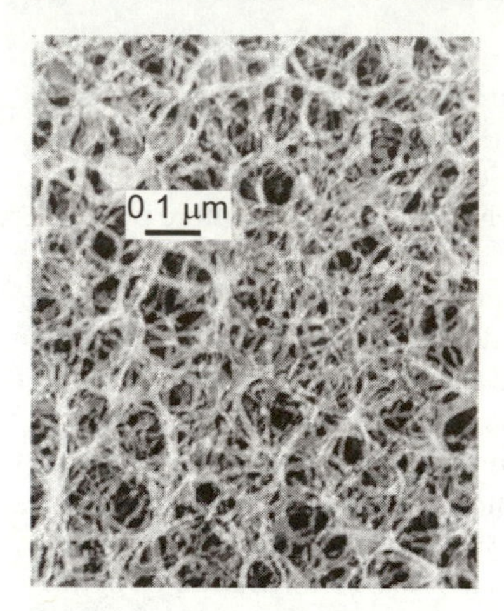

Fig. 6.10. Electron microscopy of the gelatin gel network. The chains are composed of triple helices of various lengths (10–100 nm). The concentration of gelatin is 2 %. Scale 1 cm = 100 nm (M. Djabourov, J. P. Lechaire, N. Favard, and N. Bonnet, in M. Djabourov, Contemp. Phys., 29, N°3, 273-297 (1988)).

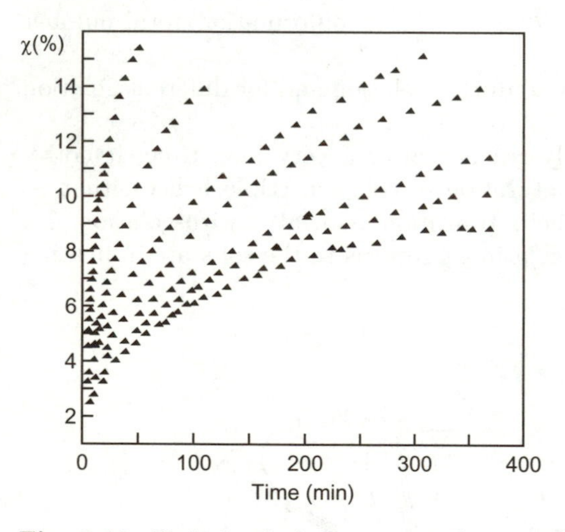

Fig. 6.11. Variation in helix content during gelation. The helix content χ and its variation with time during gelation of gelatin at different temperatures are measured (same as for Fig. 6.6) (M. Djabourov et al., *J. Phys.*, **49**, 333-343 (1988)).

(it "melts"), but the fraction of helices measured is not the same for a given temperature in heating and cooling again.

Finally, we can show that χ can be considered as the thermodynamic variable associated with gelation (the equivalent of variable p in (6.7)–(6.9)).

There is a direct correlation between shear modulus G and quantity χ.

The melting point and gelation rate of a gelatin depend strongly on the nature of the collagen and thus its origin. The hydroxyproline (an amino acid) content in the chain would seem to be the key factor here.

For example, mussels secrete a protein-like substance, byssus, which allows them to attach to a solid such as a rock. Byssus is composed of collagen, which has a relatively high melting point of approximately 90°C (the size of this collagen is of the order of 130 nm). Moreover, the problems that the use of collagens extracted from mammals (calves, sheep) following the so-called "mad cow disease" (bovine spongiform encephalopathy) pose for public health led to consideration of preparation of gelatin from fish collagens, but the gelation temperature conditions are not strictly the same as for gelatins from mammalian collagens. This question is still very open.

6.4.2 Polysaccharide Gels

Polysaccharides are natural polymers found in plants, algae, animals, and bacteria; some are electrically neutral, while others are charged. They are macromolecular compounds of glucidic nature in which the monomers are composed of one or more sugars connected together in a more or less regular sequence.

Agarose, amylose, and **amylopectin** are neutral polysaccharides that form thermoreversible gels. Agarose is one of the constituents of **agar-agar**, a gelling agent extracted from red algae of the gelidium type. Amylose and amylopectin (from the Greek word *amylos*, which means starch) form gels that play an important role in the texturing properties of starch-based food products. A solution of amylose can only be obtained at relatively high temperature (90–100°C). Amylopectin gels only at low temperature (around 1°C) and the kinetics is generally very slow (the gel times are measured in weeks).

It is necessary to report that cellulose derivatives also have gelling properties, but they are rather special: they form reversible gels on heating and then "melt" when cooled again. These are the only known gels with this characteristic.

Some gel-forming polysaccharides are electrically charged; these are **polyelectrolytes**. Carrageenans and alginates respectively extracted from red and brown algae are typical examples of such physical gels. The formation of alginate gels is strongly influenced by the presence of ions, Ca^{++} ions in particular. The electrostatic repulsion due to carboxyl groups prevents the polymer chains from coming together, and insertion of cations such as Ca^{2+} reduces the ionization and allows bridges to be formed between the chains; a structure of the "egg box" type is obtained (Rees model, Fig. 6.12). Carrageenan gels are thermoreversible while alginate gels are not.

Pectins, which are structural components of fruit peels, are also gel-forming polysaccharides whose formation can be controlled by addition of calcium ions and the pH. The basic monomer in pectins is galacturonic acid.

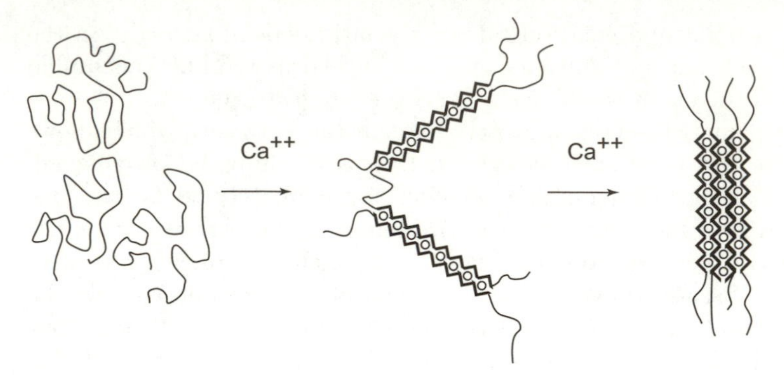

Fig. 6.12. Model of formation of an alginate gel. Ca^{++} cations favor approach of the polymer chains and are located in interstitial positions between the polysaccharide chains. They thus form a so-called "egg box" structure (Rees model).

At low pH, thermoreversible gels are obtained. Pectin gels are used in jelly type jams.

Finally, mixtures of polysaccharides also have interesting gelling properties. **Galactomannans** and **xanthan**, which do not form gels individually, gel when mixed together. Xanthan is extracted from a bacterium, *Xanthomas campestris*; since its molecules have a very low affinity for associating with each other, they are thickening agent. Galactomannans are extracted from guar and carob seeds; the monomers are sugars, galactose and mannose.

6.4.3 Modeling of the Coil ⇔ Helix Transition

As we just saw, a change in molecular conformation is the mechanism that triggers the sol/gel transition in many biopolymers. In the case of gelatin, gelation is thus initiated by a coil ⇔ helix transition.

The conformational transition in a biopolymer is encountered in many systems; it generally plays a very important role in modifying its properties. Such transitions are found in DNA (which forms a double helix) and in a protein, prion, which plays a key role in mad cow disease. These transitions occur in well-defined temperature conditions or if the quality of the solvent is altered (pH, ionic concentration). They are often considered as first order transitions because they are associated with a change in enthalpy.

In general, the appearance of a stable helical conformation in solution for proteins or polypeptides is related to the formation of hydrogen bonds which strengthen the helical structure; these bonds are established between neighboring groups (N–H and O=C, for example) in the same chain parallel to the axis of the helix (Fig. 6.13). In the case of collagen, it is in the form of a triple helix in the native state, and the breaking of the hydrogen bonds, that stabilize this conformation, causes it to pass into the coil form (the strands are randomly arranged): this is denaturation. This can be induced by altering

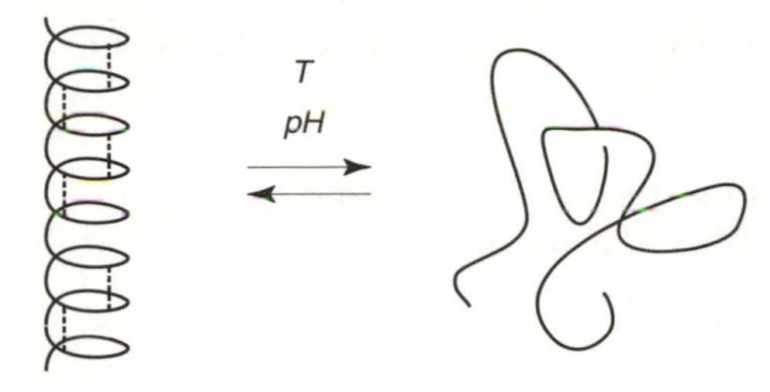

Fig. 6.13. Helix–coil transition. The hydrogen bonds (dashed lines) parallel to the axis of the helix stabilize the helical conformation. If the temperature is raised or if the pH is modified, the helix is destabilized and a coil forms. The protein is denatured.

the charges of the side groups, by varying the pH or the ionic concentration in solution, or by raising the temperature.

The coil $\Longleftrightarrow$ helix transformation, whether thermal or ionic, is in principle reversible. There are simple models that can account for this: one can predict what fraction of the monomers (or residues) are in the coil or helix form as function, for instance, of the temperature.

6.4.4 Statistical Model

To statistically describe the behavior and conformational transition of a chain, consider a chain with n monomers that are either in the **coil form** c or the **helix form** h. In principle, the conformation of each residue (helix or coil, Fig. 6.14) is known. For example, it can be considered that at least three units are necessary for accomplishing one helix turn.

A global configuration at given temperature T can thus be represented by a sequence of the type, for example, *hhccccchhhcchhhh*...

To simplify, consider two statistical hypotheses:

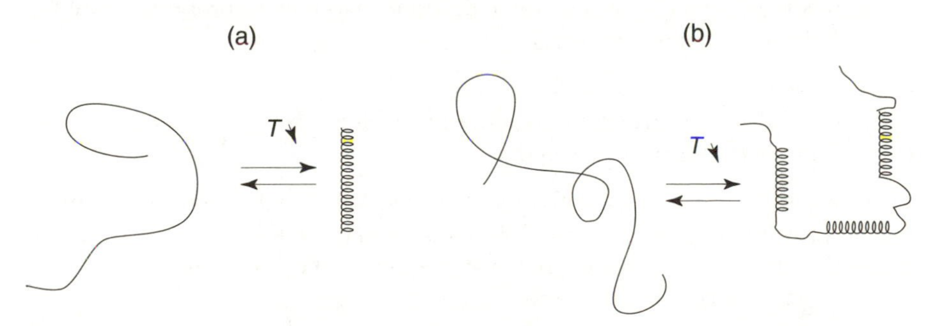

Fig. 6.14. Coil–helix transition. (**a**) Chain with two conformational states, coil and helix; (**b**) Random mixture of coil and helix states whose proportions are functions of the temperature.

- there is a mixture of states c and h in variable proportions within a chain (cf. preceding sequence): each chain is composed of n units (monomers) in helical form or coil form (Fig. 6.14b);
- each macromolecular chain is either entirely in the form of coil units (c) or entirely in the form of helical units (h). The only two possible forms will then be symbolically represented (Fig. 6.14a) as:

$$\ldots hhhhh \ldots \text{or} \ldots \text{ccccc} \ldots$$

The probability of being in this state will be defined for a chain unit in the coil state c by:

$$u = \mathrm{e}^{-G(c)/kT}/z \tag{6.10}$$

and for a unit in the helical state h by:

$$w = \mathrm{e}^{-G(h)/kT}/z \tag{6.11}$$

with

$$z = \mathrm{e}^{-G(c)/kT} + \mathrm{e}^{-G(h)/kT}$$

where $G(h)$ and $G(c)$ are the Gibbs free energy corresponding to configurations h and c.

The ratio:

$$s = w/u = \mathrm{e}^{-\{G(h)-G(c)\}/kT} \tag{6.12}$$

defines the "equilibrium constant" of the "reaction" of transformation of a coil unit into a helical unit: $c \Leftrightarrow h$.

The statistical weight u of a unit in the coil form, and thus disordered, is essentially entropic in nature (entropy S is related to rotational conformations): $u \propto \mathrm{e}^{S/k}$. On the contrary, the statistical weight w of the helical form is essentially a function of the potential energy E of stabilization of the helices: $w \propto \mathrm{e}^{-E/kT}$.

Here s is a parameter whose value determines the most probable state for a chain element at temperature T:

- $s < 1$ the coil form is favored;
- $s = 1$ both configurations are equally probable;
- $s > 1$ the helical form is favored.

Introducing the change in enthalpy ΔH associated with the coil $\Leftrightarrow$ helix transition which can be measured by calorimetry, we then assume that s is a function of the temperature that obeys a law of the Van't Hoff type:

$$\frac{\mathrm{d}\ln s}{\mathrm{d}T} = \frac{\Delta H}{kT^2} \tag{6.13}$$

We will define the transition temperature T_t by $s = 1$. By integration, we have:

$$\int_{T_t}^{T} \mathrm{d}\ln s = -\frac{\Delta H}{kT}\bigg|_{T_t}^{T} \tag{6.14}$$

or

$$\ln s = \frac{\Delta H}{kTT_t}[T - T_t] \tag{6.15}$$

Setting $\Delta T = T - T_t$, we thus have:

$$\ln s = \frac{\Delta H}{kTT_t}\Delta T \tag{6.16}$$

For the coil $\Rightarrow$ helix transition, $\Delta H < 0$; in effect, if $\Delta T > 0$, the coil form appears and is stable at high temperature ($T > T_t$).

In the case of the first hypothesis of a **random mixture of coil and helix states**, the total partition function for the chain is written as:

$$Z(n,T) = \sum_{n_c,n_h} \mathrm{e}^{-G_{tot}(n_c,n_h)/kT} \tag{6.17}$$

$G_{tot}(n_c, n_h)$ is the total Gibbs free energy of the chain in a configuration with n_c residues in the coil state and n_h in the helical state ($n = n_c + n_h$).

Using (6.10) and (6.11), (6.17) is rewritten as:

$$Z(n,T) = \sum_{n_h=0}^{\infty} \frac{n!}{n_h!(n-n_h)!} w^{n_h} u^{n-n_h} \tag{6.18}$$

or:

$$Z(n,T) = u^n \left(1 + \frac{w}{u}\right)^n = (w+u)^n \tag{6.19}$$

The helix content χ (Sect. 6.4.1) is given by: $\chi = < n_h > /n$, as:

$$< n_k > = \frac{1}{Z}\sum_{n_h} n_h w^{n_h} u^{n-n_h} \frac{n!}{n_h!(n-n_h)!} \tag{6.20}$$

we have:

$$< n_h > = \frac{w}{Z}\frac{\partial Z}{\partial w} = w\frac{\partial \log Z}{\partial w} \tag{6.21}$$

and, using $s = w/u$:

$$\chi = \frac{w}{n}\frac{\partial \log Z}{\partial w} = \frac{s}{s+1} \tag{6.22}$$

In this simple model, χ is independent of the chain length n. The dependence of χ as a function of s is shown in Fig. 6.15a. For $s = 1$, $\chi = 1/2$.

If $s < 1$ (corresponding to $\Delta T > 0$), the coil form is favored. On the contrary, for $s > 1$, we have the opposite solution: the helix form is favored $\chi \to 1$.

If we now apply the second hypothesis of the **two-state chain**, the form of $Z(n,T)$ is extremely simplified; it is written as:

$$Z(n,T) = u^n + w^n = u^n(1+s^n) \tag{6.23}$$

i.e.:

$$\chi = \frac{< n_h >}{n} = \frac{1}{n}\frac{nw^n}{u^n + w^n} = \frac{w^n}{u^n + w^n} = \frac{s^n}{1+s^n} \tag{6.24}$$

where χ is a function of n here. If $n \to \infty$, χ will vary abruptly as a step function from 0 to 1 around $s = 1$ (curve 2 in Fig. 6.15).

For $s < 1$, $\chi = 0$ if $n \to \infty$ and for $s > 1$, $\chi = 1$ if $n \to \infty$.

For small values of n, (6.24) can be rewritten as:

$$\chi = \frac{\mathrm{e}^{n\frac{\Delta H}{kT_t}\frac{\Delta T}{T}}}{1 + \mathrm{e}^{n\frac{\Delta H}{kT_t}\frac{\Delta T}{T}}} \propto \mathrm{e}^{n\frac{\Delta H}{kT_t}\frac{\Delta T}{T}} \tag{6.25}$$

If $\Delta T > 0$, there is an exponential increase in the helix content in the vicinity of the transition. The change in $\chi(T)$ for the coil $\Leftrightarrow$ helix transition in a polypeptide, poly-γ-benzyl L-glutamate, in a solvent is shown in Fig. 6.15b for different chain lengths. The transition is more pronounced the longer the chain is ($\Delta H > 0$ here because there are bonds between the solvent and the chain and helices are formed at high temprature).

For polypeptides, an additional parameter is usually included, the statistical weight of the residues in helical conformation located at the ends of the sequences.

These residues cannot establish hydrogen bonds with their neighbors, and are not favored energetically, but they control the length of the sequence of helices. The lower this statistical weight is, the more pronounced the transition is.

The preceding arguments apply to the case of gelatin, with a certain number of modifications. Helices are formed along individual chains in the gelatin. Proline-rich sequences are favorable sites for nucleation due to their

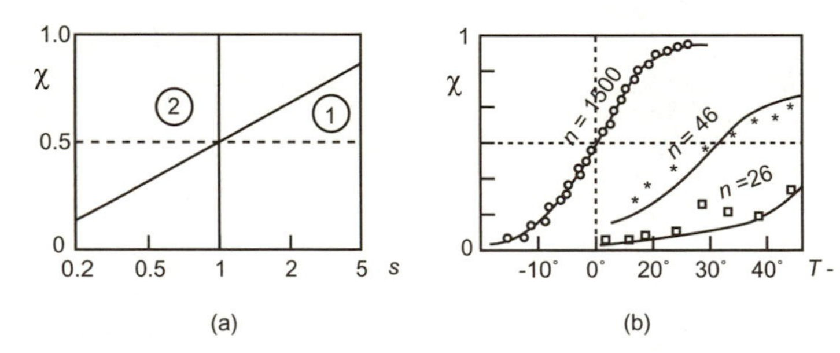

Fig. 6.15. Conformational transition. (**a**) Fraction of helices χ as a function of parameter s (logarithmic scale); 1) random mixture of coil and helical states; 2) two-state chains. (**b**) Experimental result on poly-γ-benzyl L-glutamate in a solvent which is a mixture of dichloroacetic acid (80%) and ethylene dichloride (20%). The transition is more pronounced the larger the number of units n in the chain (B. H. Zimm et al., 1959).

local rigidity. Hydrogen bonds give the gelatin chains helical rigidity, and the helical zones form interchain junctions. A chain is assumed to be involved in two junctions.

The probability f that a chain will have h units in a helical conformation at temperature T is written:

$$f(T,h) \propto \Omega \exp\left(\frac{\epsilon h}{kT}\right) \tag{6.26}$$

where ε is the binding energy corresponding to the helical unit, and Ω is the number of microscopic configuration, a quantity that is difficult to calculate. The value of h at equilibrium $h_{eq}(T)$ is calculated by finding the maximum of $f(T,h)$. We obtain $\chi_{eq}(T) = h_{eq}/N$, if N is the total number of units in the chain.

The result of numerical calculations of χ_{eq} as a function of T is shown in Fig. 6.16. It shows the formation of a loose network, but with fixed helical junctions; the kinetics of this formation is fast and is the first stage in gelation. Finally the stresses that keep the junctions fixed relax, and continuous helical sequences form; this is a slow process that theoretically ends in total restoration of collagen (Fig. 6.16b). Similar models have been developed for DNA.

The coil/helix transition in biopolymers can be studied by Rayleigh light scattering. The molecular diffusion coefficient D in the solvent (the width of the Rayleigh line is proportional to it) and the average radius $< R >$ of the coils ($D = kT/(6\pi\eta < R >)$, where η is the viscosity of the solvent) can also be measured. The transition reveals itself by a sudden increase in $< R >$ associated with a decrease in the width of the spectral line of the scattered light.

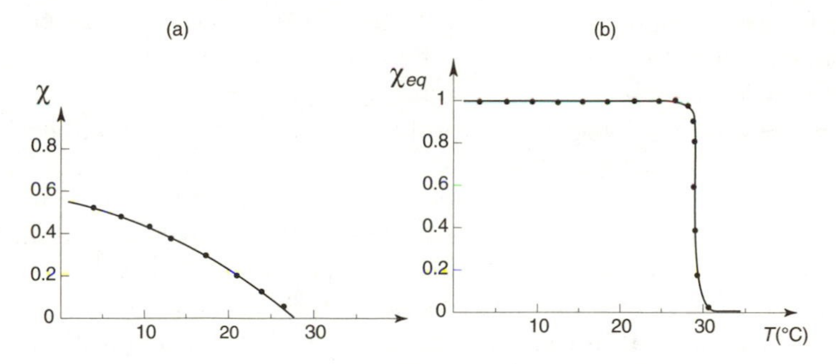

Fig. 6.16. Gelation of gelatin. Gelation occurs in two stages; (**a**) A loose network is very rapidly formed (several minutes); (**b**) After this stage, slow evolution continues (logarithmic variation). It can lead to total denaturation of the triple-helix collagen rods (M. Djabourov, *Contemp. Phys.*, **29**, 273-297 (1988)).

6.5 Main Applications of Gels and Gelation

Gels have many industrial and domestic applications. The applications of gelatin in the food industry are therefore very old.

Gels are used in the food industry for manufacturing jellies and thickeners: **texturing agents**, that is, they modify the structure of a food (a jelly or a cream is a soft substance). Gels and gelation are used in making jellies (delicatessen, jams), yogurt, ice creams, and some sauces. The molecular systems used in the food industry include gelatin, pectins (polygalacturonic acid) which form gels with sugar and give jams their texture, and polysaccharides such as carrageenans and xanthan.

Gelatin is also traditionally used in the photographic industry in the fabrication of films. Its role is to trap the silver chloride in the photographic emulsion and to prevent dispersion of the Ag^+ salt microcrystals formed after exposure to light, which would decrease the sharpness of the image.

Gels are used in the cosmetics (hydrating creams) and paint industries (manufacture of enamel paints). They play an essential role in the electrophoresis equipment currently used in biology. Gel electrophoresis allows separating fragments of DNA that can subsequently be identified. An electric field is applied to a plate where the gel has been spread and causes separation of charged fragments of a DNA fraction as a function of their size.

Gelation implies formation of a cohesive three-dimensional network associating a solute (generally a polymer) and a solvent. An important amount of solvent can be absorbed by the solute: the network swells, forming a gel (the opposite phenomenon of expulsion of part of the solvent by the gel is called **syneresis**). The swelling rates or ratio of the volumes can attain important values, which allows gels to absorb an organic or aqueous solvent. This is the principle used in "water grains" (starch grafted with polyacrylonitrile) used in horticulture for fixing important amounts of water in the soil, as well as in baby diaper pants, where the gel is a dried and crushed polyelectrolyte (for example 1 g of powder can absorb 1 kg of water by rapid swelling).

Gels are also used for their mechanical properties in very different applications: fluids for oil wells, flexible lenses, for example.

Finally, the property of gels to contract or swell in reaction to changes in their physical or chemical environment (presence of surface-active molecules and salts, for example) opens up new applications: a polyacrylamide gel contracts when exposed to an electric field (it is a polyelectrolyte and the polymer chains are electrically charged); it can thus play the role of "artificial muscle". These gels could thus be used for transforming chemical energy into mechanical energy, which could be of interest in biomedical engineering or for making sensors and mechanical devices for robots. Gels already have a classic application in pharmacology (encapsulation of drugs), but they could also have others. Indeed, since some gels are sensitive to the pH of the human body, they could release a drug in a medium with the appropriate pH: these gels contract in the basic medium of the intestine where they then release the drug. Gels that selectively absorb metal cations could also be used.

The sol ⇔ gel transition itself has many applications in materials synthesis. In geology, certain crystals can be formed from a gel phase; for example, this would be the case of quartz formed in certain conditions in silica gel (the question is still being debated). More generally, the gel phase is a medium favorable to crystal growth, since it eliminates convection on one hand (crystal growth only works by diffusion of molecules), and it decreases nucleation on the other hand (the number of microcrystals is limited). These properties of a gel thus allow obtaining growth of single crystals with a restricted number of defects. This is the principle of the **sol/gel methods** used in numerous areas of materials synthesis. These methods are also used in many other areas (Fig. 6.17): manufacture of films, aerogels, xerogels, and ceramics.

Xerogels are prepared by drying a gel: the solvent is removed by evaporation. Evaporation induces capillary pressure that causes contraction of the three-dimensional network of the gel, which tends to collapse into itself. The gel is reduced in volume by a factor of 5 to 10.

An aerogel is made from a wet gel (with solvent in the network) placed in an autoclave and dried with heat in conditions in which the solvent passes into the supercritical state. There is thus no liquid/vapor interface and no capillary forces: the gel does not collapse and retains its developed three-dimensional network. Aerogels are thus light ("airy") structures in which the

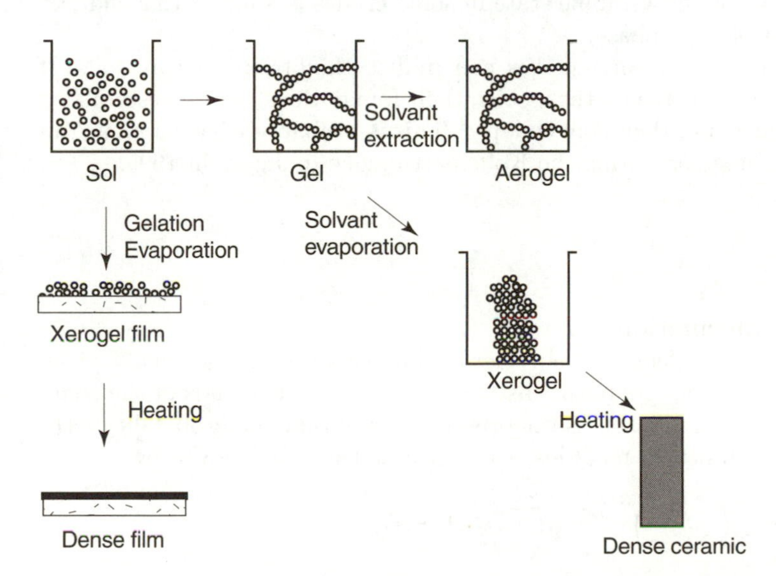

Fig. 6.17. Sol–gel methods. A gel is prepared which: is transformed into aerogel by evaporation or extraction of the solvent; is transformed into xerogel by heating. A denser material such as a ceramic can be fabricated from a xerogel.

solid volume fraction can attain 1 % of the total. This situation is similar to what is observed with dried, cracked mud.

In making an aerogel, methanol or supercritical CO_2 can be used as the solvent. Silica aerogels with a density of 200 kgm^{-3} with pores 1–10 nm in diameter can be obtained. These aerogels can be used as insulating materials, flotation materials, or for constructing solar panels. Even lighter aerogels ($\rho \approx$ 5 kgm^{-3}) can be used as acoustic isolation.

In the sol–gel method of materials synthesis, precursors are used to prepare a colloid: a metal or a metalloid surrounded by an organic (an alcohol) or inorganic ligand. For example, for an aluminum compound, $Al(C_4H_9)_3$ salts such as $Al(NO_3)_3$ are used. For silicon compounds, one can start from $Si(OC_2H_5)_4$.

Finally, we note the existence of thixotropy, which is a gel/sol or gel fluidization transition in the presence of stress (shear). The gel becomes a liquid and thus passes into the sol phase in the presence of stress. In general, a thixotropic material is a material whose viscosity is a function of the rate of variation of the stress: it decreases significantly if this rate increases.

A thixotropic system is formed of molecular aggregates, like a gel, and the thixotropy is a structural property of the material. Application of stress breaks the bonds between three-dimensional microstructures.

Thixotropy occurs in a very large number of physical gels, in clays, waxes, some liquid crystals, and some highly viscous crude oils. The presence of paraffins in the microcrystalline state in some crudes is a factor that induces the equivalent of a gel phase.

Particularly violent earthquakes can induce thixotropy in clay soils. If buildings are constructed on these soils, their foundations can be destabilized in an earthquake and they may collapse. This is probably what happened in certain coastal regions during the Kobe earthquake in Japan in 1995.

Problems

6.1. Coil Conformation

We will calculate the length of polymer chain composed of n segments which are monomers. We designate the distance between the two ends of the chain by $\boldsymbol{r}$, where one end is taken as the origin. Assume that the probability $w(\boldsymbol{r})$ that vector $\boldsymbol{r}$ will have a modulus between $\boldsymbol{r}$ and $\boldsymbol{r} + \mathrm{d}\boldsymbol{r}$ is given by:

$$w(\boldsymbol{r})\,\mathrm{d}\boldsymbol{r} = \left(\frac{3}{2nl^2\pi}\right)^{3/2} \exp\left(\frac{-3r^2}{nl^2}\right) 4\pi r^2 \mathrm{d}r$$

where l is the length of a segment.

1. Calculate the mean chain length and compare it with the length of a hypothetically linear chain. What conclusion can you draw from this?
2. Application: $n = 10^3$ and $l = 0.3\text{nm} - n = 10^4$, $l = 0.3\text{nm}$.

6.2. Conformation of a Protein

A biological polymer is composed of N segments which are monomers that can assume two conformations α or β with energy E_α and E_β and respectively corresponding to lengths a and b. Tension X is applied to both ends of the chain.

1. Calculate the chain partition function.
2. Determine its length as a function of X. An expansion limited to the first order in X is performed.

6.3. Sol/gel Transition and Percolation

Gelation is modeled with a percolation process (Sect. 6.3.1). Assuming that there is site percolation, the probability that a site will belong to some cluster, that is, that a site will be occupied, is designated by p. $n_s(p)$ is the probability of having clusters at s sites. Assume that in the vicinity of the critical threshold p_C corresponding to gelation, $n_s(p_C) \propto s^{-\tau}$ with $\tau = 2.5$ and one has $n_s(p)/n_s(p_C) \propto \exp(-cs)$, where $c = (p - p_C)^2$.

1. Calculate p and the characteristic quantity S_m of the size of a cluster.
2. Show that S_m diverges at p_C and give the value of the critical exponent.

6.4. Gelation and Property of Water

Liquid water with a network of hydrogen bonds between the molecules is compared to a system undergoing gelation. Show in a simple way that the anomalous density of water can be explained with this model.

7 Transitions and Collective Phenomena in Solids. New Properties

7.1 Transitions with Common Characteristics

Melting and sublimation of a solid are two important phase transitions, but they are not the only changes of state in solids. Indeed, the appearance of a magnetic phase in a nonmagnetic solid when its temperature is reduced below a certain point (called the Curie temperature) is another type of important phase transition (paramagnetic/ferromagnetic transition), investigated in particular by P. Curie and P. Weiss. The transition from a conducting to a superconducting state (superconductivity) and ferromagnetism reflect the appearance of new properties of the solid state, implying electron coupling.

In general, we note that two concepts play a major role in phase transitions: the order parameter and symmetry breaking (Sect. 1.2.3). The order parameter is a thermodynamic average which can have several values in certain conditions. If we describe magnetization of a system with the **Ising model**, magnetization can have two values $\pm \boldsymbol{M}$ in the absence of a magnetic field below critical temperature T_C (corresponding to the Curie temperature). The free energy or free enthalpy of the system has two minima of the same height separated by a maximum at its center for the two values of the order parameter $\pm \boldsymbol{M}$; this corresponds to the magnetization and it is the average value of the spin S_i^z associated with each lattice point (individual microscopic magnetization corresponds to each spin) and which assumes values of $\pm 1/2$ in this model (Fig. 7.1b).

When the temperature of a magnetic system in a zero field is decreased from temperature $T > T_C$, the system will pass into a state where the magnetization is still zero when it ends up at $T < T_C$. This does not correspond to the lowest free energy, and it is thus totally unstable: the least perturbation will induce reorganization of spins so that they will all be oriented "up" (right part of the free energy well) or "down" (left part). However the partial spin orientation (up/down) would correspond to the formation of magnetic domains that require supplementary interface energies which increase the free energy, such state is not possible. The final choice determining the orientation of magnetization is accidental (or the result of a small perturbation in the system), and the final result is no longer symmetric despite the fact that the free energy is symmetric. Magnetization should be invariant with respect

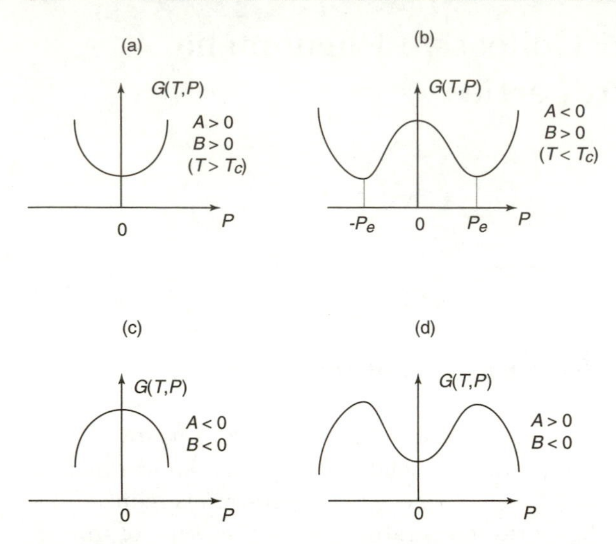

Fig. 7.1a–d. Free energy $G(T, P)$. G is expressed as a function of an order parameter P, magnetization M, for example (with no field). Diagrams (**a**) and (**b**) correspond to a second order phase transition. These curves can be obtained with a Landau expansion of type (7.53).

to rotation around an axis having its orientation: this is called **symmetry breaking**.

By analogy, this concept of an order parameter can be used for the liquid/gas transition, but the situation is slightly different because the liquid and the gas do not have the same symmetry. The number of particles is generally fixed and constitutes a restriction imposed on the system, which will be in a mixed state with a meniscus separating the liquid and vapor phases. It will pass from one homogeneous phase to another depending on the available volume.

In the case of a binary liquid/gas system, two variables are involved, the density (or volume) and the concentration. The coexisting states are determined with a tangent plane common to the two surfaces representing the free energy that establishes the connecting line between these two states, characterized by two coordinate points (v_1, x_1), (v_2, x_2). The length of the line connecting these two points, or concentration $x = v_{Gas}/(v_{Liq} + v_{Gas})$ can be selected as the order parameter in this case. Near the liquid/gas critical point, another possible choice is the ratio $(\rho_{Liq} - \rho_{Gas})/\rho_C$, where ρ_{Liq} is the density of the liquid and ρ_{Gas} is the density of the gas phase and ρ_C is the density at the liquid/gas critical point.

Phase transitions are classified in two broad categories (Sect. 1.2.3), first order transitions (associated with latent heat) on one hand, and second order transitions (without latent heat) on the other. If we consider specifically the second order transitions, they are characterized by particular properties of their thermodynamic potentials at the transition point: certain second derivatives of thermodynamic potentials with respect to state variables approach infinity or zero at the transition point. It has also been observed that thermodynamic quantities such as the specific heats, compressibility κ_T for a fluid, magnetic susceptibility χ_T for a ferromagnetic, and the dielectric constant for a ferroelectric will approach infinity around the critical point (Fig. 7.2). As for the order parameters associated with these transitions (magnetization, electrical polarization, $(\rho_{Liq} - \rho_{Gas})/\rho_C$ for the liquid/gas transition), they approach zero at the critical point. Moreover, the laws of the behavior of the equivalent thermodynamic quantities characterizing these systems (for example, κ_T and χ_T) are often very similar, which was far from the case of first order phase transitions. We say that critical phenomena (in the vicinity of a second order transition) have a **universal character**.

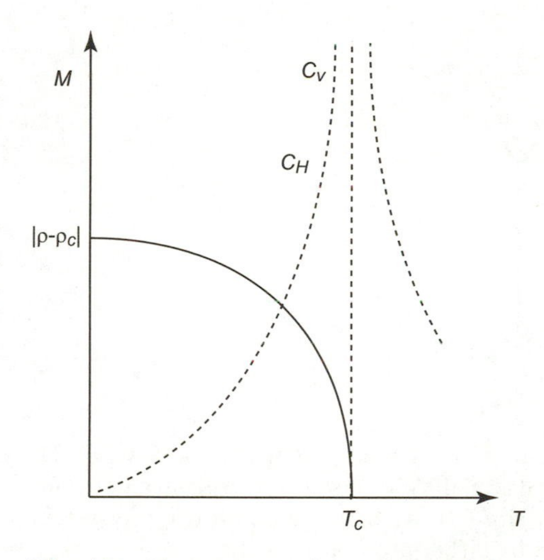

Fig. 7.2. Temperature dependence of the order parameters. The temperature dependence of the magnetization M and $|\rho - \rho_C|$ and the specific heats C_V and C_H as a function of T in the vicinity of T_C are shown.

7.2 The Order–Disorder Transition in Alloys

In 1934, Bragg and Williams described a transition observed in alloys of the Cu–Zn type (brass), manifested by the way in which these groups of atoms are ordered in a crystal lattice. Using X-ray spectra, it was shown that below a certain transition temperature, the atoms are ordered in two interpenetrating sublattices: all of the zinc atoms are in one lattice, and all of the copper atoms are in the other. The ordered lattice has cubic symmetry.

At the transition temperature, a specific heat peak characteristic of second order phase transitions is observed. The theoretical description they gave of this phenomenon was used as a standard for studying numerous transition phenomena: magnetism, superconductivity, liquid crystals, etc.

Assume that there is an equal number of different atoms A and B on one hand, with two types of sites α and β on the other. Each α site is uniquely surrounded by β sites and vice versa (Fig. 7.3). Identical sites (α or β) are the next nearest neighbors. If the interaction energy between two different neighboring atoms V_{AB} is lower than the average interaction energy between atoms of the same species $(V_{AA} + V_{BB})/2$, the system should be completely ordered at very low temperatures, where all A atoms occupy α sites and all B atoms occupy β sites, because this arrangement corresponds to the lowest energy for the system. At high temperature, the system becomes increasingly disordered.

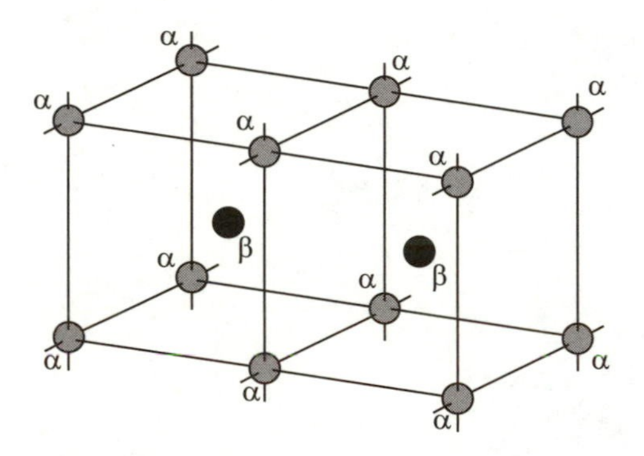

Fig. 7.3. Order–disorder transition in an alloy. A binary alloy AB (50% A–50% B) forms an ordered system at low temperature: all A atoms occupy a given site (α, for example) and B atoms occupy another site (β). At high temperatures, the system is disordered, and each site is occupied by either atom arbitrarily.

We will designate the probability that an α site will be occupied by an A atom by p_α; $w_\alpha = 1 - p_\alpha$ is the probability that they will be occupied by a B atom; p_β, w_β are the probabilities associated with β sites. We then introduce an order parameter P representing the long-range order:

$$P = \frac{p_\alpha - \frac{1}{2}}{\frac{1}{2}} \tag{7.1}$$

(P is defined identically on β sites). In the state of total disorder, $p_\beta = p_\alpha = w_\alpha = w_\beta = 1/2$ and $P = 1$.

If N is the total number of sites and assuming that there are N/2 α sites and N/2 β sites, the respective number of A atoms on a α site, N_A^α and B atoms on a β site, N_B^β can be written using (7.1):

$$N_A^\alpha = (1+P)\frac{N}{4}, \quad N_B^\beta = (1+P)\frac{N}{4} \tag{7.2}$$

The numbers of A atoms on a β site, N_A^β, and B atoms on an α site, N_B^α, are given by:

$$N_A^\beta = (1-P)\frac{N}{4}, \quad N_B^\alpha = (1+P)\frac{N}{4} \tag{7.3}$$

If z is the coordination number, that is, the number of β sites surrounding an α site and if we assume that the probability of having pairs AA, BB, and AB on neighboring sites is the product of the probabilities of occupation of these sites, then we will have:

$$N_{AA} = zp_\alpha(1-p_\beta)N/2 = zN(1-P^2)/8 = N_{BB} \tag{7.4}$$

where N_{AA} and N_{BB} are the numbers of pairs AA and BB. Setting $z = 8$ for a f.c.c. lattice, we will then have:

$$N_{AA} = N_{BB} = N(1-P^2) \tag{7.5}$$

Moreover, the number of pairs AB and BA is given by:

$$N_{AB} = [p_\alpha p_\beta + (1-p_\alpha)(1-p_\beta)]z(N/2) \tag{7.6}$$

that is:

$$N_{AB} = 2N(1+P^2) \tag{7.7}$$

The internal energy U of the system is determined by the number of nearest neighbors which we calculated.

We assume that only the interactions between nearest neighbors count and that they are independent of the state of the environment of the atoms. Then:

$$U = N_{AA}V_{AA} + N_{BB}V_{BB} + N_{AB}V_{AB} \tag{7.8}$$

U is written:

$$U = U_0 + NP^2V \tag{7.9}$$

where

$$U_0 = N(V_{AA} + V_{BB} + 2V_{AB}) \tag{7.10}$$

$$V = 2V_{AB} - V_{AA} - V_{BB} \tag{7.11}$$

This method is related to the Weiss theory of ferromagnetism, which hypothesizes the existence of a molecular field; it excludes any kind of correlation with the global environment of each atom. We must then calculate the free energy F of the system by first calculating the entropy. The number of configurations is counted, that is, the number of ways A and B atoms can be distributed in α and β sites. The number of configurations W is given by:

$$W = \left[\frac{(N/2)!}{N_A^\alpha! N_B^\alpha!}\right]\left[\frac{(N/2)!}{N_A^\beta! N_B^\beta!}\right] \tag{7.12}$$

that is

$$W = \frac{(N/2)!}{[\frac{(1+P)}{4}N]![\frac{(1-P)}{4}N]!}\frac{(N/2)!}{[\frac{(1-P)}{4}N]![\frac{(1+P)}{4}N]!} \tag{7.13}$$

The entropy of configuration S of the solid is correlated with W by the Boltzmann equation: $S = k \log W$. Using the Stirling equation for $N!$ in (7.13), S is written:

$$S/k = (N/2)\{2\log 2 - N[(1+P)\log(1+P) + (1-P)\log(1-P)]\} \tag{7.14}$$

The free energy $F = U - TS$ is then written:

$$F = U_0 + VNP^2 - (kTN/2)\{2\log 2$$
$$-[(1+P)\log(1+P) + (1-P)\log(1-P)]\} - TNS_0 \tag{7.15}$$

where S_0 is the entropy of the A and B atoms in their lattice.

At thermodynamic equilibrium, F should be minimum. This minimum corresponds to the values of P that are solutions of the equation:

$$\frac{\partial F}{\partial P} = 2NPV + \frac{NkT}{2}\log\frac{1+P}{1-P} = 0 \tag{7.16}$$

Other than the obvious solution $P = 0$, (7.16) has two other real solutions that can be obtained graphically using a limited expansion:

$$2NPV + NkT\,[P + (1/3)P^3 + \ldots] = 0 \tag{7.17}$$

Defining the critical temperature T_C by $kT_C = -2V$, we see that for $T > T_C$, $P = 0$ is the only possible solution. For $T < T_C$, there are the following two possible solutions in the vicinity of T_C.

$$P = \pm\sqrt{3\frac{T_c - T}{T}} \approx \pm\sqrt{3\frac{T_c - T}{T_c}} \tag{7.18}$$

The order parameter $P(T)$ has an infinite slope at T_C.

This calculation holds only if $P \ll 1$ so that a series expansion of (7.16) can be performed (that is, for $(T_C - T)/T_C \ll 1$). This treatment corresponds to the Bragg–Williams method.

The order parameter P is a simple function of $(T_C - T)$ in the vicinity of temperature T_C and vanishes at T_C. It is represented by a parabola with a tangent vertical to T_C which has the shape in Fig. 7.2. By calculating the change in entropy at the transition (for $T = T_C$), it is possible to verify that there is no latent heat associated with the transition: it is second order.

It is also possible to directly expand the function F from (7.15); it is written as:

$$\frac{F}{kN} = \frac{F_0}{kN} + (T - T_c)\frac{P^2}{2} + T\frac{P^4}{12} \tag{7.19}$$

This again assumes $P \ll 1$ and thus $|T_C - T|/T_C \ll 1$. F_0 is the contribution of the degrees of freedom that are not related to the configuration and interactions of A and B atoms to the free energy. An expansion of a thermodynamic potential such as F as a function of order parameter P is called a **Landau expansion**.

Second order phase transitions have been experimentally demonstrated in alloys, particularly by specific heat measurements. This is the case for the Cu–Zn alloy, where the specific heat tends to diverge at T_C ($T_C = 742$ K), and the law of variation of P deviates from (7.18).

In some alloys of the A_3B type (for example, $AuCu_3$, where $T_C = 640$ K), the transition is first order.

7.3 Magnetism

7.3.1 Characterization of Magnetic States

In general, three types of magnetism are distinguished: **diamagnetism**, **paramagnetism**, and **ferromagnetism**. There are also several ferromagnetic states: ferromagnetism per se, antiferromagnetism, ferrimagnetism, and helimagnetism. Only ferromagnetism is really associated with phase transitions.

Diamagnetism appears when there is a current loop in a material formed by electrons rotating around the nucleus. It directly results from Lenz's law that the induced moment opposes the magnetic field. This is manifested by the fact that the susceptibility χ_T defined by $\boldsymbol{M} = \chi_T \boldsymbol{H}_0$ (where $\boldsymbol{M}$ is the magnetization and $\boldsymbol{H}_0$ is the applied magnetic field) is negative. The effect is small and independent of the temperature.

Paramagnetism originates in the electron spins of the system. The spins, associated with individual magnetic moments μ, tend to align in a certain direction but this tendency is opposed by thermal motion. This form of magnetism is thus a function of the temperature. In strong magnetic fields, the susceptibility at fixed temperature is no longer constant because the magnetization tends to become saturated when all of the spins and thus their individual magnetic moments are aligned.

In ferromagnetic systems, there can be a macroscopic magnetic moment (that is, magnetization) even without a magnetic field. They exhibit spontaneous magnetization, at least in certain temperature conditions; this arises from coupling of spins localized at different sites in the solid lattice. Spin coupling and the structure of the crystal determine the type of magnetic state observed.

There can thus be a state in which all spins are aligned in the same direction, and this corresponds to **ferromagnetism**. They can also form two groups of spins, each distributed in a sublattice but aligned in antiparallel directions, where the resulting macroscopic magnetization is zero: this

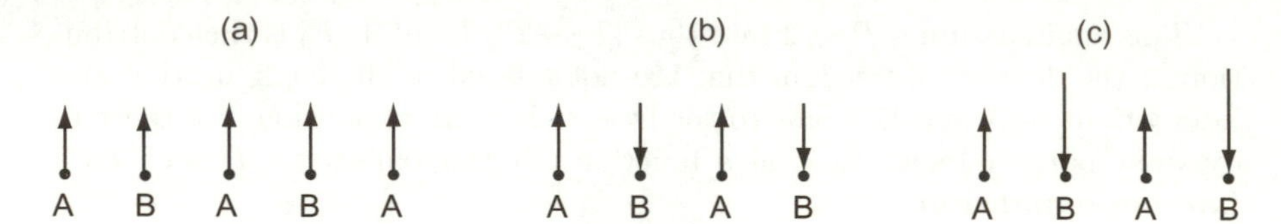

Fig. 7.4. Magnetic systems. The three types of magnetic systems are indicated: (**a**) ferromagnetic solid: all individual magnets have the same orientation; (**b**) antiferromagnetic solid: the individual magnets are antiparallel on sites A and B; (**c**) ferrimagnetic solid: the individual moments are antiparallel but unequal, and the resultant moment is zero.

is the **antiferromagnetism**. There is also the situation where the spins in each of the sublattices do not have the same value; if they are all oriented in antiparallel directions, the macroscopic magnetization is not zero: this is **ferrimagnetism**. The spins can also be oriented in the same direction but arranged in helices: this is **helimagnetism** (Fig. 7.4).

Metals such as Fe, Ni, and Gd are typically ferromagnetic systems, and a compound such as $CrBr_3$ is also. The ferromagnetic–paramagnetic transition temperature is the Curie temperature. In general, the antiferromagnetic/paramagnetic transition temperature is the Néel point. These temperatures are generally lower than the Curie temperature (610 K for MnO, 24 K for $FeCl_3$). Ferrites, whose general formula is $MO–Fe_2O_3$, where M is a divalent metal, are typical ferromagnetic compounds; Fe_3O_4 is the best known example. It is necessary to note that only ions that have a permanent magnetic moment can produce magnetic compounds. The ions concerned are those with an incomplete electron shell (The 3 d-shell for the series Fe, Co, Ni ..., the 4 f-shell for rare earths such as Gd, Dy ...).

7.3.2 The Molecular Field Model

The simplest model for describing a ferromagnetic system and the magnetic phase transition is the **molecular field** model. It has a strong similarity to the method used in the physics of liquids for obtaining an equation of state such as the van der Waals equation: it consists of replacing all interactions undergone by a particle in a solid (atoms, molecules, electrons) by a single field called the molecular field.

There are several quantum models that can be used for treating magnetism. For a ferromagnetic system, the **Heisenberg** model is the most general. It is assumed that a spin $\boldsymbol{S}_i$ is located on each lattice site (i). An interaction energy is associated with the quantum observables $\boldsymbol{S}_i$, represented by the following **Hamiltonian**:

$$\mathcal{H} = - \sum_{<i,j>} J_{ij} S_i S_j \tag{7.20}$$

J_{ij} is the coupling constant between neighboring spins or exchange energy. In fact, this model is simplified by assuming that each atom has a spin S_i associated with a magnetic moment $\mu_0 = geh/4\pi mc$ (g is the Landé factor) that can have two values ± 1, that is, in only two directions, up or down. This spin variable is designated by σ_i for each of the N lattice sites ($i = 1, 2 \ldots N$). ($\sigma_i = 1$ corresponds to spin orientation up, while $\sigma_i = -1$ corresponds to orientation down). We will assume $J_{ij} = J$. The situation where all spins point in the same direction corresponds to $J > 0$, and the situation where the spins are antiparallel corresponds to $J < 0$. The interaction energy of the system is written:

$$\mathcal{H} = -\sum_{<i,j>} J\sigma_i\sigma_j \tag{7.21}$$

The notation $< i, j >$ means that the summation is only over neighboring sites. This model is called the **Ising model**. For $J > 0$, the system will be ferromagnetic, and for $J < 0$, it is antiferromagnetic.

To completely calculate the thermodynamic potentials of the system and determine the magnetization, it is necessary to introduce the expression giving $\mathcal{H}$ in the total partition function by taking the sum over all configurations. The complete calculation without approximations has only been performed in the case of a two-dimensional system by Onsager. An approximation previously proposed by P. Weiss, called the self-consistent field method, can be used: in the expression of $\mathcal{H}$, spins σ_j can be replaced by their average value $< \sigma >$, then it is calculated with the partition function. In the presence of an external magnetic field $\boldsymbol{H}$, we write:

$$\mathcal{H} = -\sum_{<i,j>} J\sigma_i < \sigma_j > -\mu_0 \mathbf{H} \sum_i \sigma_i$$

$$\mathcal{H} = -zJ < \sigma > \sum_i \sigma_i - \mu_0 \boldsymbol{H} \sum_i \sigma_i \tag{7.22}$$

$$\mathcal{H} = -[zJ < \sigma > +\mu_0 \boldsymbol{H}] \sum_i \sigma_i \tag{7.23}$$

where z is the number of nearest neighbors. This is like a system of independent spins exposed to an effective spins exposed to an effective field $\boldsymbol{H}_{ef} = zJ < \sigma > /\mu_0 + \boldsymbol{H} = \boldsymbol{H}_W + \boldsymbol{H}$, where $\boldsymbol{H}_W$ is often called the **Weiss field** (P. Weiss advanced the first theory of magnetism). We thus come to the classic calculation of the magnetization of a system of independent spins in the presence of the effective field $\boldsymbol{H}_{ef}$. Since each spin can have two values ± 1, partition function Z for the individual spins is written:

$$Z = \mathrm{e}^{\mu_0 H_{ef}/kT} + \mathrm{e}^{-\mu_0 H_{ef}/kT} \tag{7.24}$$

We can easily calculate $< \sigma_i > = < \sigma >$ with this distribution function or, which is equivalent, $M = N\mu_0 < \sigma >$, which is the total magnetization of the system. This is the crucial stage of the method.

Knowing that $F = -kT \ln Z$, $M = -(\frac{\partial F}{\partial H})_T$, we find:

$$M = N\mu_0 \tanh\left\{\frac{\mu_0}{kT}\left(H + zJ < \sigma > /\mu_0\right)\right\} \tag{7.25}$$

We can set $M_\infty = N\mu_0$, and this quantity is the saturation magnetization. (7.25) is thus rewritten:

$$M = M_\infty \tanh\left\{\frac{\mu_0 H}{kT} + \frac{zJM}{kTM_\infty}\right\} \tag{7.26}$$

We say that (7.26) is a self-consistent equation because it expresses M or $< \sigma >$ with an equation which is a function of M.

The general solution of (7.26) can be obtained graphically. If the magnetic field is zero ($\boldsymbol{H} = 0$), we obtain M by looking for the intersections of the two curves:

$$X = \frac{\mathbf{M}}{\mathbf{M}_\infty} = \tanh x \quad X = \frac{kT}{zJ}x = \frac{T}{T_c}x$$

where $T_C = zJ/k$. The solutions obtained are shown in Fig. 7.5.

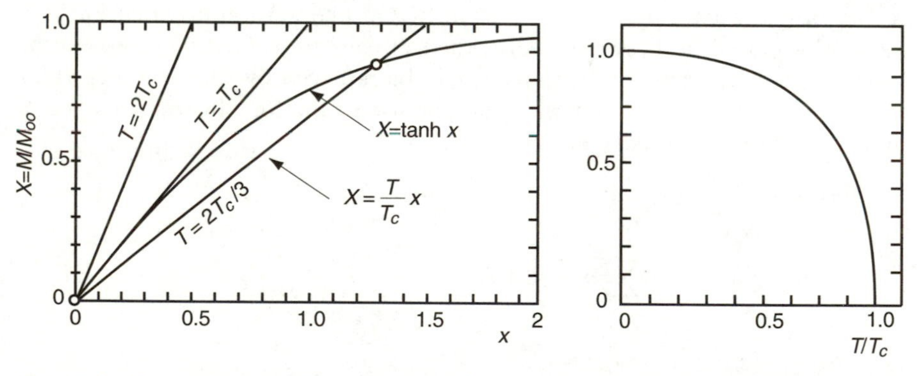

Fig. 7.5. Magnetization. The graphic solution of (7.26) for $H = 0$ is shown. For $T < T_C$, solution M is the intersection of the line $X = xT/T_C$ with the curve $\tanh x$.

For $T < T_C$, there is a solution $x \neq 0$ (in addition to $x = 0$, which is always a solution) which is a function of the temperature. If T approaches zero, x will approach 1, which corresponds to saturation magnetization. This solution vanishes at $T = T_C$. If the external field is no longer zero, magnetization no longer appears at $T = T_C$ and there is no longer a critical temperature because there is constant magnetization. At temperature T_C (Curie temperature), there is a **ferromagnetic–paramagnetic phase transition**.

Equation (7.25) can also be rewritten by solving it again for calculating the exponentials of function tanh and by taking the logarithms. We then obtain the new equation:

$$\frac{1}{2}kT \log \frac{1+ <\sigma>}{1- <\sigma>} = zJ <\sigma> + \mu_0 H \tag{7.27}$$

With no external field, an equation similar to (7.16) obtained in treating the order/disorder transition is obtained. $zJ = -2V$ is the coupling constant and $<\sigma>$ plays the role of order parameter. We will set $zJ/k = T_C$.

A solution can also be found for $<\sigma>$ and thus for M by limited expansion of (7.27) in the vicinity of T_C. We will find an expression for $<\sigma>$ similar to the expression for P given by (7.19). The magnetization $M(T)$, which is the order parameter, is reduced to zero with an infinite slope at T_C.

If we return to (7.26), which gives the magnetization and is derived with respect to H in the vicinity of T_C (thus for $M \approx 0$) and taking the limit for $H = 0$, we can easily find the form of the magnetic susceptibility $\chi_T = (\partial M/\partial H)_T$:

$$\chi_T \propto \frac{1}{T - T_c} \tag{7.28}$$

This expression is none other than the Curie–Weiss law for a ferromagnetic. It only holds in the vicinity of the Curie point.

7.3.3 Bethe Method

The molecular field model does not take into account the correlations between neighboring sites. Bethe introduced a method that takes into account the short-range order, that is, the number of neighboring spins for a spin localized on a site and having the same orientation. He found that this short-range order exists above the critical point, contrary to the long-range order (macroscopic magnetization) which is only nonzero below the critical point. The Bethe method (also called the Bethe–Peierls method) is equivalent to the quasi-chemical Guggenheim method and can be considered the first approximation of the so-called Kikuchi cluster variation method.

This approximation in a way is reduced to considering a subsystem of the spin system "immersed" in all of the spins and is constituted of a given spin and its nearest neighbors γ ($\gamma = 2$ for a linear system, $\gamma = 4$ for a plane square lattice, etc.).

The fraction of up spins is designated by x_1 and the fraction oriented down is designated by x_2. We thus have $x_1 + x_2 = 1$. We will introduce a second set of probabilities for a pair of spins: y_1 is the probability that the two neighboring spins will be both up, y_2 is the probability that one of the neighboring spins will be up and the other down, and y_3 is the probability that both spins will be down. Since the overall probability is normalized to one, we have: $y_1 + 2y_2 + y_3 = 1$. The factor 2 accounts for the fact that there are two possible situations for spins of opposite orientation. The probability that one of the spins will be up, regardless of the orientation of the other, is $x_1 = y_1 + y_2$. We have $x_2 = y_2 + y_3$ for the probability that the spin is down. The sum of these two probabilities is normalized and their difference corresponds to the average magnetization or the long-range order: $s = x_1 - x_2$. We will assign an energy to each of the three states of the pair: $\varepsilon_1 = -J$ corresponds to the pair were both the spins are up, $\varepsilon_2 = J$ is the energy for the pair in which the spins are antiparallel, and $\varepsilon_3 = -J$ is the energy for the pair with both spins down. J is the coupling constant. For $J > 0$, parallel orientations are favored.

To obtain a simple expression for the entropy, we note that the total number of W of ways of placing the spin pairs in the lattice is given by the number of combinations:

$$W = \frac{N_p!}{[N_p y_1]![N_p y_2]!^2[N_p y_3]!} \tag{7.29}$$

where $N_p = N\gamma/2$ is the number of pairs, N is the number of lattice sites, and γ is the number of nearest neighbors.

Using the relation $S = k \ln W$ and Stirling's equation, we find:

$$S_{pair}/k = -\frac{N\gamma}{2}(y_1 \ln y_1 + 2y_2 \ln y_2 + y_3 \ln y_3) \tag{7.30}$$

The entropy of the individual sites must also be taken into consideration:

$$S_{site}/k = -N(x_1 \ln x_1 + x_2 \ln x_2) \tag{7.31}$$

However, a fraction of this entropy has already been included into the entropy of the pairs and it is thus necessary to correct this over-count of γN sites. The internal energy U is given by the expression:

$$U = \frac{\gamma N}{2}(\epsilon_1 y_1 + 2\epsilon_2 y_2 + \epsilon_3 y_3) \tag{7.32}$$

Hence, the complete expression for the free energy per site related to $\beta = 1/kT$:

$$\begin{aligned} \varPhi \equiv \frac{\beta F}{N} &= \frac{\gamma}{2}\beta(\epsilon_1 y_1 + 2\epsilon_2 y_2 + \epsilon_3 y_3) + \frac{\gamma}{2}(y_1 \ln y_1 + 2y_2 \ln y_2 + y_3 \ln y_3) \\ &+(1-\gamma)(x_1 \ln x_1 + x_2 \ln x_2) - \beta\lambda(y_1 + 2y_2 + y_3 - 1) \end{aligned} \tag{7.33}$$

The Lagrange multiplier λ was introduced to maintain the normalization condition. Minimizing (7.33) with respect to the three variables and taking into consideration the conditions $x_1 = y_1 + y_2$ and $x_2 = y_2 + y_3$, we find equations of the following type for y_1:

$$y_1 = (x_1 x_1)^{\overline{\gamma}} \mathrm{e}^{-\beta\epsilon_1} C_1 \tag{7.34}$$

where C_1 is a constant. In general y_1 can be put in the form $y_1 = E_i/Z$ $(i = 1, 2, 3)$, where:

$$
\begin{aligned}
E_1 &\equiv (x_1 x_1)^{\overline{\gamma}} \exp(-\beta\epsilon_1) \\
E_2 &\equiv (x_1 x_2)^{\overline{\gamma}} \exp(-\beta\epsilon_2) \\
E_3 &\equiv (x_2 x_2)^{\overline{\gamma}} \exp(-\beta\epsilon_3)
\end{aligned} \tag{7.35}
$$

where constants C_1, C_2, and C_3 are obtained by writing the normalization conditions $y_1 + 2y_2 + y_3 = 1$ and with

$$
Z = E_1 + 2E_2 + E_3 \quad \text{and} \quad \overline{\gamma} = (\gamma - 1)/\gamma \tag{7.36}
$$

Equation (7.35) can be used to express point variables x_1 and x_2 in terms of themselves. We still have a self-consistency condition:

$$
x_1 = \frac{E_1 + E_2}{Z} \quad \text{and} \quad x_2 = \frac{E_2 + E_3}{Z} \tag{7.37}
$$

For convenience, the following ratio can be introduced:

$$
\frac{x_1}{x_2} = \frac{E_1 + E_2}{E_2 + E_3} \tag{7.38}
$$

using (7.35) and setting:

$$
z \equiv \left(\frac{x_1}{x_2}\right)^{\overline{\gamma}} \quad \text{and} \quad t \equiv \exp(-2\beta J) \tag{7.39}
$$

Equation (7.38) can be written:

$$
z^{1/\overline{\gamma}} = \frac{z^2 + zt}{zt + 1} \tag{7.40}
$$

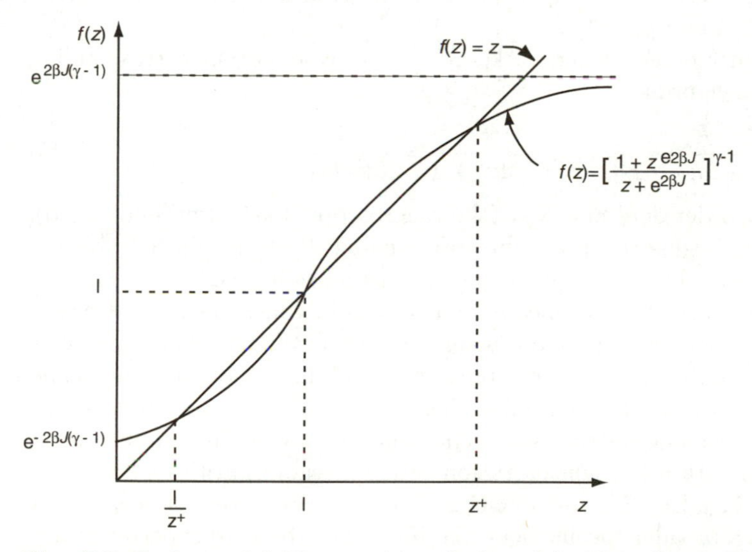

Fig. 7.6. Bethe model. The graphic solution of (7.41) is shown. The intersections of $f(z)$ and line z give the solutions z_+ and $z_- = 1/z_+$ corresponding to the orientations of opposite spins.

Dividing the two terms by z, we obtain:

$$z^{1/(\gamma-1)} = (\frac{z+t}{zt+1}) \quad \text{or} \quad z = (\frac{z+t}{zt+1})^{\gamma-1} \equiv f(z) \tag{7.41}$$

This equation must be solved with a graphic method such as the Weiss method (Fig. 7.6). The curve representing function $f(z)$ is S-shaped, and its intersection with line z gives the solutions of the equation. At low temperature, there are three intersections corresponding to the roots z_-, $z_0 = 1$ and z_+. The symmetry of function $f(z)$ imposes $z_- = 1/z_+$, which corresponds to the same solution, but where the orientation of all spins is reversed. These two solutions only exist if the slope of the tangent to function $f(z)$ for $z = 1$ is greater than 1. Slope c of the tangent to $f(z)$ is equal to:

$$c = (\gamma - 1)\left(\frac{t+z}{zt+1}\right)^{\gamma-2} \frac{1-t^2}{(zt+1)^2} \tag{7.42}$$

For $z = 1$:

$$c = (\gamma - 1)\frac{1-t^2}{(1+t)^2} \tag{7.43}$$

Critical temperature T_C corresponds to $z = 1$, that is:

$$\frac{2J}{kT_C} = \ln(\frac{\gamma}{\gamma - 2}) \tag{7.44}$$

This result shows that a one-dimensional system ($\gamma = 2$) has no critical point. For $T > T_C$, the only solution corresponds to $z = 1$. There is no long-range order. At a very high temperature, the distribution of the pairs is given by $y_1 = y_3 = 1/4$ and $y_2 = 1/4$.

In the vicinity of T_C, we can take $z = 1$ and calculate y_1 corresponding to the short-range order:

$$y_1 = \frac{E_1}{E_1 + 2E_2 + E_3} = \frac{z^2}{z^2 + 2zt + 1} = \frac{1}{2(t+1)} \tag{7.45}$$

Long-range order does not exist (the macroscopic magnetization is zero), but local order begins to appear in approaching T_C from above, which is proportional to y_1. For $z \neq 1$, graphic solutions must be used.

Finally, there are also intermediate situations in which the spins are coupled by an exchange interaction but where there is nevertheless no long-range order: these are **spin glasses**. The spins are frozen in random directions and we have the equivalent of a molecular glass (Chap. 5). Such a system can be described with a model of the Ising type. Such behaviors are characteristic of metal alloys with a low concentration of magnetic impurities such as the alloys: Au_{1-x}, Fe_x, Eu_x Sr_{1-x}, and Cu_{1-x} Mn_x. A very distinct susceptibility maximum and a broader specific heat maximum are observed experimentally. A magnetic atom can be coupled with its neighbors simultaneously by magnetic and antiferromagnetic interaction; in this case, there is **frustration**.

7.3.4 Experimental Results

The theories of second order phase transitions allow expressing the thermodynamic quantities X in the vicinity of a transition point (called **critical point**) with laws of the general form:

$$X \propto \epsilon^{x} \tag{7.46}$$

where $\varepsilon = |T - T_C|/T_C$ is postulated and x is called the **critical exponent** of the quantity considered.

When an order parameter can be defined, as in the case of an order/disorder transition and magnetism, the associated critical exponent is designated by β: $M \propto \varepsilon^{\beta}$ (with no external magnetic field).

Moreover, we associate:

- exponents γ (for $T > T_C$) and γ' $(T < T_C)$ for the magnetic susceptibility χ_T (and compressibility κ_T for a fluid); $\chi_T \propto \varepsilon^{-\gamma}$;
- exponents $\alpha(T > T_C)$ and $\alpha'(T < T_C)$ are associated with the specific heat C_H and C_V, $C_V \propto \varepsilon^{-\alpha}$.

In the presence of an external magnetic field and on the critical isotherm $(\varepsilon = 0)$, we also have a critical exponent defined by $H \propto M^{\delta}$.

The molecular field models (of the Weiss or Bethe type) predict $\beta = 1/2, \gamma = 1, \delta = 3$ for magnetic transitions and a discontinuity for the specific heat (corresponding to $\alpha = 0$).

The magnetic transition has been the subject of a very large number of experimental studies for more than a century. J. Hopkinson was the first to observe the disappearance of magnetization of iron at a high temperature in 1890, and the first systematic studies of the transition in the vicinity of the transition temperature were conducted by Pierre and Jacques Curie at the end of the 19th century. The magnetization and magnetic susceptibility of a material can be measured directly or indirectly. The magnetization of a magnetic system can be measured by NMR (nuclear magnetic resonance) and the magnetic susceptibility can be measured by neutron scattering; they have a magnetic moment that can be coupled with the magnetic moment of the magnetic material; they are sensitive to fluctuations in magnetization in the vicinity of T_C which are directly linked with κ_T. In the same way, a light scattering experiment can be conducted in transparent materials (MnF_2, which is antiferromagnetic, for example).

In the 1960s, it was found that the predictions of the molecular field models were no longer verified in magnetic materials in the immediate vicinity of T_C (a similar conclusion was obtained for fluids in the vicinity of the liquid/gas critical point). The experimental results converge toward the following values of the critical exponents: $\beta = 0.36, \gamma = 1.33, \alpha = 0.003, \delta = 4.5$. About the same values are found for the critical exponents in fluids in the vicinity of the liquid/gas critical point.

The differences with the predictions of the classic models are thus relatively important. The validity of molecular field models is consequently very limited.

7.4 Ferroelectricity

7.4.1 Characteristics

Spontaneous electrical polarization P appears in some crystals below a specific temperature of the material. This is called **ferroelectricity** by analogy with the appearance of constant magnetization in a ferromagnetic. A ferroelectric solid by definition has a constant macroscopic electric dipole moment in the absence of an electric field. At high temperatures, the individual electric dipoles of the ions are randomly oriented in the solid (paraelectric phase), then when the temperature is decreased, they undergo a transition to a temperature which is also called the Curie temperature: in the absence of an electric field, the individual dipoles are oriented in a common direction and the solid becomes ferroelectric; it has undergone a paraelectric/ferroelectric transition. For a material to be ferroelectric, it is necessary for the centers of gravity of the positive and negative electric charges to be distinct and thus for the crystal to have no center of symmetry.

The ferroelectric/paraelectric phase transition can be either first- or second-order. Application of an electric field can modify the order of the transition: it is always first order in the presence of an electric field.

It is also necessary to note the existence of materials called electrets, which are organic substances, generally waxes, that have a constant electric dipole moment formed when they solidify in the presence of an electric field. The wax molecules in effect have an electric dipole moment that can be oriented by an electric field. This polarization can persist for years after solidification, but the system is metastable. Electrets can also be obtained by ion implantation.

In some crystals, the distribution of the electric charges and the magnitude of the individual electric dipoles of the lattice cells are very sensitive to a change in the temperature, heating, for example, and permanent electrical polarization P can appear during variation of the temperature: this is **pyroelectricity**. Any ferroelectric is a pyroelectric, but the opposite is not true (tourmaline, for example, is a nonferroelectric pyroelectric).

Dielectric polarization can also be obtained by compressing a crystal: on application of external stress C, electrical polarization P appears. Moreover, application of an electric field E causes elastic deformation e. This is called **piezoelectricity**. It was discovered in 1880 by Pierre and Jacques Curie in a quartz crystal. Voigt gave the phenomenological theory of piezoelectricity. There is a linear correlation between the quantities;

$$P = \varepsilon E + dC \quad e = dE + LC \tag{7.47}$$

where ε is the dielectric constant, d is the piezoelectric coefficient, and L is the elastic coefficient.

We emphasize that although all ferroelectrics are piezoelectrics, a solid can be piezoelectric without necessarily being ferroelectric (this is the case of quartz, for example). Application of external stress induces electrical polarization in the crystal by shifting the centers of gravity of the electric charges.

A situation similar to the one encountered in magnetic materials is also found, where each sublattice has electrical polarization but has the opposite sign: the resulting macroscopic polarization is zero. The solid is **antiferroelectric**.

In conclusion, we emphasize that there is an essential difference between ferroelectricity and ferromagnetism: ferromagnetism arises from the interaction between distinct individual magnetic dipoles, while the polarization of a ferroelectric is an effect inherent in the entire structure of the crystal.

7.4.2 The Broad Categories of Ferroelectrics

There are two broad categories of ferroelectric materials:

- those for which the ferroelectric/paraelectric phase change is associated with an **order/disorder transition**;
- those for which the transition is induced by a **displacive transition** that alters the crystal structure.

Potassium dihydrogen phosphate, KH_2PO_4 (KDP), is the classic example of a ferroelectric associated with an order–disorder transition at a temperature of 123 K (the transition is called weakly first order because the latent heat is very low). The order/disorder transition is triggered by movements of hydrogen atoms which, inserted in the system of PO_4 tetrahedrons, can occupy two equilibrium positions in the vicinity of each tetrahedron (Fig. 7.7). At high temperatures, these two positions are occupied with equal probability: we have a disordered state without polarization. Below the critical temperature T_C, the hydrogens become "blocked" at specific sites: order appears with nonzero electrical polarization.

Displacive transitions are characteristic of perovskites of the general formula ABO_3 (Sect. 1.4.1). Barium titanate, $BaTiO_3$, has a perovskite structure. Above 120°C, the Ti^{4+} ion is simultaneously located at the center of a cube whose vertices are occupied by Ba^{2+} ions and an octahedron formed by O^{2-} ions (Fig. 7.8). In this structure, the crystal is paraelectric because the centers of gravity of the negative and positive charges coincide. However, if the temperature is lowered below a temperature T_C, the Ti^{4+} ions will move in the same direction, the direction of axis z, causing the appearance of spontaneous electrical polarization P, and the O^{2-} ions move in the opposite direction. The crystal has undergone a cubic/quadratic transition at $T_C = 120°C$ (the transition is first order) associated with a paraelectric/ferroelectric transition. The $BaTiO_3$ crystal undergoes other structural

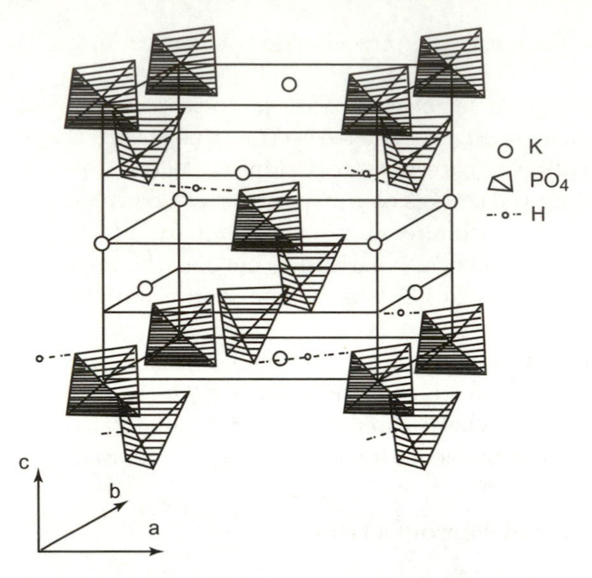

Fig. 7.7. Order/disorder ferroelectric. The lattice of KH_2PO_4 is shown. The para-electric/ferroelectric transition is triggered by movements of hydrogen atoms which can occupy two equilibrium positions between two neighboring tetrahedrons.

transitions at a lower temperature but retains its property of being ferroelectric.

There are more complex displacive transitions which can be associated with rotations of oxygens and associated octahedrons in perovskites. This is the case in $SrTiO_3$, but the transition is not associated with the appearance of ferroelectricity ($T_C = 99.5$ K).

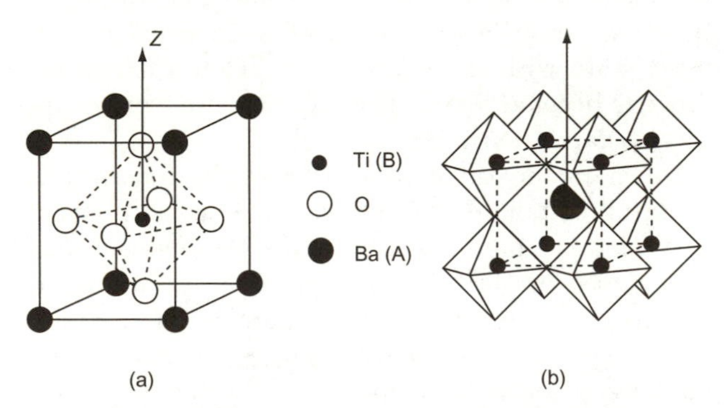

Fig. 7.8. Ferroelectric of the perovskite type. These solids have the general formula ABO_3. Shown: (**a**) the unit cell of $BaTiO_3$; (**b**) the three-dimensional lattice of BaO_6 octahedrons. The transition is triggered by movement of Ti^{4+} ions in the z-direction, while the O^{2-} ions move in the opposite direction.

Transition temperatures T_C vary within a relatively wide range: 760 K for $KNbO_3$, 271 K for $CaB_3O_4(OH)_3 \cdot 3H_2O$ (the transition is second order), 123 K for KH_2PO_4, 794 K for $NaNbO_3$, and 148 K for $NH_4H_2PO_4$, which are antiferroelectrics.

Recent (1999) electric potential measurements on thin films of crystalline ice deposited between 40 and 150 K on platinum surfaces revealed the existence of ferroelectricity in ice.

7.4.3 Theoretical Models – the Landau Model

If we return to the situation of a crystal like $BaTiO_3$ in the vicinity of its transition temperature ($T > T_C$), we find that the Ti^{4+} ions move a short distance. Vibration modes are associated with these motions, and the system oscillates with a frequency ω; these vibrations are the normal lattice vibration modes (Sect. 2.4.1). When the temperature decreases and approaches T_C, the frequency of a certain mode progressively decreases and reaches zero at T_C. This is called a **soft mode**. The existence of a soft mode is characteristic of structural transitions. The square of the frequency ω^2 of this mode (which is a mode corresponding to optical phonons) is proportional to the restoring force of the Ti^{4+} ions, it becomes zero when there is a phase transition, and the lattice deformation due to ion movement becomes permanent.

In the case of ferroelectricity, which is associated with an order/disorder transition, it is necessary to consider that the positive H^+ ion can occupy two equivalent energy positions located at distance d from the center of the cell which thus has a local electric dipole moment $\pm P$ as a function of the position of the ion. The potential energy has the form of the double well, as shown in Fig. (7.9). The average polarization $< p >$ is given by the relation:

$$< p > = \frac{p\mathrm{e}^{pE_l/kT} - p\mathrm{e}^{-pE_l/kT}}{\mathrm{e}^{pE_l/kT} + \mathrm{e}^{-pE_l/kT}} \tag{7.48}$$

where E_l is the local field, the sum of two terms: the external electric field E and the internal field created by the polarization $\lambda P/\varepsilon_0$, where ε_0 is the dielectric constant, P is the polarization, and λ is the polarization constant. Assuming that there are N cells in the crystal, we have $P = N < p >$.

The maximum saturation electrical polarization is $P_S = Np$ and we easily obtain from (7.48):

$$\frac{P}{P_S} = \tanh x \quad \text{with} \quad x = \frac{pE}{kT} + \frac{T_C}{T}\frac{P}{P_S} \tag{7.49}$$

taking $kT_C = \lambda N p^2/\varepsilon_0$. This equation is again self-consistent and is equivalent to the equations found for the order/disorder (7.16) or magnetic (7.26) transition. A numerical solution can be found by plotting the $\tanh x$ curve as a function of x and finding its intersection with the line representing:

$$\frac{P}{P_S} = \frac{T}{T_C}x - \frac{pE}{kT_c} \tag{7.50}$$

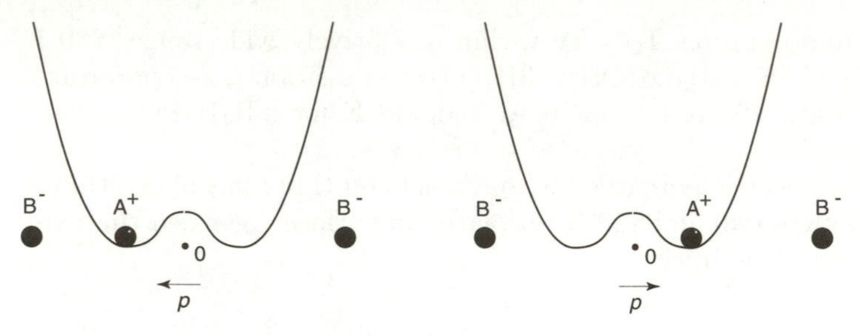

Fig. 7.9. Polarization associated with an order–disorder transition. The positive ion A^+ (a proton, for example) can occupy two equilibrium positions between two neighboring ions B^-. They correspond to polarization $\pm p$ in each of these positions.

For $E = 0$, $x = 0$ is still a solution, the only one when $T > T_C$. For $T < T_C$, there is a second solution, $x \neq 0$ which gives the polarization P. For $E \neq 0$, there is always a solution (we have a graph completely analog to the one in Fig. 7.5). By performing an expansion limited to the first order in the vicinity of T_C, we can easily show that:

$$\frac{P}{P_S} \approx x = \frac{pE}{kT} + \frac{T_C}{T}\frac{P}{P_S} \tag{7.51}$$

This leads to:

$$P = \frac{Np^2}{kT(T - T_C)}E \tag{7.52}$$

We find the Curie law for the dielectric constant:

$$\varepsilon = (\frac{P}{E}) = (\frac{C\varepsilon_0}{|T - T_c|})$$

where $C = Np^2/\varepsilon_0 k$ is the Curie constant.

It has been observed that the free energy of a system obtained with a molecular field model can be expanded for the order parameter (7.17), either for the order/disorder transition or for the magnetic transition. In 1936, Landau and then Tisza generalized the expressions of this type. The Landau model applies particularly well to the ferroelectric transition.

With Landau, let us assume that there is an order parameter P for the system studied, which is the electrical polarization in the case of a ferroelectric. The free enthalpy in the absence of an external field E is a function $G(T, p, P)$. In the vicinity of the critical point, we will assume, and this is the second hypothesis, that a Taylor expansion can be performed for $G(T, p, P)$ as a function of P in the vicinity of T_C. For reasons of symmetry in the case of a ferroelectric (but also a ferromagnetic and the order/disorder transition), the expansion can only contain even terms in P (no situations $\pm P$ are favored). We can then write:

$$G(T, p, P) = G_0 + \frac{1}{2}A(p, T)P^2 + \frac{1}{4}B(p, T)P^4 \ldots \tag{7.53}$$

G_0 is a function of the contribution of the other degrees of freedom, but also of T and p. Coefficients $A, B \ldots$ in the expansion are functions of p, T, but we will assume here that they vary slightly with p. We can study the variation of the potential $G(T, p, P)$ as a function of P as a function of the sign of coefficients A and B in (7.53). Figure 7.1 shows that diagrams a and b correspond to a second order phase transition. For $T > T_C$, the order parameter at equilibrium corresponding to the minimum of potential G is zero, and it is nonzero for $T < T_C (\pm P_e)$.

There will be a phase transition corresponding to continuous change of P at temperature T_C such that $A(T_C) = 0$. At $T = T_C$, the concavity of the curve changes sign. The solution $P = 0$ is no longer stable and corresponds to a maximum of G. Such a transition is only obtained if B is positive. Landau simply chose to take $A = a(T - T_C)$. We will assume that B is constant in the vicinity of T_C and set $B = b$.

We can then rewrite (7.53):

$$G(T, p, P) = G_0 + \frac{a}{2}(T - T_C)P^2 + \frac{b}{4}P^4 \tag{7.54}$$

At equilibrium, the order parameter is the solution of the equation $(\partial G / \partial P) = 0$. We easily find the following solutions:

$$P_e = 0 \quad \text{if} \quad T \leq T_C, \qquad P_e = \pm \left[\frac{a(T_C - T)}{b}\right]^{1/2} \quad \text{if} \quad T < T_C \tag{7.55}$$

A simple calculation of the entropy shows that the latent heat is zero at the transition, which is characteristic of a second order phase transition and that there is discontinuity of the specific heat C_E at T_C (Fig. 7.10 and Problem 7.1).

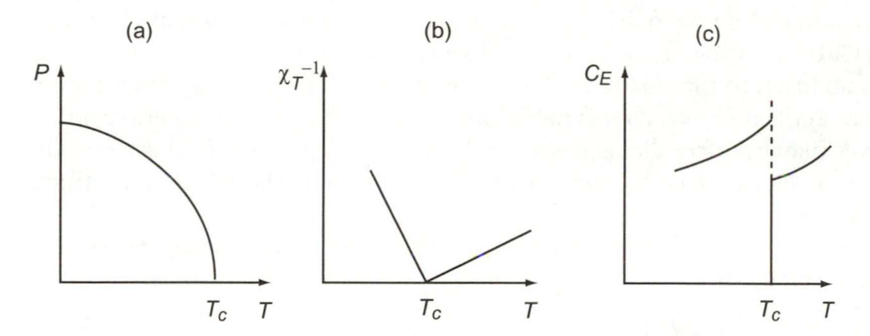

Fig. 7.10. Predictions of the Landau model. The model predicts the critical exponents (**a**) $\beta = 1/2$, (**b**) $\gamma = \gamma' = 1$, (**c**) discontinuity of the specific heat.

In the presence of an external electric field E, the thermodynamic potential should be written $\psi = G - EP$, where G is given by the Landau expansion (7.54). The dielectric constant ε_T is defined by $1/\varepsilon_T = (\frac{E}{P})_T$. At equilibrium, the electrical polarization P obeys the equation with $E = 0$, and we have from (7.54) and (7.55):

$$\varepsilon_T^{-1} = a(T - T_C) \text{ if } T \geq T_C,\ \varepsilon_T^{-1} = 2a(T_C - T) \text{ if } T \leq T_C \qquad (7.56)$$

The Curie law is indeed found with this model.

The Landau model is in fact a molecular field model since a limited expansion of G or F of type (7.54) can be obtained from the expressions for F calculated with the Bragg–Williams and Ising models.

In the case of ferroelectrics, the results of the measurements of order parameter P, dielectric constant ε, and specific heat C_E are compatible with the predictions of these models. The critical exponents $\beta = 1/2$ and $\gamma = 1$ are found by experimentation up to values of $|T - T_C|/T_C$ of the order of 10^{-3}–10^{-4}. Discontinuity of the specific heat is also found.

7.5 Superconductivity

7.5.1 A Complex Phenomenon

Superconductivity consists of the appearance of a new state in a conducting solid in which the electric resistivity vanishes (equivalently, the conductivity becomes infinite) at temperature T_C. The phenomenon was discovered by Kamerlingh Onnes in 1911 in mercury (the transition temperature is 4.15 K). It was then demonstrated in a certain number of metals such as aluminum, subsequently in alloys at very low temperatures. $T_C = 15K$ for Nb_3Sn and 23.2 K for Nb_3Ge. The discovery of superconductivity in copper oxides of the Ba–La–Cu–O type by Bednorz and Müller in 1986 opened a new stage in the study of this phenomenon (the transition temperature is 32 K). Superconductivity was very quickly discovered in other phases, compounds of the $YBa_2Cu_3O_{7-x}$ type ($T_C = 91$ K), and similar families.

In addition to the existence of zero electric resistivity, the superconductor is characterized by specific magnetic properties: placed in a magnetic field, it behaves like a perfect diamagnetic, and magnetic induction $\boldsymbol{B}$ is zero inside a solid in the superconducting state. This property is the **Meissner effect** (Fig. 7.11).

This happens as if there were magnetization $\boldsymbol{M}$ inside the superconducting material so that:

$$\boldsymbol{B} = \mu_0(\boldsymbol{H} + \boldsymbol{M}) \qquad (7.57)$$

A magnetic field $\boldsymbol{H}$ can also destroy superconductivity if it is higher than a critical value $\boldsymbol{H}_C$ which is a function of the temperature. The Meissner effect leads to distinguishing two types of superconductors:

- Superconductors of the **first kind or type I**: $\boldsymbol{B}$ is strictly zero in the solid for $\boldsymbol{H} < \boldsymbol{H}_C(\mathbf{T})$. If $\boldsymbol{H} > \boldsymbol{H}_C(\mathbf{T})$, the material stops being superconducting and $\boldsymbol{B}$ penetrates it (Fig. 7.11).

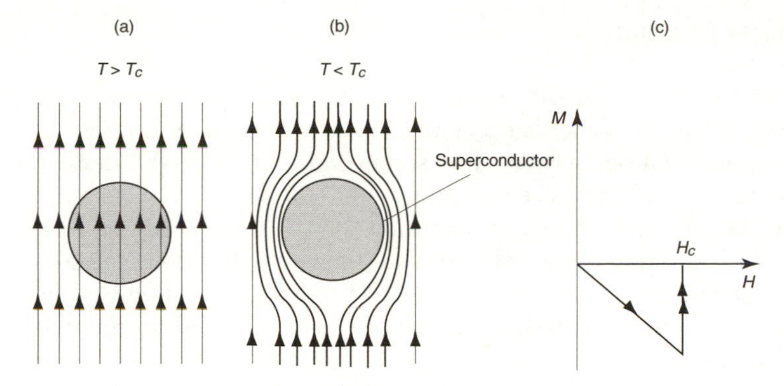

Fig. 7.11. Type I superconductor. (**a**) $T > T_C$, the solid is a normal conductor, and the flux lines penetrate it. (**b**) In the superconducting state ($T < T_C$), the flux lines do not penetrate, $\boldsymbol{B} = 0$, in the solid. (**c**) If $\boldsymbol{H} > \boldsymbol{H}_C$, the solid stops being superconducting.

- In superconductors of the **second kind or type II**, there is a hybrid structure: a) for $\boldsymbol{H} < \boldsymbol{H}_{C1}(\mathbf{T})$, $\boldsymbol{B} = 0$ and there is a strict Meissner effect; b) for $\boldsymbol{H}_{C1}(\mathbf{T}) < \boldsymbol{H} < \boldsymbol{H}_{C2}(\mathbf{T})$, the superconductor is in a mixed state, the induction only partially penetrates in the form of quantified flux tubes, and we have a partial superconducting state; c) if $\boldsymbol{H} > \boldsymbol{H}_{C2}$, the superconducting state totally vanishes (Fig. 7.12).

The value of the critical magnetic fields depends on the material of thc superconductor. In general, $\boldsymbol{H}_C$ is relatively low for type I superconductors (corresponding to inductions less than 0.1 T for elements such as V, Pb, Hg, In, Ti, etc.). This is a serious handicap to their practical use. On the other hand, the critical fields $\boldsymbol{H}_{C2}$ are much higher for type II superconductors (they attain 25 T for Nb_3Sn) which are usually metal alloys. The normal state/superconducting state transition is of **second order**.

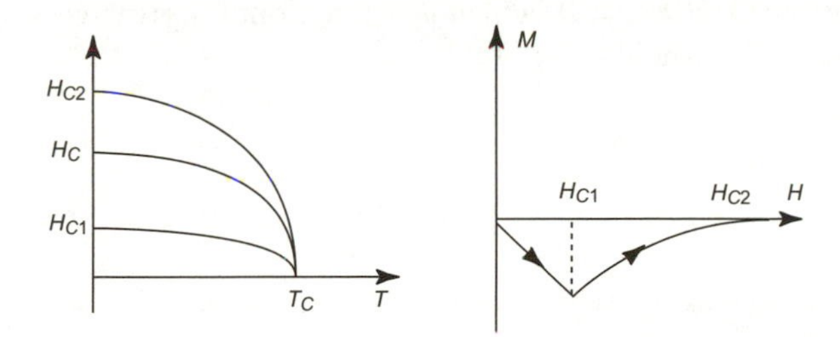

Fig. 7.12. Type II superconductor. If $\boldsymbol{H} < \boldsymbol{H}_{C1}(T)$, $\boldsymbol{B} = 0$ in the superconductor, the situation is similar to the one in Fig. 7.11b; for $\boldsymbol{H}_{C1}(T) < \boldsymbol{H} < \boldsymbol{H}_{C2}(T)$, the flux lines only partially penetrate; if $\boldsymbol{H} > \boldsymbol{H}_{C2}(T)$, the solid is no longer superconducting.

7.5.2 Theoretical Models

The first theory for describing the superconducting state was proposed by London in 1935. Of a phenomenological nature, it consisted of adding an equation giving the relation between the superconducting current density and the magnetic field to Maxwell's equations.

Landau and Ginzburg proposed an equation for the free energy in 1950 by introducing an order parameter that would allow giving a complete description of the phenomenon as a function of the temperature. If n_S is the number of electrons in a superconducting state, n_S is associated with an effective wave function Ψ such that $|\Psi|^2 = n_S$. Using a Taylor expansion for F_S similar to (7.54) but introducing an energy term that accounts for the heterogeneity of the order parameter in the system (similar to the term introduced by van der Waals in capillary theory and the term in the model of spinodal decomposition, (2.60)), we then have:

$$F_S = F_N + \frac{1}{2}\int_V \left[\alpha(T - T_C)|\Psi|^2 + \frac{\beta}{2}|\Psi|^4 + \gamma|\nabla\Psi|^2\right]\mathrm{d}\boldsymbol{r} \tag{7.58}$$

where F_N corresponds to the free energy of the normal part of the solid.

In the absence of an external magnetic field and assuming a homogeneous superconductor, (7.58) is rewritten:

$$F_S = F_N + V\left[\frac{\alpha(T - T_C)}{2}|\Psi|^2 + \frac{\beta}{4}|\Psi|^4\right] \tag{7.59}$$

where T_C is the normal state/superconducting state transition temperature and α and β are two parameters assumed to be independent of T.

At equilibrium, the classic results of the Landau model are found:

$$|\Psi|^2 = (T_c - T)\frac{\alpha}{\beta} \quad \text{if} \quad T < T_C, \qquad |\Psi|^2 = 0 \quad \text{if} \quad T > T_C \tag{7.60}$$

Taking (7.58) again, we can show (from the Schrödinger equation) that $\gamma = h^2/8\pi^2 m$, where m is the mass of the electron and h is the Planck constant, and write the condition $(\partial F/\partial\Psi) = 0$, which is a functional derivative. Function $\Psi(r)$ varies in space. We then have:

$$\alpha(T - T_C)|\Psi| + \beta|\Psi|^3 - \frac{h^2}{8\pi^2 m}\nabla^2\Psi = 0 \tag{7.61}$$

with the condition $\boldsymbol{n}.\nabla\Psi = 0$, which ensures that there is no current flux through the surface.

For simplification, let us take a function Ψ varying in only one direction: z, (7.61) becomes:

$$-\frac{h^2}{8\pi^2 m}\,\frac{1}{\alpha(T - T_C)}\,\frac{\mathrm{d}^2\Psi}{\mathrm{d}z^2} + |\Psi| + \frac{\beta}{\alpha(T - T_C)}|\Psi|^3 = 0 \tag{7.62}$$

The coefficient of the first term should have the dimensions of the square of length $\xi^2(T)$. $\xi(T)$ is the Landau–Ginzburg **coherence length**:

$$\xi(T) = \left(\frac{h^2}{8\pi^2 m\alpha T_C}\right)^{\frac{1}{2}} \left(\frac{T_C}{|T_C - T|}\right)^{\frac{1}{2}} = \xi(0)\left(\frac{T_C}{|T_c - T|}\right)^{\frac{1}{2}} \tag{7.63}$$

The coherence length diverges at T_C. For $T > T_C$, $\xi(T)$ corresponds to the length, or distance, over which the order parameter exists in the material; for $T < T_C$, ξ is the distance over which there is changing in the order parameter with respect to its equilibrium value.

The penetration thickness δ of an external field in a superconductor can be calculated: all currents and external magnetic fields are expelled from the solid except in a thin layer of thickness $\delta < \xi$.

The Landau–Ginzburg model gives a good description of the behavior of a superconductor, its magnetic aspects in particular.

In 1957, Bardeen, Cooper, and Schrieffer advanced a microscopic theory of superconductivity, called BCS, based on the hypothesis of the existence of electron pairs in the superconducting state. These paired electrons of opposite spins are called **Cooper pairs**. The BCS theory assumes that there is an attractive interaction between the conducting electrons in the solid which compensates the Coulomb repulsive force. This interaction is indirect; when an electron moves through the lattice it deforms it, the second electron is coupled to the first one via this deformation (this via the electron–phonon interaction). Cooper pairs move in the crystal lattice without any energy losses to the lattice ions: resistivity vanishes in the superconducting phase. This suggests that the stronger the electron–lattice interaction, the greater the chances that the material will be superconducting. This explains why metals which are good electrical conductors (like Cu and Ag) are not superconducting while Pb and Sn are. Moreover, at high temperature, the vibration amplitude of the lattice ions is high and does not allow electron–lattice coupling and thus the creation of Cooper pairs to take place. This could explain why superconductivity is found at low temperature except for the new copper oxide superconductors. In the BCS model, the critical temperature T_C can be calculated:

$$T_C = 1.14\Theta_D \exp\left(\frac{-1}{UD(\varepsilon_F)}\right)^{\frac{1}{2}} \tag{7.64}$$

where Θ_D is the Debye temperature, U is the interaction energy, and $D(\varepsilon_F)$ is the electron density on the Fermi level. T_C is a function of the isotope mass of the superconductor via Θ_D, an experimentally verified result. There is also a coherence length $\xi(T)$ which has the same form as (7.63), but the energy of a Cooper pair and the energy of two unpaired electrons are separated by a gap $\Delta(T)$ which is a function of the temperature and vanishes at T_C. Gap $\Delta(T)$ varies like the order parameter for $T < T_C$ and is zero for $T > T_C$:

$$\Delta(T) \propto (T_C - T)^{1/2} \tag{7.65}$$

The new superconductors of the YBaCuO type have the advantage of a relatively high transition temperature (higher than the boiling point of liquid nitrogen, 77 K). Superconducting compounds of the BiSrCaCuO and

TlSrCaCuO type have been synthesized with critical temperatures near 125 K. The "record" for T_C up to now is for a copper oxide with Ba, Hg, and Ca atoms ($T_C = 134$ K). These materials, called **high-temperature superconductors**, are of type II and have a special crystallographic structure: plane CuO structures alternate with layers in which atoms such as Ba and Y are inserted (Fig. 7.13). The situation is much more complicated in this type of material and it is still not certain whether the BCS model is pertinent for explaining the existence of the superconducting state. The Cu_2O planes develop local antiferromagnetic correlations induced by magnetic interactions between copper ions, in this case called diluted quantum antiferromagnetism. The dilution comes from dopants such as barium which carry mobile charges to the copper oxide layers. For a certain initial critical concentration, the system will undergo a magnetic phase transition and will become antiferromagnetic. There is thus competition between magnetic order and superconductivity in this type of material which must be understood to satisfactorily explain the electron coupling mechanism (and thus the formation of possible Cooper pairs) and the superconducting transition.

Finally, it is necessary to note that superconductivity has been observed in fullerene compounds of the K_3C_{60} type ($T_C = 18$ K), in fullerene doped

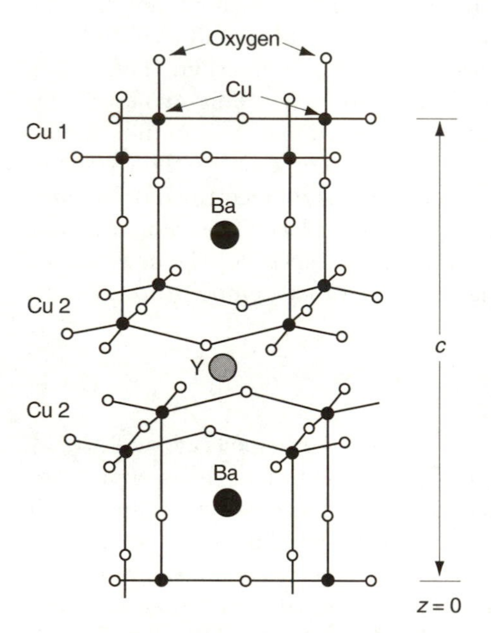

Fig. 7.13. Crystal structure of YBaCuO superconductor. The CuO plane structures play an essential role in the superconducting property of these copper oxides (Ourmazd et al., *Nature*, **325**, 309 (1987), copyright MacMillan Magazines Limited).

with tribromomethane ($T_C = 117K$) and in magnesium diboride MgB_2 at $T_C = 39K$.

In type II superconductors, the role of the flux lines that penetrate the material seems important, particularly in high-temperature superconductors. We will discuss this problem in Chap. 9 (Sect. 9.2.5).

7.6 Universality of Critical Phenomena

7.6.1 Critical Exponents and Scaling Laws

The phase change theories such as the van der Waals model for fluids, the Bragg–Williams model for the order/disorder transition, the Weiss model for magnetism, and the general Landau theory are so-called molecular field theories. The critical exponents (defined, for example, by (7.46)) predicted by these theories which hold for second order transitions are identical, even when different physical states are involved. We thus find: $\beta = 1/2$ for the critical exponent of the order parameter whether it is magnetization, electrical polarization, or the quantity $(\rho - \rho_C)/\rho_C$ in a fluid in the vicinity of the critical point, $\gamma = 1$ for the critical exponent of the generalized susceptibility. The interaction forces involved in very different physical systems are clearly not of the same nature, but the critical phenomena, that is, the behavior of physical quantities in the vicinity of a critical point, have a universal character. This is due to the theoretical treatment method used: interactions between particles (atoms, molecules, ions, spins) are replaced in all of the systems considered by an average field acting locally on each particle. This distinguishes second order phase transitions from first order transitions which do not have this characteristic of universality: when an exponent can be defined for certain quantities such as the order parameter (liquid crystals, for example), they differ greatly from one system to the other.

The experimental situation is a little less clear, even if the values of the experimentally found exponents for very different systems are very close or identical. Exponent β is equal to 0.33 for ferromagnetic/paramagnetic, helium normal state/superfluid state, and liquid/gas transitions near the critical point. Universality is again verified, but with different values of the critical exponents than those predicted by molecular field theories.

We find in fact that molecular field theories produce satisfactory results when temperature T is relatively far from T_C, with the exception of ferroelectrics and superconductors, where it still holds up to the immediate vicinity of T_C. The principal shortcoming of the molecular field theories is to hide all the effects of fluctuations in the macroscopic quantities and not to take into account the existence of short-range order in the material. In a magnetic system, the magnetic contribution to the specific heat C_H does not strictly vanish at T_C, contrary to the predictions of the Weiss model (Fig. 7.14).

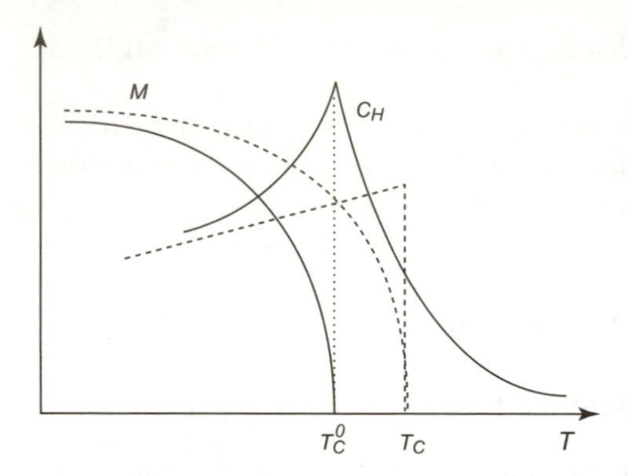

Fig. 7.14. Influence of fluctuations. The dashed curves represent the predictions of the molecular field theories for the magnetization M and specific heat C_H and the solid curves correspond to the experimental results. The magnetic contribution to C_H does not strictly vanish at T_C^0 and short-range order persists for $T > T_C^0$.

The magnetic interaction has a coherence length ξ which allows conserving nonzero local magnetization for $T = T_C + \delta T$ ($\delta T > 0$) and $\xi \to \infty$ when $T \to T_C$, as demonstrated with the Landau–Ginzburg model which takes into account these local fluctuations of the order parameter by construction.

The Landau model clearly demonstrates this universality of the phenomena since it assumes the existence of a single form for the potential (this universality is found with the law of **corresponding states** in fluids), but it only very reflects reality imperfectly. This led physicists to look for other models to account for these phenomena.

The hypothesis that lead to progress was formulated by Domb, Kadanoff, and Widom: they assumed that near a critical point, the thermodynamic quantities would obey **scaling laws**. If we return to (7.27), for example, the order parameter $<\sigma>$ is given by the expression:

$$\frac{1}{2}\log\frac{1+<\sigma>}{1-<\sigma>} - \frac{T_C}{T}<\sigma> = \frac{\mu_0 H}{kT} \tag{7.66}$$

This expression is the equation of state for the magnetic system. Expanding (7.66) in series and setting $\varepsilon = (T - T_C)/T_C$, $x = \mu_0 H/kT$, we have:

$$<\sigma>\varepsilon + \frac{<\sigma>^3}{3} = x \tag{7.67}$$

This equation has a **scaling property** with respect to $<\sigma>$, ε and x because if these quantities are replaced by $<\sigma>' = \lambda <\sigma>, \lambda^2\varepsilon = \varepsilon', \lambda^3 x = x'$, we find that (7.67) retains the same form and is written:

$$<\sigma>'\varepsilon' + \frac{<\sigma>'^3}{3} = x \tag{7.68}$$

The equation of state is invariant for an appropriate change in scale. This property can be generalized to all physical systems in the vicinity of a second order transition. This is once more a manifestation of the universal character of these transitions. Scaling laws are also found in fluid mechanics.

In general, it is believed that the thermodynamic potentials, G, for example, can be in the form of the sum of two terms in the vicinity of a critical transition: one part G_S which has singularity at T_C and another which does not. Remember that near the critical point, short-range order appears (the coherence length $\xi(T)$ defined above diverges at T_C), expressed itself by the existence of this singularity in G_S. The scaling property of (7.68) is thus generalized to the potential G_S, which is a function of ε and $\boldsymbol{H}$ in the presence of an external field $\boldsymbol{H}$:

$$\lambda G_S(\varepsilon, H) = G_S(\lambda^n \varepsilon, \lambda^m H) \tag{7.69}$$

where G_S is by definition a homogeneous function of the first degree with respect to the variables and n and m are two integers.

All critical exponents of the system can easily be calculated as a function of n and m with this hypothesis (Problem 7.4). In this way we show that:

$$\beta = \frac{1-m}{n}, \quad \delta = \frac{m}{1-m}, \quad \gamma = \gamma' = \frac{2m-1}{n}, \quad \alpha = \alpha' = 2 - \frac{1}{n} \tag{7.70}$$

Eliminating m and n between these relations, we obtain new equalities between the critical exponents:

$$\alpha + 2\beta + \gamma = 2 \quad \text{(Rushbrooke)}, \quad \alpha + \beta(\delta + 1) = 2 \quad \text{(Griffiths)}$$
$$\gamma = \beta(\delta - 1) \quad \text{(Widom)} \tag{7.71}$$

7.6.2 Renormalization Group Theory

The preceding considerations can be summarized as follows: when approaching a critical point, the large-wavelength fluctuations of the order parameter tend to dominate the properties of the system in the critical region, and they can be related to a scale which is schematically the scale of the range of these fluctuations. All of these quantities obey a scaling law that allows establishing relations of type (7.71) between the critical exponents. Another consequence of this conclusion is that the values of the exponents will not be a function of the detailed atomic structure of the system but will be determined by the long-range fluctuations. This is the basic hypothesis of **renormalization group** theory.

Kadanoff advanced the fundamental hypothesis that the effect of fluctuations of the order parameter in the vicinity of the critical point will not be altered if the observation scale is changed if a scale larger than the distance between lattice sites and smaller than its size is used. On this basis, Wilson introduced a calculation technique in 1970 called the renormalization group, borrowed from particle physics, which can be used to calculate the critical exponents. We will only give the principle of the calculation here, reporting the more complete details in appendix C.

If we take the example of ferromagnetism, the Ising Hamiltonian representing the energy of the system is written:

$$\mathcal{H}_0 = -J_0 \sum_{i,j} \sigma_i \sigma_j - \mu_0 H_0 \sum_i \sigma_i \tag{7.72}$$

where σ_i and σ_j are spin operators, H_0 is the external field, and J_0 is the coupling constant. The first summation is taken on the nearest neighbors, and the second summation is taken on all spins. The principle of the method consists of decoupling the singular part of magnetic origin F_S with critical behavior at T_C and the regular part which has no anomalies and corresponds to other degrees of freedom of the energy in calculating the free energy $F = F_S + F_R$. We have:

$$\mathrm{e}^{-F/kT} = \left[\sum_{(\Omega_0)} \mathrm{e}^{-\beta \mathcal{H}_0} \right] \mathrm{e}^{-F_R/kT} \tag{7.73}$$

(Ω_0) designates all spin configurations. The Kadanoff method is then applied: if we take a two-dimensional system in the summation Σ_{ij} of (7.72), for example, only one out of two spins will be taken into consideration (the black points in Fig. 7.15). It will then be necessary to recalculate the reduced variables J_0/kT and $\mu_0 H_0/kT$ to account for the fact that only one out of two sites is summed, and a new effective Hamiltonian with a new regular part F_{R1} is obtained.

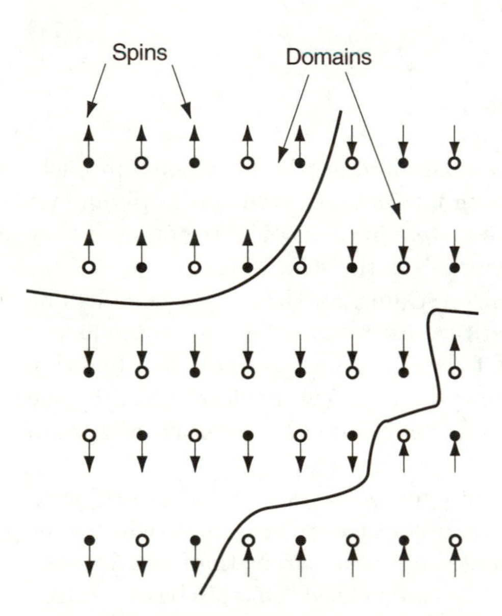

Fig. 7.15. Renormalization group. Each site i bears a spin associated with a magnetic moment. Instead of summing on all sites i in (7.72), the summation will only be performed on one every other site: this is the principle of renormalization group theory. Then an iteration is performed.

$$e^{-F/kT} = \Big[\sum_{(\Omega_1)} e^{-\beta\mathcal{H}_1}\Big]\ e^{-F_{R1}/kT} \tag{7.74}$$

(Ω_1) represents the new configuration.

This process can be iterated by doubling the scale each time; this operation is called decimation and it is accompanied by a renormalization of the variables. The residual spins play the role of a medium which transmits an effective interaction to the remaining spins included in these summations. The critical point will then be an end point or fixed point for which no scale change will modify the effective Hamiltonian, and this is its definition. We then find a scaling law for the thermodynamic potential of type (7.69).

We can then show that all critical exponents of the system can be calculated with finite expansions as a function of parameters $y = 4 - d$ and $1/n$, where d is the **dimension of the system** ($d = 3$ for most systems) and n is the dimension of the **order parameter** ($d = 2$ for the Ising model).

Wilson's theory allows calculating the critical exponents as a function of d and n using different realistic forms for the Hamiltonian. The predictions that it gives (Table 7.1) are in very good agreement with the experimental findings and the theory thus has a universal character.

Table 7.1. Critical exponents predicted by Renormalization Group Theory.

Exponents	α, α'	β	γ, γ'	δ
Mean field theory	0	0.5	1	3
Ising model $d = 2$	0	1/8	7/4	15
Ising model $d = 3$	0.12	0.31	1.25	5
Heisenberg model $d = 3$	- 0.07	0.35	1.4	

7.7 Technological Applications

The technological applications of phase transition phenomena in solids are very numerous. We will leave aside here all of those that directly concern metallurgy and do not involve solidification or solid/solid transitions of the type encountered with martensitic steels and those that we have already treated. We will only give a brief overview of the perspectives opened up by the new materials.

Let us begin with magnetism, whose technological applications have been known for a very long time (the first application was the compass, invented by the Chinese). The uses of magnetic materials in electrical engineering are classic and in the past twenty years have been extended to NMR (nuclear magnetic resonance) imaging which requires high magnetic fields. There are schematically two types of magnetic materials: soft magnetic materials for which the coercive field necessary for eliminating induction in the material is weak and the hysteresis cycle is narrow; hard magnetics, which have wide hysteresis cycles associated with a strong coercive field and high residual magnetization.

Alloys of the Fe–Ni type, called Permaloys, are soft magnetics. Hard magnetic materials are used as permanent magnets; they are often alloys of Fe, Co, Ni, Al (with Cu, Nb, Ti additives). They are formed by spinodal decomposition of a strongly magnetic phase (FeCo) and a weakly magnetic phase (NiAl). Other types of new magnetic alloys have also been developed, in particular, cobalt-rare earth alloys of the $SmCo_5$ type; they are used in miniaturized circuits. The heat treatments for all of these magnetic alloys play an essential role (in order to obtain uniform grain size).

New magnetic materials have been developed in recent years, in particular, amorphous magnetic alloys of the $Fe_{80}B_{20}$, $Fe_{40}Ni_{40}B_{20}$ type, etc. The metalloid content is between 15 and 25%. These alloys, generally in the form of thin films and ribbons, have weak coercive fields and are used in low-power devices, particularly audio and video recording systems. These are the materials used in DVD disks.

Magnetic materials have the property of **magnetostriction**: they may change their length when magnetized (there is coupling between elastic energy and magnetic energy).

Magnetization induces lattice distortion. Magnetostriction is utilized in strain detection devices and transductors (to transmit stresses). An alloy such as $TbO_{0.3}Dy_{0.7}Fe_2$ is an example of a magnetostrictive material. More recently, **magnetoresistive** materials with nonlinear coupling between the electrical resistance and the magnetic field have been developed. This effect can be very important and can lead to gigantic magnetoresistances in transition metal oxides like $La_{1-x}Mn_{1-x}O_3$, which is a conducting ferromagnetic, for example. These materials are particularly used in magnetic reading and recording devices (disk heads).

Ferroelectrics are classically used in a large number of electrical and electromechanical devices. Capacitors utilize ferroelectrics of the $BaTiO_3$ type, for example. The piezoelectric properties of materials such as quartz, ceramics of the PZT type ($PbZr_{1-x}Ti_xO_3$) or polymers such as PVF_2 (polyvinylidene fluoride) are used in ultrasound wave generators and detectors (sonar and hydrophones, medical devices for destroying renal calculi (kidney stones), for example), but also in microphones and accelerometers. The electro-optical applications of ferroelectric materials have also expanded significantly because they are nonlinear materials: the frequency of the electromagnetic wave that passes through them may be changed (a green light wave can be transformed into a blue wave), as well as the propagation velocity of the light. $NaNbO_3$ and PZT ceramics, possibly doped with lanthanum, are used for their electro-optical properties in different optical telecommunications systems. Pyroelectric materials have important applications in infrared imaging (local temperature variations that modify the polarization are detected).

Finally, it is necessary to note that the use of ferroelectric films in data storage memories has been considered. This takes advantage of the capacity of ferroelectric microdomains to very rapidly change their polarization orientation. The best candidates are perovskites of the PZT or $SrBiO_2Ta_2O_9$ type.

The technological interest of superconductivity is obviously potentially very great since it offers the prospect of using electrical conducting wires with no losses due to the Joule effect. Superconducting magnetic coils (Nb–Ti, Nb_3–Sn, and Nb–Zr) are already being constructed, particularly in NMR since they allow obtaining intense magnetic fields with no energy losses (inductions of 20–30 T can be obtained). Such magnetic windings (cryocables) are also used in particle accelerators or machines of the tokamak type for controlled fusion. Type II superconductors with a high critical field $\boldsymbol{H}_{C2}$ are used, since the field limits the use of cryocables (in the same way, the limiting factor is the critical current density).

Superconductors could also be used in zero-loss power transmission lines, electric motors, and alternators. In the last case, the improvement in energy efficiency (several percent) is not as important as the reduction in weight and volume since the bulkiness of the motors could be reduced by a factor of 3 to 5.

The new high-temperature superconductors of the YBaCuO type also have very interesting prospects, but one of the unresolved obstacles to their use remains their forming: it is necessary to obtain wires capable of supporting high current densities and these materials, which are often ceramics, are fragile since they are brittle. Other perspectives might be opened by materials as magnesium diboride and fullerene compounds. The applications of these materials in the form of films, for example, in microelectronics or in magnetic field detectors (Squid or superconducting quantum interference device) based on the Josephson effect would pose fewer problems and will undoubtedly be a future trend.

Problems

7.1. Predictions of Molecular Field Models

1. Using a molecular field model, for example, the Landau model, calculate the entropy for the system at equilibrium. Determine the change in entropy ΔS at temperature T_C. What conclusion can you draw from this?
2. Calculate the constant-field specific heat C_E for $E = 0$ for such a system. Show that there is discontinuity for C_E at T_C.

7.2. Order/Disorder Transition with pressure
We are investigating the influence of pressure on an order/disorder transition in a binary alloy of simple cubic crystal symmetry. We assume that the internal energy of the alloy per site has the form

$$U = U_0(a) + P^2 V(a)$$

where a is the distance between two nearest neighboring sites. $U_0(a)$ is the potential energy of the crystal and P is the order parameter. A pressure p is applied to the material.

1. Calculate the potential $G(T, p, P, a)$.
2. Write the equilibrium conditions in the presence of p. If the transition is still considered second order, calculate coordinates (p_c, T_c) of the transition.
3. Assuming that the transition remains second order, taking the value of a in the vicinity of a_0, its value at the transition, with $p = p_c$ and $T \approx T_c$. Postulating:

$$U'(a) = U_0'(a_0) + U_0''(a_0)\delta a$$

and

$$V(a) = V(a_0) + V'(a_0)\delta a$$

determine δa and P at equilibrium. Show that there is a temperature T_c^* beyond which the transition stops being second order.

7.3. Modeling a Structural Transition
Consider a displacive phase transition in a solid triggered the displacement of an ion of mass m of coordinate q in a given direction. Assume that this ion is subject to a potential

$$V(q) = V_0 + \frac{1}{2}\alpha q^2 + \frac{1}{4}\beta q^4 \quad \text{with} \quad \alpha = a(T - T_C)$$

where T_C is the transition temperature.

1. Write the equation of motion. Designate the damping coefficient by γ.
2. Setting $q = q_e + \delta q(t)$, where q_e is the equilibrium position, write the equation for $\delta q(t)$. Determine q_e.
3. Neglecting m, calculate $\delta q(t)$. Give the expression for the characteristic damping time. What happens at T_C?

7.4. Universality and Critical Exponents
Assume that the Gibbs function G for a magnetic system obeys the equation $\lambda G(\varepsilon, H) = G(\lambda^n \varepsilon, \lambda^m H)$, where H is the magnetic field and

$$\varepsilon = \frac{T - T_C}{T_C}$$

and n and m are two integers.

1. Differentiating the equation for λG with respect to H, define a scaling law for $M(\varepsilon, 0)$.
2. Select $\lambda^n \varepsilon = \mathrm{C^{te}}$, express β as a function of m and n.
3. Find a scaling law for $M(0, H)$ from the derivative of λG with respect to H, introducing the critical exponent δ defined by

$$H \propto |M|^\delta$$

   and selecting $\lambda^m H = \mathrm{C^{te}}$, determine δ as a function of m.
4. Differentiating twice with respect to H, define a scaling law for χ_T, taking $\lambda^n \varepsilon = \mathrm{C^{te}}$, determine critical exponent γ.

7.5. Piezoelectricity

Consider the phase transition in a ferroelectric crystal. It also has the property of being piezoelectric: the deformation x is associated with the electrical polarization and we assume that the corresponding mechanical energy is

$$\frac{1}{2} c^p x^2 - \frac{1}{2} \gamma x P^2$$

where c^p is the elastic constant, γ is the electrostriction constant, and P is the polarization.

1. Write an equation giving the free enthalpy G.
2. Determine the stress X in the solid and the value of x for $X = 0$.
3. Rewrite G as a function of P and determine the expression for the dielectric constant.

8 Collective Phenomena in Liquids: Liquid Crystals and Superfluidity

8.1 Liquid Crystals

8.1.1 Partially Ordered Liquid Phases

A crystal melts when the thermal energy which tends to create disorder, becomes greater than the intermolecular interaction energy which stabilizes the periodic structure of the crystal, ensuring its cohesion. The structure is then broken up and the molecular positional order is destroyed. The molecules are then free to move randomly. On the contrary, if the molecules are rod-shaped, a different process can be observed. At a certain temperature (melting point of the crystal), the thermal energy can be sufficient to destroy the positional order, but still insufficient to oppose the intermolecular forces responsible for the orientational order. A **mesomorphic phase** is then obtained: the molecules can conserve a preferred orientation within the liquid. Finally, at a higher temperature (melting point of the mesomorphic phase), this ordered liquid phase will give rise to the isotropic phase of the normal liquid, when the thermal energy will be sufficient to overcome the contribution of the potential energy which favors alignment of the molecules. In crystals formed from quasispherical molecules, each molecule occupies a well-defined place in the lattice; the centers of gravity of the molecules are located in a three-dimensional periodic lattice – the molecules have positional order. In a crystal, rod-shaped molecules also have positional order, but they also have the same direction at any point; there is in addition orientational order. This positional order breaking induces the appearance of the liquid crystal phase.

Mesomorphic phases, commonly called liquid crystals, are thus intermediate phases between the liquid phase and the solid phase. Several thousand organic compounds have now been identified as presenting mesomorphic phases. As a function of the molecular geometry, a system initially in the crystalline state can pass through one or more mesomorphic phases before being transformed into an isotropic liquid. The transitions can be induced either by varying the temperature (thermotropic mesomorphism) or with a solvent (lyotropic mesomorphism). We are only interested in thermotropic liquid crystals.

The discovery of liquid crystals was only possible after the development of techniques of microscopic analysis, that is, in the 19th century. In 1853 the German physicist Rudolph Virchow was undoubtedly the first to observe a liquid crystal phase, myelin, a substance that surrounds the nerves and is contained in Schwann's sheath.

The German physicist Lehmann, who was working on melting, identified the liquid crystal state in 1888. Reinitzer began to characterize this product by studying its melting point and noted an unusual phenomenon: this compound seemed to have two melting points. On heating, the crystal melted at 145.5°C to form a cloudy liquid; subsequently, the liquid suddenly cleared at 178.4°C. On cooling, the opposite phenomena occurred. While crystals with a three dimensional structure, well-defined on the atomic scale, are often birefringent, on the contrary, liquids whose molecules are randomly distributed are characterized by a single refractive index. Lehmann and Reinitzer, surprised to observe birefringence in the intermediate liquid phase of cholesteryl benzoate, decided to call this phase the **liquid crystal** phase. This term was later taken up by G. Friedel.

Vorländer showed that materials exhibiting a liquid crystal phase are constituted of rod-shaped molecules, while molecules of spherical shape or configuration do not produce this type of phase.

A very large variety of intermediate phases has been observed up to now. Mesomorphic phases can be observed when the molecules have a certain rigidity and elongated or flat shapes. Liquid crystals are generally distinguished by taking into account the order that characterizes them. We note that **plastic crystals** are also an intermediate phase between the crystalline and liquid states (Sect. 1.4.1).

8.1.2 Definition of Order in the Liquid Crystal State

Two types of order should be considered: positional order and orientational order.

◇ *Positional Order*

Positional order can be characterized by the particle density correlation function, defined by:

$$F(\boldsymbol{x}-\boldsymbol{x}') = <\rho(\boldsymbol{x})\rho(\boldsymbol{x}')> \tag{8.1}$$

where $\rho(\boldsymbol{x})$ is the local particle density, and $< >$ is the ensemble average. In any homogeneous phase, $<\rho(\boldsymbol{x})\rho(\boldsymbol{x}')>$ is a function of $\boldsymbol{x}-\boldsymbol{x}'$ only.

In a crystal, $<\rho(\boldsymbol{x})\rho(\boldsymbol{x}')>$ is a periodic function of $\boldsymbol{x}-\boldsymbol{x}'$, where the periods are the basis vectors, $\{\boldsymbol{a}_i\}$ of the lattice.

In an isotropic liquid, on the contrary, $<\rho(\boldsymbol{x})\rho(\boldsymbol{x}')>$ is only a function of $|\boldsymbol{x}-\boldsymbol{x}'|$ and attains a constant value equal to $<\rho>^2$ when $|\boldsymbol{x}-\boldsymbol{x}'|$ becomes larger than several intermolecular distances. The range of values of $|\boldsymbol{x}-\boldsymbol{x}'|$ for which $<\rho(\boldsymbol{x})\rho(\boldsymbol{x}')> - <\rho>^2 \neq 0$ defines the *correlation length* ξ. In a normal liquid, ξ is independent of the direction considered, since the system is isotropic. In the mesomorphic phase, we will see that other situations are found.

◊ *Orientational Order*

For rod-shaped molecules, the direction of orientation of the molecule $\boldsymbol{a}$, is the direction of the axis of the rod representing the molecule. If they have a preferred axis of orientation $\pm\boldsymbol{n}$, the orientation distribution function can be defined by $f(\theta,\varphi)$, where θ and φ are the Euler angles in the reference system (x,y,z), where the z-axis is parallel to $\boldsymbol{n}$. The axis $\boldsymbol{a}$ of a molecule is then defined by:

$$a_x = \sin\theta\cos\phi, \quad a_y = \sin\theta\sin\phi \quad \text{and} \quad a_z = \cos\theta \tag{8.2}$$

$f(\theta,\phi)\mathrm{d}\Omega$ is then the probability of finding molecules with orientation **a** in a elementary solid angle $\mathrm{d}\Omega = \sin\theta\mathrm{d}\theta\mathrm{d}\varphi$ in the vicinity of the (θ,ϕ)-direction.

Given the cylindrical symmetry, $f(\theta,\phi)$ should be independent of φ so that $f(\theta,\phi) = f(\theta)$. Moreover, the equivalence between $\boldsymbol{n}$ and $-\boldsymbol{n}$ implies that $f(\theta) = f(\pi-\theta)$. If the molecular axes are randomly distributed (isotropic orientation), then $f(\theta) = (1/4\pi)$.

The degree of orientation can be characterized by a parameter $< s >$, defined by the following expression:

$$< s >= \frac{1}{2} < 3\cos^2\theta - 1 > \tag{8.3}$$

where $< >$ is the ensemble average over all molecules and θ is the deviation angle of the principal axis of the molecule under consideration with respect to $\boldsymbol{n}$, the preferred axis of orientation of the surrounding molecules. It follows that:

$$< s >= \frac{1}{2}\int(3\cos^2\theta - 1)f(\theta)\sin\theta\mathrm{d}\theta \tag{8.4}$$

Note that in a system where there is no preferred orientation (isotropic orientation), $< s >= 0$. If the molecules are all oriented along $\boldsymbol{n}$ (perfect alignment), then $< s >= 1$.

8.1.3 Classification of Mesomorphic Phases

The liquid crystal phases obtained are obviously a function of the molecular unit. Phases whose molecules have elongated shapes on one hand, and phases obtained with discoid molecules on the other hand are generally distinguished.

a) Elongated and Rigid Molecules

Three broad classes are distinguished: **nematics, cholesterics, smectics**.

Nematic phases are constituted of rod-shaped molecules which are identical to their mirror image (achiral molecules); two examples of the molecules

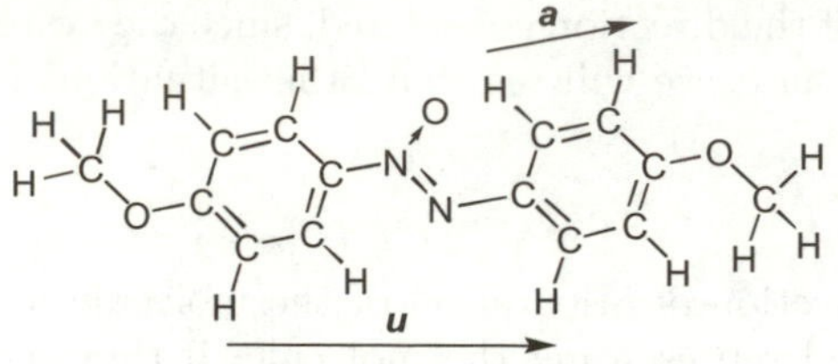

4,4'-dimethoxyazoxybenzene (para-azoxyanisole, PAA)
T_{KN} = 118 °C ; T_{NI} = 135.5 °C

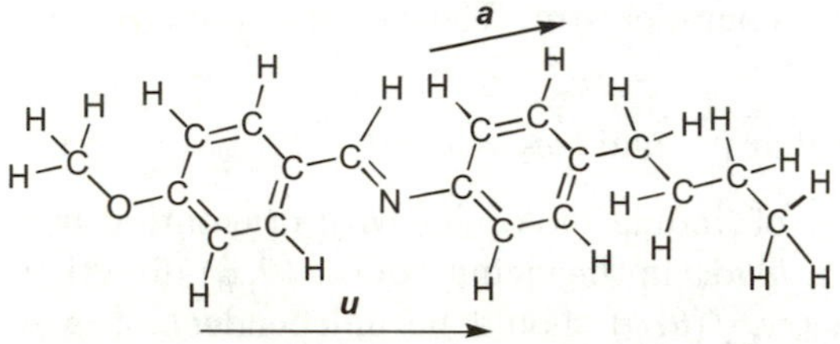

N-(4-methoxybenzylidene)-4'-butylaniline (MBBA)
T_{KN} = 22 °C ; T_{NI} = 47 °C

Fig. 8.1. Two examples of nematogenic molecules. T_{KN} is the solid–nematic transition temperature. T_{NI} is the nematic–isotropic liquid transition temperature. Two types of groups of protons are distinguished in these molecules: The protons in type I are bound to the rigid part of the molecule (proton pairs bound to rings) and the protons in type II belong to methyl groups whose reorientational movements are almost free.

are shown in Fig. 8.1: 4,4′-dimethoxyazoxybenzene (*p*-azoxyanisole, PAA) and N-(4-methoxybenzylidene)-4′-butylaniline (MBBA).

In the absence of stress, the nematic phase should be single domain with a constant and arbitrary directional vector $\boldsymbol{n}$. However, in practice, a nematic phase is composed of **domains**, each characterized by a directional vector $\boldsymbol{n}$, which are the result of external stresses (effects of walls or interfaces).

In each domain, the molecular alignment is shown in all macroscopic tensor properties: index, permeability, magnetic susceptibility. From the optical point of view, a nematic domain is a uniaxial medium with the optical axis along $\boldsymbol{n}$. The difference between the refractive indexes measured with polarizations parallel and perpendicular to $\boldsymbol{n}$ is important ($n_{\parallel} - n_{\perp} = 0.2$ for PAA). In all cases encountered, the domains of nematic liquid crystals seem to have the rotational symmetry around the $\boldsymbol{n}$-axis.

The degree of alignment of the molecules along $\boldsymbol{n}$ is characterized by order parameter $< s >$ defined by (8.4). The states of the directional vector $\boldsymbol{n}$ and $-\boldsymbol{n}$ are indistinguishable. If the molecules have a permanent dipole moment, there will be just as many dipole moments oriented in one direction as in the opposite direction, so that the medium is not ferroelectric. The positional order of the molecules is close to the order of an isotropic liquid; it is thus

Fig. 8.2. Nematic phase. In each monodomain, the molecules are oriented in a preferred direction n, but the positional order of the molecules is close to the order in a normal liquid.

characterized by a very short-range density correlation. However, small-angle X-ray analysis shows that the correlation length $\xi_{\|}$ along orientation axis $\boldsymbol{n}$ is very different from correlation length $\xi_{\perp}$ in a direction perpendicular to $\boldsymbol{n}$.

Nematics flow like liquids, and the shear is seldom higher than ten times the viscosity of water; for para-azoxyanisole (PAA), the viscosity is 10^{-2} dekapoise (10^{-3} dekapoise for water).

Cholesteric phases are composed of rod-shaped molecules which are not identical to their mirror image (chiral molecules); the term "cholesteric" is due to the fact that the phase was first observed in derivatives of cholesterol (Fig. 8.3a); cholesteric liquid crystals not derived from cholesterol were subsequently demonstrated, for example, act-amyl-*p*-(-4-cyanobenzylideneamino)-cinnamate (Fig. 8.3b). Within a domain of such a phase, the axis of orientation of the molecules is not constant; it progresses helically in the material (Fig. 8.4) so that if we designate the axis of the helix by $\mathbf{O}\boldsymbol{z}$, $\boldsymbol{n}$ is defined in reference system $Oxyz$ by:

$$n_x = \cos(q_0 z + \varphi), \quad N_y = \cos(q_0 z + \varphi), \quad n_z = 0 \tag{8.5}$$

Note that the axis of the helix, Oz and angle φ are both arbitrary; in practice, they are imposed by external stresses (effects of walls or interfaces). The structure is thus periodic along $\mathbf{O}\boldsymbol{z}$ but as in a nematic, states $\boldsymbol{n}$ and $-\boldsymbol{n}$ are equivalent. For this reason, Λ, the spatial period of the helical structure, is equal to one-half of the helical pitch:

$$\Lambda = \frac{\pi}{|q_0|} \tag{8.6}$$

Typical values of Λ are much larger than the size of the molecules; it is of the order of 3000 Å. On the contrary, Λ is of the order of magnitude of the optical wavelengths so that this uniaxial medium can be made to interact with a light beam and the value of Λ can be obtained by studying the Bragg scattering. q_0 can be positive or negative (right-handed or left-handed helix); its value is a function of the temperature.

Smectics include a vast family of mesophases with laminar structures. These molecules also have a rod-shaped structure; they have a preferred axis of orientation. The layer thickness is well defined and can be measured by X-ray diffraction and it is approximately equal to either one or two times the

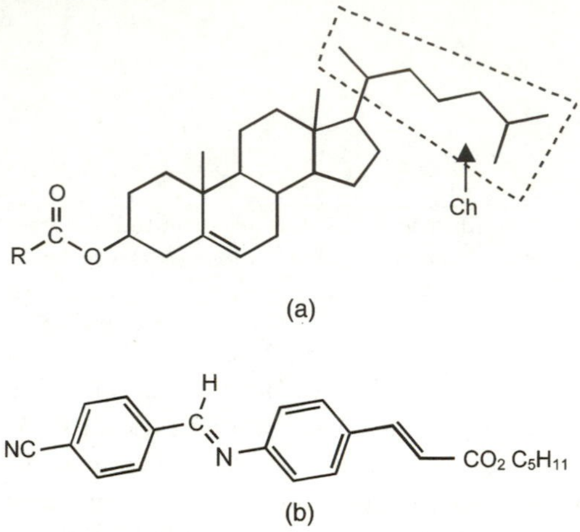

Fig. 8.3. Two examples of molecules representing a cholesteric phase. (**a**) Cholesterol esters giving rise to a cholesteric phase where Ch is a saturated hydrocarbon chain; cholesteryl nonanoate obtained for R = $CH_3(CH_2)_7$ and Ch = C_8H_{17} has a cholesteric phase between 78.5°C and 91°C; (**b**) act-amyl-*p*-(4-cyanobenzylideneamino)-cinnamate, which forms a cholesteric phase between 95°C and 105°C.

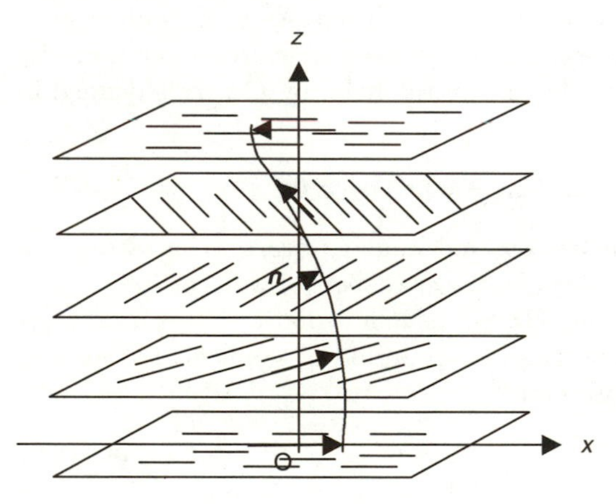

Fig. 8.4. Arrangement of molecules in a cholesteric phase. Directional vector $\boldsymbol{n}$ describes a helix of axis Oz.

length of the molecules. An impressive number of different smectic structures has already been identified, and one probably knows only a limited number of the groups of this family. We will only consider two types of smectics here: smectic A and smectic C, but many other interesting categories have been documented (smectics B, F, I, H, ...). Figure 8.5 shows two types of molecules with smectic phases.

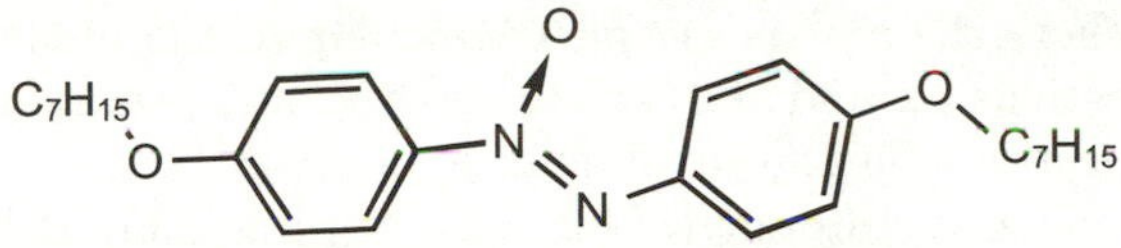

4,4'-diheptoxyazoxybenzene (HOAB)

$T_{KSC} = 74.5\,^{\circ}\mathrm{C}$; $T_{SCN} = 95.5\,^{\circ}\mathrm{C}$; $T_{NI} = 124\,^{\circ}\mathrm{C}$

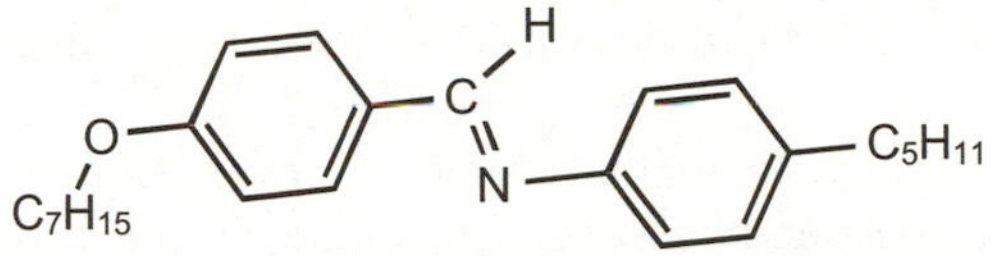

N-(4-heptoxybenzylidene)-4'-butylaniline (HBBA)

$T_{KSG} = 23\,^{\circ}\mathrm{C}$; $T_{SGSB} = 58\,^{\circ}\mathrm{C}$; $T_{SBSC} = 64\,^{\circ}\mathrm{C}$; $T_{SCSA} = 68\,^{\circ}\mathrm{C}$;
$T_{SAN} = 80\,^{\circ}\mathrm{C}$; $T_{NI} = 83\,^{\circ}\mathrm{C}$

Fig. 8.5. Two examples of molecules with smectic phases. T_{KSG} is the "solid–smectic G" transition temperature, T_{SGSB} is the "smectic G–smectic B" transition temperature, T_{SBSC} is the "smectic B–smectic C" transition temperature, T_{SCSA} is the "smectic C–smectic A" transition temperature, and T_{SAN} is the "smectic A–isotropic liquid" transition temperature.

The molecular arrangement in a **smectic A** is shown in Fig. 8.6a. Its structure is characterized as follows:

- The molecules are arranged in successive layers of the same thickness. Their centers of gravity do not have long-range order so that each layer can be considered as a two-dimensional liquid.
- The thickness of each layer is close to one or two times the length of the molecule depending on the type of molecule.
- In each layer, the molecules have a preferred orientation $\boldsymbol{n}$ perpendicular to the plane of the layer.

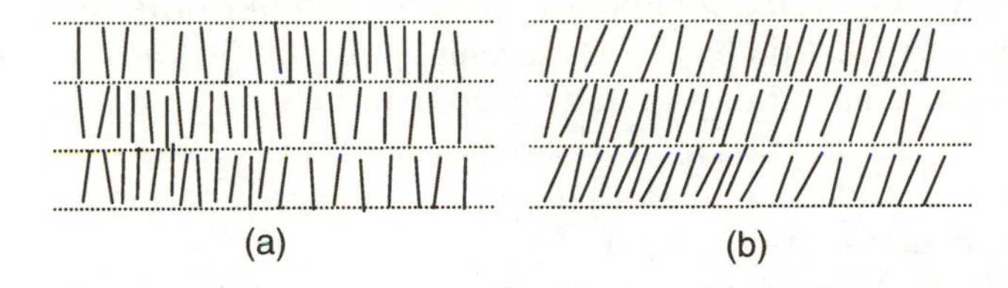

Fig. 8.6. Molecular arrangements in smectic phases. The molecules are arranged in layers of the same thickness. In smectic A phases (**a**), the preferred orientation axis of the molecules is perpendicular to the layers. In smectic C phases (**b**), the preferred orientation axis is tilted at angle θ with respect to the normal to the layers.

The system is uniaxial, where the axis is Oz perpendicular to the plane of layers with equivalent directions Oz and $-Oz$, as in the case of a nematic. In principle, no permanent electric polarization should be observed.

The molecular arrangement in a **smectic C** is shown in Fig. 8.6b. It has the following two characteristics:

- As in the case of smectic A, the molecules are arranged in successive layers of the same thickness and the centers of gravity of the molecules exhibit no long-range order, so that each layer can be considered as a two-dimensional liquid.
- The molecules in each layer have the same preferred orientation $\boldsymbol{n}$ which forms an angle θ with the perpendicular to the plane of the layer; the medium is biaxial and corresponds to monoclinic symmetry (symmetry group C_{2h}).

Consider an orthogonal reference system $Oxyz$ such that Oz is perpendicular to the planes of the layers and Oy perpendicular to directional vector $\boldsymbol{n}$. The medium is invariant:

- under reflection in a mirror Ozx;
- under rotation by angle π around axis Oy,

and there is an inversion point i. Such a medium cannot be ferroelectric given its symmetry.

The molecular arrangement just described and corresponding to a smectic C phase is only obtained if the molecules are optically inactive (achiral molecules or racemic mixture).

If the molecules are chiral or if chiral (optically active) molecules are added to a smectic C phase, a distortion of the structure is then observed. Although the directional vector still has a constant direction in each layer, it undergoes slight angular displacements around Oz from layer to layer so that a helical configuration is globally obtained, called C*. In this new medium, the monoclinic symmetry is probably lost since it is no longer possible to find invariance in mirror reflection and there is no longer any inversion center.

However, in each layer, C_2 symmetry applies (invariance in mirror reflection Ozx and invariance by rotation by angle π around axis Oy); it leads to cancellation of the electric polarization along Oz and Ox but does not permit its cancellation along Oy; an electric moment should thus appear along Oy, that is, in a direction perpendicular to the normal of the layer and in the direction of preferred orientation in the layer.

In the smectic C* phase, the electric moment undergoes the same angular displacement around Oz as directional vector $\boldsymbol{n}$ from layer to layer; helical electric polarization is obtained, where the axis of the helix is perpendicular to the planes of the layers. In smectic liquid crystals, there is thus a remarkable correlation between chirality and ferroelectricity.

b) Discotic Molecules

The two main classes are columnar phases (D) and nematics (N_D).

Since 1961, it has been possible to obtain mesomorphic phases from discotic molecules (flat disk-shaped molecules). In fact, the first systems studied were mixtures obtained by pyrolysis of petroleum pitches. It was only in 1978 that well-characterized discotic molecules capable of producing mesomorphic phases could be obtained (example in Fig. 8.7). Synthesis of a large variety of molecules revealed the existence of several phases:

In columnar phases (D), the discotic molecules are more or less uniformly stacked to form columns which either have no positional correlation between them (phase D) or are arranged in a regular geometric system which can be hexagonal (phase D_h), rectangular (phase D_r), etc. To account for the more or less regular character of stacking of the molecules in the columns, we will denote a hexagonal system of columns in which the molecules are relatively regularly stacked by D_{ho} (using o for order) and a hexagonal system of columns in which the molecules are irregularly stacked (correlation over several molecules) by D_{hd} (using d for disorder). Note that it is always difficult to distinguish between D_{ho} and D_{hd} since the correlation length of the positions along the columns is always finite.

In nematic phases (N_D), the centers of gravity of the discotic molecules are randomly arranged, but their plane is preferably oriented perpendicular to a preferred direction that defines directional vector $\boldsymbol{n}$.

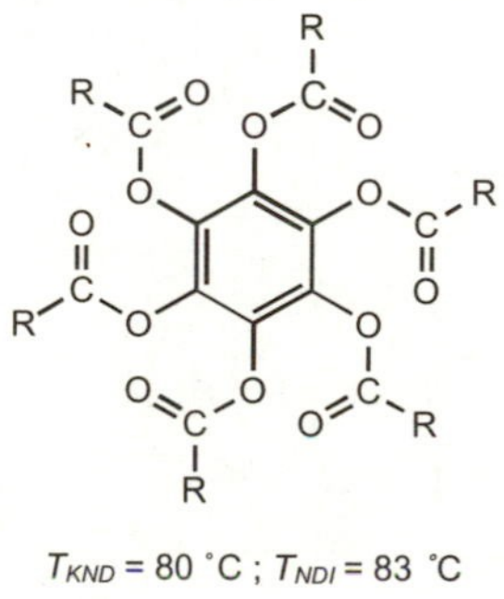

T_{KND} = 80 °C ; T_{NDI} = 83 °C

Fig. 8.7. Example of a disc-shaped molecule (discotic molecule): T_{KND} is the "solid–nematic D" transition temperature; T_{NDI} is the temperature of the "nematic D–isotropic liquid".

8.1.4 The Nematic Phase and Its Properties

We will only consider the nematic phases composed of elongated molecules here.

The nematic phase has weaker symmetry than the isotropic liquid: spherical symmetry for the isotropic phase (normal liquid); axial symmetry for the nematic liquid. The nematic phase is *more ordered* than the isotropic phase. This notion of order can be made more precise by defining:

- the direction of the preferred orientation of the axis of the molecules $\boldsymbol{n}$;
- the degree of alignment of the molecules along $\boldsymbol{n}$ defined by alignment order parameter $< s >$.

a) Measurement of Order Parameter $< s >$ in the Nematic Phase

We want to define the order parameter selected to be able to easily link it with the anisotropies of the measurable physical quantities.

◇ *Diamagnetic Anisotropy*

Let us select a Cartesian reference system fixed in space, Oxyz, with the axis Oz parallel to the nematic axis, $\boldsymbol{n}$. We will designate the number of molecules per unit of volume by n.

The principal diamagnetic susceptibilities of a molecule corresponding to the principal axes of the molecule are designated η_1, η_2, and η_3, with $\eta_2 = \eta_3$ to account for the cylindrical symmetry of each molecule.

Consider one of the molecules in the nematic phase shown in Fig. 8.8. Its axis of revolution i forms angle θ with respect to axis Oz. The other orthonormal axes of the molecule, $\boldsymbol{j}$ and $\boldsymbol{k}$, are selected so that $\boldsymbol{k}$ is a vector perpendicular to plane $(\boldsymbol{i}, \mathrm{O}\boldsymbol{z})$.

If a low-amplitude magnetic field $\boldsymbol{H}$ is applied along the preferred orientation axis Oz, the induced magnetic dipole moment of this molecule, μ, is:

$$\mu = H(\eta_1 \cos\theta \boldsymbol{i} + \eta_2 \sin\theta \boldsymbol{j}) \tag{8.7}$$

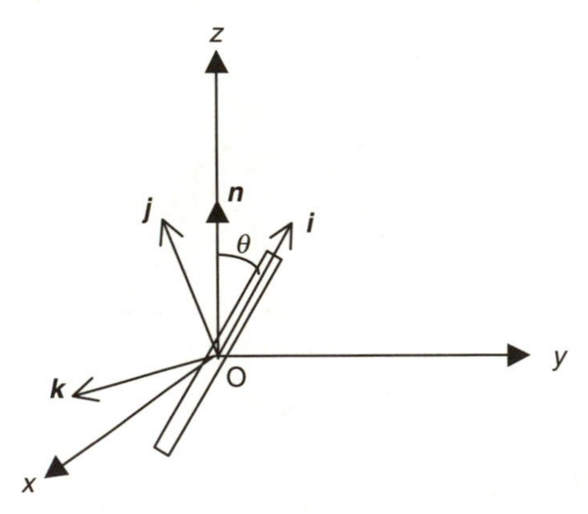

Fig. 8.8. Nematic molecule in the laboratory reference system. The laboratory reference system Oxyz is selected so that Oz is parallel to directional vector $\boldsymbol{n}$. The reference system of the molecule (i, j, k) is selected so that i is the axis of the rod and $\boldsymbol{k}$ is perpendicular to plane (i, n).

The component along Oz of μ,designated as μ_z, is equal to:

$$\mu_z = H(\eta_1 \cos^2\theta + \eta_2 \sin^2\theta) = H[(\eta_1 - \eta_2)\cos^2\theta + \eta_2] \tag{8.8}$$

Now consider a nematic sample; its magnetization component (magnetic moment per unit of volume) will be:

$$M_z = n < \mu_z >= nH[(\eta_1 - \eta_2) < \cos^2\theta > +\eta_2] \tag{8.9}$$

Knowing that $< \cos^2\theta >= 1/3\,(2 < s > +1)$, we finally obtain:

$$M_z = nH[(\eta_1 - \eta_2)\frac{1}{3}(2 < s > +1) + \eta_2] \tag{8.10}$$

that is:

$$M_z = \chi_{mz} H \tag{8.11}$$

where:

$$\chi_{mz} = n(\overline{\eta} + \frac{2}{3}\eta_a < s >) \tag{8.12}$$

with $\overline{\eta} = (\eta_1 + 2\eta_2)/3$ and $\eta_a = \eta_1 - \eta_2$.

If a magnetic field $\boldsymbol{H}$ is now applied perpendicular to the preferred orientation axis, Oz, for example along Oy, the induced dipole moment of this molecule, μ, is:

$$\mu = H(\eta_1 \sin\varphi \sin\theta\, \boldsymbol{i} - \eta_2 \sin\varphi \cos\theta\, \boldsymbol{j} - \eta_2 \cos\varphi\, \boldsymbol{k}) \tag{8.13}$$

The component of μ along Oy, designated μ_y, is equal to:

$$\mu_y = nH(\eta_1 \sin^2\varphi sin^2\theta + \eta_2 sin^2\varphi \cos^2\theta + \eta_2 \cos^2\varphi) \tag{8.14}$$

Magnetization M_y along Oy is thus:

$$M_y = nH(\eta_1 < \sin^2\varphi >< \sin^2\theta > +\eta_2 < \sin^2\varphi >< \cos^2\theta >$$
$$+\eta_2 < \cos^2\varphi > \tag{8.15}$$

and knowing that $< \sin^2\varphi >=< \cos^2\varphi >= 1/2$, we obtain:

$$M_y = nH[\frac{1}{3}\eta_1 + \frac{2}{3}\eta_2 - \frac{1}{3}(\eta_1 - \eta_2) < s >] \tag{8.16}$$

that is:

$$y = \chi_{my} H \quad \text{and} \quad \chi_{my} = n[\overline{\eta} - \frac{1}{3}\eta_a < s >] \tag{8.17}$$

where we set:

$$\overline{\eta} = \frac{1}{3}(\eta_1 + 2\eta_2) \quad \text{and} \quad \eta_a = \eta_1 - \eta_2 \tag{8.18}$$

A similar result is obviously obtained for χ_{mx}. Combining these results, we obtain:

$$< s >= \frac{\chi_{my} - \chi_{mz}}{n(\eta_1 - \eta_2)} \tag{8.19}$$

$\eta_1 - \eta_2$ can be obtained from measurements of the susceptibilities of the solid crystal if its structure is known. Measurement of the anisotropy $\chi_{mz} - \chi_{my}$ in the nematic phase can be used to calculate the value of $< s >$.

Similarly, $< s >$ can be correlated with the optical anisotropy: birefringence and circular dichroism.

◇ *Nuclear Magnetic Resonance (NMR)*

The order parameter $< s >$ can also be obtained by proton NMR analysis. As an example, consider the case of PAA, whose formula is given in Fig. 8.1. We can distinguish in this molecule:

- type I protons, four pairs of protons bound to the rigid part of the molecule (ring protons); each pair is the site of an induced magnetic dipole interaction between protons. It is important to note that the $\boldsymbol{a}$ axis of the interacting proton pairs is approximately parallel to the $\boldsymbol{u}$ axis of the rigid part of the molecule.
- type II protons, two triples of protons (methyl groups). Each triple is the site of induced dipole interactions modulated by the free reorientational movement of these groups.

As we will see later, in the presence of high magnetic induction $\boldsymbol{B}$ ($>$ 0.1 Tesla), the preferred orientation axis $\boldsymbol{n}$ of the molecules is aligned along $\boldsymbol{B}$ with no significant change in order parameter $< s >$. We will henceforth note the direction of $\boldsymbol{B}$ and $\boldsymbol{n}$ by Oz.

Taking into consideration the magnetic dipole interaction between the spins of the two protons in the molecule, we can show that the system has several nuclear transitions associated with magnetic resonance lines. If μ_0 is the magnetic permeability and γ is the gyromagnetic ratio of the protons ($\gamma = 2.6751 \times 10^8\ \mathrm{T}^{-1}\ \mathrm{s}^{-1}$), they will correspond to frequencies ω such as:

$$\omega = \gamma \left[B \pm \frac{\mu_0}{4\pi} \frac{\hbar\gamma}{d^3} \frac{3}{2} < s > \right] \tag{8.20}$$

where d is the distance between the protons of a pair.

In the isotropic phase ($< s > = 0$), a single resonance line is predicted, while in the nematic phase, where $< s > \neq 0$, a doublet will be observed. The difference in frequency $\Delta\omega$ between the two lines is equal to:

$$\Delta\omega = 3 \frac{\mu_0}{4\pi} \frac{\hbar\gamma^2}{d^3} < s > \tag{8.21}$$

The order parameter $< s >$ can thus be determined directly from the difference $\Delta\omega$.

For type II protons, there will be a single resonance line regardless of the nematic order: $\omega = \gamma B$.

The overall spectrum is thus composed of a central line with frequency $\omega = \gamma B$ and a doublet whose frequencies are given by $\omega = \gamma \left[B \pm \frac{\mu_0}{4\pi} \frac{\hbar\gamma}{d^3} \frac{3}{2} < s > \right]$.

The frequency difference $\Delta\omega$ between the two components of the doublet is proportional to the order parameter:

$$\Delta\omega = 3\frac{\mu_0}{4\pi}\frac{\hbar\gamma^2}{d^3} < s > \tag{8.22}$$

$< s >$ is thus directly determined from $\Delta\omega$ if d, the distance between the type I protons, is known.

Figure 8.9 shows the simulation of a NMR proton spectrum when PAA is in the nematic phase. The evolution of $< s >$ as a function of the temperature can be determined from the evolution of $\Delta\omega$.

Finally, Fig. 8.10 shows the measurements of order parameter $< s >$ performed in MBBA with different techniques: Raman, NMR, birefringence, diamagnetic anisotropy. Very good consistency of the results is obtained.

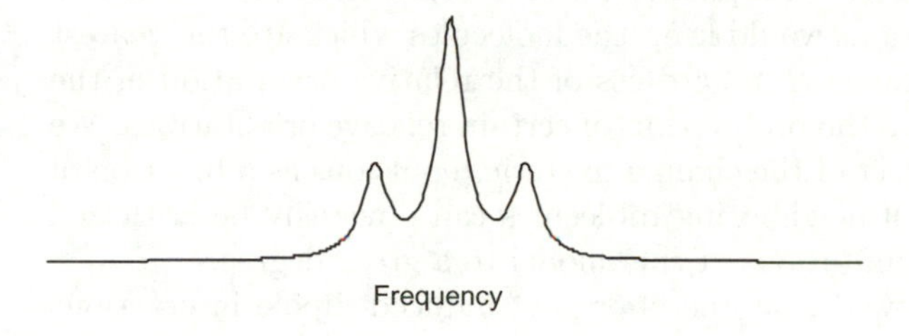

Fig. 8.9. NMR line of a nematic. The overall spectrum is composed of a central line corresponding to type I protons and a doublet corresponding to type II protons, characterized by frequency difference $\Delta\omega = 3\frac{\mu_0}{4\pi}\frac{\hbar\gamma^2}{d^3} < s >$.

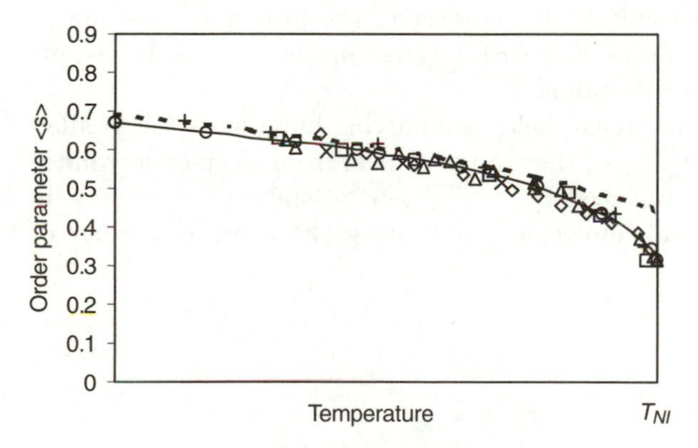

Fig. 8.10. Measurements of order parameter $< s >$ performed with different methods. (○ Raman, ⊓ NMR, × and + birefringence, △ magnetic anisotropy). The dashed curve corresponds to the Maier–Saupe theory (1958 and 1960) discussed in Sect. 8.1.4b; the solid curve corresponds to an improvement in this theory by Humphries, James, and Luckhurst (1972).

b) Statistical Models of Nematic Order – Isotropic Liquid–Nematic Transition

We will present a first model, which is a theory of the mean field type with ion interaction in s^2 proposed by Maier and Saupe in 1960.

Assume that the molecules are in the shape of rigid rods of cylindrical symmetry (Fig. 8.2). In the nematic phase characterized by order parameter $< s >$, the energy of interaction of a molecule i with its near neighbors is a function of: – its orientation with respect to the average direction of orientation of the surrounding molecules – order parameter $< s >$, characterizing the surrounding molecules. The interactions to consider are a priori: induced dipole interactions (van der Waals attractive forces); repulsive steric interactions. These two types of interaction should obviously take into account the rigid and cylindrical character of these molecules.

Maier and Saupe proposed simple modeling of intermolecular interactions, assuming that the joint effect of repulsive (steric) interactions and attractive (induced dipole) interactions would keep the molecules which are the nearest neighbors at a fixed distance R, regardless of the relative orientation of the molecules, thus neglecting the prohibition of certain relative orientations. We note in passing that neglect of the change in steric repulsions as a function of the relative orientation of neighboring molecules can obviously be criticized and should explain the limitations of this model to a great degree.

Maier and Saupe showed that the energy of induced dipole interactions of a molecule i with its m_i nearest neighbors is written as follows:

$$H_{int,i} = -\frac{b}{R^6} - \frac{a}{R^6} \sum_{k=1}^{m_i} s_k s_i \tag{8.23}$$

with a and $b > 0$ and $s_i = 1/2\ (3\cos^2\theta_i - 1)$. θ_i characterizes the deviation of the orientation of a molecule i with respect to the preferred direction of orientation in the nematic phase (Fig. 8.8). Interactions between more distant molecules are neglected in this model.

Introducing the so-called "mean field" simplifying hypothesis (also called molecular field) in which $\sum_{k=1}^{m_i} s_i$, the sum of s_k over m_i molecules surrounding the molecule i will be replaced by $< \sum_{k=1}^{m_i} s_k >$ where $< >$ represents an ensemble average over all molecules. Indicating the average number of nearest neighbors by m,

$$< \sum_{k=1}^{m_i} s_k >= m < s >$$

so that:

$$H_{int,i} \cong -\frac{b}{R^6} - \frac{a}{R^6} m < s > s_i \tag{8.24}$$

where the "mean field ", which is coupled to s_i, is represented by the term $(a/R^6)m < s >$. Introducing the molar volume, V, proportional to R^3, the interaction energy becomes:

$$H_{int,i} \cong -\frac{B}{V^2} - \frac{A}{V^2} < s > s_i \tag{8.25}$$

The partition function $Z_{int,i}$ relative to this interaction is then written:

$$Z(T, < s >)_{int,i} = \exp\left(\frac{1}{kT}\frac{B}{V^2}\right) \int_0^{\pi} \mathrm{d}\theta_i \sin\theta_i \exp\left(\frac{A}{kT}\frac{1}{V^2} < s > s_i\right) \tag{8.26}$$

knowing that $< s_i >$ must satisfy the following constraint $< s_i >=< s >$. Introducing:

$$\alpha = \frac{1}{kT}\frac{A}{V^2} < s > \tag{8.27}$$

the last condition is written:

$$\frac{\int_0^{\pi} \mathrm{d}\theta_i \sin\theta_i s_i \exp(\alpha s_i)}{\int_0^{\pi} \mathrm{d}\theta_i \sin\theta_i \exp(\alpha s_i)} = \alpha\frac{kT\ V^2}{A} \tag{8.28}$$

and setting $u = \cos\theta_i$ and:

$$I(\alpha) = \frac{\int_0^1 \mathrm{d}u \frac{1}{2}(3u^2 - 1)\exp\left[\alpha\frac{1}{2}(3u^2 - 1)\right]}{\int_0^1 \mathrm{d}u \exp\left[\alpha\frac{1}{2}(3u^2 - 1)\right]} = \alpha\frac{kT\ V^2}{A} \tag{8.29}$$

we then have the equation:

$$\frac{I(\alpha)}{\alpha} = \frac{kT\ V^2}{A} \tag{8.30}$$

This allows obtaining the function $\alpha(T)$ and consequently correlating order parameter $< s >$ with temperature T. We find that $< s >= 0$ $(\alpha = 0)$ is a solution of this equation regardless of T.

The non zero solutions of (8.30) are the y axis of the intersections of the curve $y = \frac{I(\alpha)}{\alpha}$ with the line $y = \frac{kT\ V^2}{A}$ (Fig. 8.11). For $\frac{kT\ V^2}{A} > 0.2228$, there is only one solution, $\alpha = 0$. For $\frac{kT\ V^2}{A} < 0.22283$, three solutions are obtained, including the solution corresponding to $\alpha = 0$. Finally, for $\frac{kT\ V^2}{A} = 0.2228$, there are only two solutions: $\alpha = 0$ and $\alpha = 1.453$.

The change of the average energy $< \Delta E >$, free energy ΔF, and entropy ΔS associated with dependence of the order parameter from 0 to $< s >$ are respectively equal to:

$$< \Delta E >= -\frac{A}{V^2} < s >^2 \tag{8.31}$$

$$\Delta F = -kT\{\log[Z(T, < s >)_{int}] - \log[Z(T, 0)_{int}]\} \tag{8.32}$$

$$\Delta S = \frac{< \Delta E >}{T} - \frac{\Delta F}{T} = -\frac{A}{TV^2} + k\{\log[Z(T, < s >)_{int} - \log[Z(T, 0]_{int})\} \tag{8.33}$$

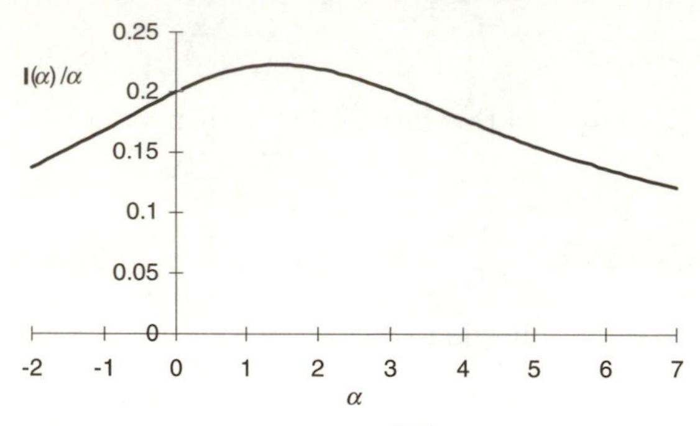

Fig. 8.11. Dependence of $\frac{I(\alpha)}{\alpha}$ as a function of α.

Determination of the transition temperature between nematic liquid and isotropic liquid requires knowing the chemical potentials of these two phases and the transition temperature corresponding to equilibrium of pressures, temperatures, and chemical potentials.

Now consider one mole of nematic liquid; it contains N_0 molecules with mN_0 interacting close pairs.

Per mole, the changes in energy, $\Delta < \mathcal{E} >$, entropy $\Delta\mathcal{S}$, free energy $\Delta\mathcal{F}$, and free enthalpy $\Delta\mathcal{G}$ associated with order parameter $< s >$ are then written as:

$$\Delta\mathcal{E} = -\frac{1}{2}N_0 < E >= -\frac{1}{2}N_0\frac{A}{V^2} < s >^2 \tag{8.34}$$

$$\Delta\mathcal{S} = N_0\Delta S = N_0\left\{-\frac{A}{TV^2} < s >^2 + k\log[z(T, < s >)_{int}]\right\} \tag{8.35}$$

with:

$$\begin{aligned} z(T, < s >)_{int} &= \frac{Z(T, < s >)_{int}}{Z(T, < 0 >)_{int}} \\ &= \frac{1}{2}\int_0^\pi \mathrm{d}\theta_i \sin\theta_i \exp\left\{\frac{1}{kT}\frac{A}{V^2} < s > s_j\right\} \end{aligned} \tag{8.36}$$

so that:

$$\begin{aligned} \Delta\mathcal{F} &= \Delta < \mathcal{E} > -T\Delta\mathcal{S} \\ &= N_0\left\{\frac{A}{2V^2} < s >^2 - kT\log[z(T, < s >)_{int}]\right\} \end{aligned} \tag{8.37}$$

and:

$$\Delta\mathcal{G} = N_0\Delta\mu = \Delta\mathcal{F} - p\Delta\mathcal{V} \tag{8.38}$$

where $\Delta\mathcal{V}$ is the change of the molar volume associated with order parameter $< s >$.

At low pressure, the term $p\Delta\mathcal{V}$ can generally be neglected, so that:

$$\Delta\mu \cong \frac{A}{2V^2} < s >^2 - kT \log[z(T, < s >)_{int}] \tag{8.39}$$

At the "nematic–isotropic" transition temperature, the coexistence of nematic and isotropic phases implies the equality of the chemical potentials of the two phases ($\Delta\mu = 0$), that is, for $\alpha \neq 0$, using (8.27) and (8.39):

$$2\frac{\log[z(\alpha)_{int}]}{\alpha^2} = \frac{kT\,V^2}{A} \tag{8.40}$$

The value of α, the solution of the preceding system of equations, must satisfy the following equation:

$$\frac{I(\alpha)}{\alpha} = 2\frac{\log[z(\alpha_{int})]}{\alpha^2} \tag{8.41}$$

Equations (8.29), (8.36), and (8.41) allow obtaining the temperature T_{NI} and the value of the order parameter in the nematic phase at the transition, $< s >_{NI}$.

The graphic solution of the preceding equation is obtained by looking for the point of intersection (different from $\alpha = 0$) of the curves $y = I(\alpha)/\alpha^2$ and $y = 2\log[\mathrm{z}(\alpha)_{int}]/\alpha^2$.

Figure 8.12 shows that the solution corresponds to $\alpha = 1.96$ and $y = 0.220$, that is:

$$\frac{A}{kT_{NI}\,V^2} < s_{NI} >= 1.96 \quad \text{and} \quad \frac{kTN_{NI}\,V^2}{A} = 0.220$$

where T_{NI} is the isotropic–nematic transition temperature and $< s_{NI} >$ is the value of the order parameter in the nematic phase at temperature T_{NI}.

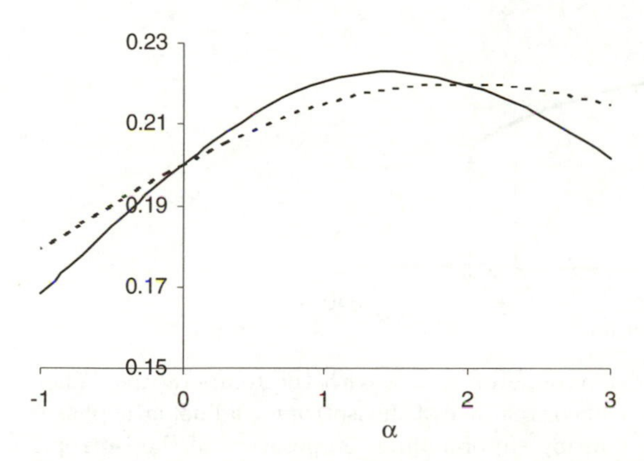

Fig. 8.12. Graphic solution of (8.41). The intersections of the curves $I(\alpha)/\alpha$ (solid line) and $2\log(z_{int})/\alpha^2$ (dashed line) correspond to the points ($\alpha = 0$; $\frac{I(\alpha)}{\alpha} = 0.2$) and ($\alpha = 1.96$; $\frac{I(\alpha)}{\alpha} = 0.220$).

Hence:

$$< s_{NI} >= 0.429 \quad \text{and} \quad T_{NI} = 0.220 \frac{A}{kV^2}$$

At constant T and p, the limits of existence of the isotropic and nematic phase correspond to:

$$\frac{\partial^2 \mathcal{G}}{\partial^2 < s >} = 0 \tag{8.42}$$

$\mathcal{G}$ is approximately equal to $\mathcal{F}$, i.e.:

$$\mathcal{G} = N_0 \left\{ -\frac{B}{V^2} + \frac{A}{2V^2} < s >^2 - kT \log[z(T, < s >)_{int}] \right\} \tag{8.43}$$

so that at constant T and p, using (8.27), the condition $\frac{\partial^2 \mathcal{G}}{\partial^2 < s >} = 0$, which defines the stability limit, is written as:

$$\frac{kT\ V^2}{A} = \frac{\partial^2 \log[Z(\alpha)_{int}]}{\partial \alpha^2} \tag{8.44}$$

The stability limit of the nematic phase ($\alpha \neq 0$) is obtained for:

$$\alpha = \frac{A}{kT\ V^2} < s >= 1.45 \quad \text{and} \quad \frac{kT\ V^2}{A} = 0.22283$$

which corresponds to the following physical state:

$$< s_N^* >= 0.323 \quad \text{and} \quad T_N^* = 0.2228 \frac{A}{kV^2}$$

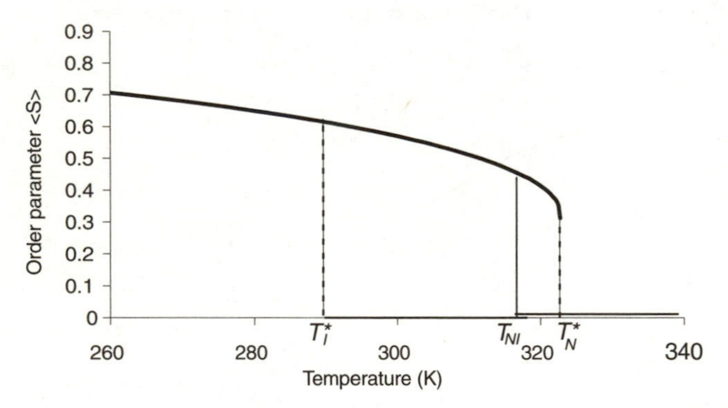

Fig. 8.13. Variation of the order parameter $< s >$ with the temperature. When $A/V^2 = 2.10^{-20}$, the temperature of coexistence of the isotropic and nematic phases is equal to $T_{NI} = 318.8$ K, the limiting supercooling temperature of the isotropic phase is $T_I^* = 290$ K and the limiting superheating temperature of the nematic phase is $T_N^* = 323$ K. The hysteresis observed is in fact very low (a few degrees).

As for the stability limit of the isotropic phase ($\alpha = 0$), it will be obtained for:

$$\frac{kT\ V^2}{A} = 0.2$$

The corresponding physical state is defined by $< s_I^* >= 0$ and $T_1^* \approx 0.2A/(kV^2)$.

The curve in Fig. 8.13 represents the dependence of the order parameter $< s >$ in the neighborhood of the transition temperature T_{NI}^* in the case where $A/V^2 = 2.10^{-20}$ J.

We thus have the following situation:

- for $T < T_I^*$, the nematic phase is stable and the isotropic phase is unstable;
- for $T_I^* < T < T_{NI}$, the nematic phase is stable and the isotropic phase is metastable;
- for $T_{NI}^* < T < T_I^*$, the isotropic phase is stable and the nematic phase is metastable;
- for $T > T_N^*$, the isotropic phase is stable and the nematic phase is unstable.

The Maier–Saupe model is a molecular field model. In conclusion, we can summarize the results as follows:

- At temperature T_{NI}, the value of the order parameter in the nematic phase is not dependent on the choice of A/V^2 and the applied pressure: $< s_{NI} >= 0.429$. This seems to be well verified, even though the experimental values found for $< s_{NI} >$ are much lower: $< s_{NI} >_{exp} \cong 0.3$. Figure 8.10 allows comparing the results of this theory with the experimental results obtained with different techniques. Substantial improvements in this theory by Humphries, James, and Luckhurst (1972) now allow obtaining excellent agreement.
- The phase transition temperature, $T_{NI} = 0.220A/(kV^2)$, is proportional to A/V^2. We assumed A/V^2 to be constant, but this parameter is probably a function of the temperature, since it is supposed to account both for van der Waals interactions and steric repulsions, which are a function of the temperature. Its dependence on the temperature, difficult to predict, should significantly modify the variation of $< s >$ as a function of the temperature.

Finally, we determine the latent transition heat (Problem 8.1) such that: $L/T_{NI} = 0.418R = 3.46$ Jmole $^{-1}$K^{-1}. This value is not in very good agreement with the experimental results obtained, as Table 8.1 shows.

The Maier–Saupe model is based on simplifying hypotheses: it describes the intermolecular interactions by only partially taking into account the molecular structure; the "mean field" approximation neglects of the fluctuations of each molecular environment, since each molecule is assumed to be placed in the same average environment (8.24).

Table 8.1. Experimental results for thermodynamic parameters characterizing the isotropic liquid–nematic transition.

Nematics	T_{NI}(K)	L(J/mole)	L/T_{NI}
PAA	408.6	570	1.39
CBOOA	373.13	1100	2.95
4-4′-diheptyloxyazoxybenzene	397.33	1050	2.64
TBAA	509.63	750	1.47

Moreover, it is desirable to have a general model of the "nematic–isotropic liquid" transition which would allow determining the correlations of the behavior of materials exhibiting such a transition; the ideal would be to have a simple phenomenological theory capable of explaining the general evolution of the physical properties of these materials without having to take either the details of the molecular interactions or the details of the molecular structure into consideration. Using this type of strategy, de Gennes proposed a simple model that generalizes the method proposed by Landau (Sect. 7.4.3) to predict the development of the order parameter in the vicinity of the phase transition temperature: the Landau–de Gennes model.

It is assumed that the free energy F is an analytical function of the order parameter. In our case, this order parameter is $< s >$, which characterizes the degree of alignment of the molecules. One makes a Taylor expansion of F as a function of the order parameter $< s >$ in the vicinity of $< s >= 0$. In the absence of any external electric or magnetic field, this expansion is written a priori:

$$F = F_0 + K(T) < s > + \frac{1}{2}A(T) < s >^2 + \frac{1}{3}B(T) < s >^3 + \frac{1}{4}C(T) < s >^4 + \mathcal{O}(< s >^5) \tag{8.45}$$

However, we have to know the temperature dependence of the coefficients of this expansion; nevertheless, referring to molecular theories, we expect a slow change of these coefficients.

The states of equilibrium of the system must satisfy the following equations:

$$\frac{\partial F}{\partial < s >} = 0 \quad \text{and} \quad \frac{\partial^2 F}{\partial^2 < s >} > 0 \tag{8.46}$$

that is:

$$K(T) + A(T) < s > + B(T) < s >^2 + C(T) < s >^3 + \mathcal{O}(< s >^4) = 0$$

$$A(T) + 2B(T) < s > + 3C(T) < s >^2 + \mathcal{O}(< s >^3) > 0$$

For the isotropic phase, $< s >= 0$, to be stable for all temperatures above T_{NI}, it is necessary to assume that $K(T) = 0$. Moreover, if we only consider a fourth-order expansion in the expression of $F(< s >)$, the equilibrium conditions can be written as:

$$< s > [A(T) + B(T) < s > +C(T) < s >^2] = 0$$

$$A(T) + 2B(T) < s > +3C(T) < s >^2 > 0$$

It is necessary for $C(T) > 0$ in order that the equilibrium values are stable. We eliminated the term $< s >$ in the expansion of F; do we have grounds for eliminating the term $< s >^3$, as in the case of a "para-ferroelectric" transition or a "para-ferromagnetic" transition?

If $B(T)$ were zero, the equilibrium conditions would become:

$$< s > [A(T) + C(T) < s >^2] = 0$$
$$A(T) + 3C(T) < s >^2 > 0$$

The solutions would be: $< s >= 0$ for $A(T) > 0$ and $< s >^2 = -A(T)/C(T)$ for $A(T) > 0$. The transition for $A(T) = 0$ would be second order (with no discontinuity of $< s >$ at the transition temperature) and in the ordered phase ($< s > \neq 0$), states $< s >$ and $- < s >$ would be equally probable, which is unacceptable. In effect, the solution $< s > > 0$ corresponds to the preferred alignment of the molecules along one axis, while the solution corresponding to $< s > < 0$ corresponds to preferred arrangement of the molecules perpendicular to the axis. States $< s >$ and $- < s >$ thus represent different molecular arrangements which are not connected by any symmetry. There is thus no reason for these two types of alignment to have the same free energy. It is necessary for $B(T)$ to be nonzero and to give preference to the stability of the solution $< s > > 0$, corresponding to the preferred alignment of the molecules along one axis, and it is necessary for $B(T) < 0$.

Assume that $C(T)$ and $B(T)$ are not a function of the temperature and set: $B(T) = -B_0$, $C(T) = C_0$ with B_0 and $C_0 > 0$. In addition, take:

$$A(T) = \alpha(T - T^*) \quad \text{with} \quad \alpha > 0 \tag{8.47}$$

• For $A(T) < 0$, i.e., $T - T^* < 0$, the values of $< s >$ corresponding to equilibrium states are solutions of the equations $\frac{\partial F}{\partial < s >} = 0$ and $\frac{\partial^2 F}{\partial^2 < s >} > 0$, we then obtain:

$$< s >_1 = \frac{B_0 + \sqrt{B_0^2 - 4\alpha(T - T^*)C_0}}{2C_0} \tag{8.48}$$

and

$$< s >_3 = \frac{B_0 - \sqrt{B_0^2 - 4\alpha(T - T^*)C_0}}{2C_0} \tag{8.49}$$

Noting that $F(< s >_1) < F(< s >_3)$, we immediately find that $< s >_1$ corresponds to a stable state and $< s >_3$ corresponds to a metastable state.

• For $0 < A(T) < B_0^2/(4C_0)$, i.e., $(T - T^*) < (B_0^2/4\alpha C_0)$, the equilibrium states correspond to:

$$< s >_1 = \frac{B_0 + \sqrt{B_0^2 - 4\alpha(T - T^*)C_0}}{2C_0} \quad \text{and} \quad < s >_2 = 0 \tag{8.50}$$

- For $0 < (T - T^*) < 2B_0^2/(9\alpha C_0)$, $F(< s >_1) < F(< s >_2)$: $< s >_1$ defines a stable state while $< s >_2$ corresponds to a metastable state.
- For $2B_0^2/(9\alpha C_0) < (T - T^*) < B_0^2/(4\alpha C_0)$, $< s >_1$ corresponds to a metastable state and $< s >_2$ corresponds to a stable state.
- The equality, $F(< s >_1) = F(< s >_2)$, is obtained for temperature T_{NI},

$$T_{NI} = T^* + \frac{2B_0^2}{9\alpha C_0} \tag{8.51}$$

the temperature of coexistence of the nematic phase ($< s >_1 = 2B_0/(3C_0)$) with the isotropic phase ($< s >_2 = 0$).

Temperatures $T_I^* = T^*$ and $T_N^* = T^* + B_0^2/(4\alpha C_0)$ respectively correspond to the stability limits of the isotropic and nematic phases $(\partial^2 F/\partial^2 < s >) > 0$.

• For $A(T) > B_0^2/(4\alpha C_0)$, i.e., $(T - T^*) < B_0^2/(4\alpha C_0)$, only the isotropic phase ($< s >_2 = 0$) is stable.

The dependence of $F(s) - F_0$ for different values of $T - T^*$ is shown in Fig. 8.14. Figure 8.15 shows the change of the equilibrium values of s as a function of $T - T^*$.

In conclusion, we emphasize that the application of this model is well adapted to describing pretransition effects in the isotropic phase. In the nematic phase, where $< s >$ is greater than 0.3, one should not be limited to a fourth-order expansion of F in order to appropriately predict the behavior of $< s >$ in this phase.

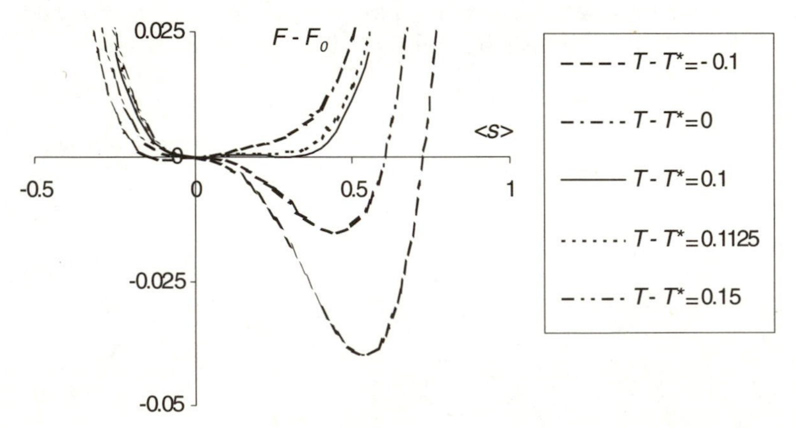

Fig. 8.14. Variation of $F(s) - F_0$ with s for different temperatures $T - T^*$.

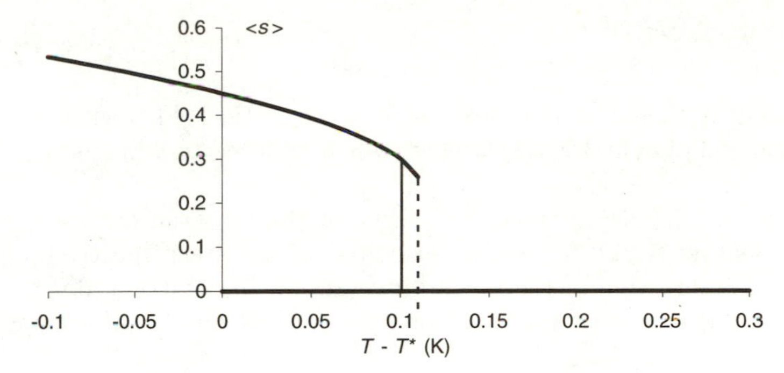

Fig. 8.15. Equilibrium values of s as a function of $T - T^*$.

c) Magnetic Birefringence in the Isotropic Phase

The Landau–de Gennes model, with an external field, allows writing the free energy per mole in the isotropic phase as:

$$F - F_0 = \frac{1}{2}\alpha(T - T^*) < s >^2 + \frac{1}{3}B_0 < s >^3 + \frac{1}{4}C_0 < s >^4 + N_0 W(< s >) \tag{8.52}$$

where $W(< s >)$ is the average potential orientation energy per molecule for the molecules exposed to the external field and N_0 is Avogadro's number.

If an external magnetic field is applied along the Oz-axis, we saw in (8.12) that the magnetic susceptibility of the nematic medium will be a function of the order parameter $< s >$:

$$\chi_{mz} = n(\overline{\eta} + \frac{2}{3}\eta_a < s >) \tag{8.53}$$

The average magnetic interaction energy per unit of volume is then written:

$$w(< s >) = -\frac{1}{2}\mu_0(1 + \chi_{mz})H^2 = -\frac{1}{3}\mu_0 n\eta_a < s > H^2 + K \tag{8.54}$$

where K is constant if the magnetic field is stationary.

Since the order induced by the magnetic field, in this initially isotropic material is low (typically: $< s > \approx 10^{-5}$), we can write the expression for the free energy F by neglecting the terms $< s >^3, < s >^4, \ldots$, so that:

$$F - F_0 = \frac{1}{2}\alpha(T - T^*) < s >^2 - \frac{1}{3}N_0\mu_0\eta_a < s > H^2 \tag{8.55}$$

At equilibrium, $\frac{\partial F}{\partial < s >} = 0$, that is:

$$\alpha(T - T^*) < s > - \frac{1}{3}N_0\mu_0\eta_a \; H^2 = 0 \tag{8.56}$$

from which:

$$< s >= \frac{1}{3} \frac{N_0 \mu_0 \eta_a}{\alpha(T - T^*)} H^2 \tag{8.57}$$

Such a material, isotropic in the absence of a magnetic field, becomes weakly anisotropic and of cylindrical symmetry when exposed to a magnetic field $\boldsymbol{H}$.

The optical index and dielectric susceptibility of the material are then characterized by tensors $\overline{\overline{\boldsymbol{n}}}$ and $\overline{\overline{\chi_e}}$, whose principal values are respectively n_1 and χ_{e1} along the direction of the magnetic field, and n_2 and χ_{e2} along a perpendicular axis. Here n_i and χ_{ei} ($i = 1$ or 2) obey the Clausius–Mosotti relation:

$$\frac{n_i^2 - 1}{n_i^2 + 2} = \frac{1}{3}\chi_{ei} = \frac{1}{3}\frac{N_0}{V_M}\alpha_i \tag{8.58}$$

where α_i is the average polarizability of the molecules along direction i and V_M is the molar volume. We then obtain:

$$\frac{3(n_1^2 - n_2^2)}{(n_1^2 + 2)(n_2^2 + 2)} = \frac{1}{3}\frac{N_0}{V_M}(\alpha_1 - \alpha_2)$$

that is, by setting: $n = (n_1 + n_2)/2$ and $\Delta n = n_1 - n_2$ and for weak anisotropy index:

$$\frac{\Delta n}{n} = \frac{1}{2}\frac{(\alpha_1 - \alpha_2)(n^2 + 2)^2}{9}\frac{N_0}{n^2 V_M} \tag{8.59}$$

Now $(\alpha_1 - \alpha_2)$ is correlated with the anisotropy of the molecule α_a and with the order parameter $< s >$ by the relation $\alpha_1 - \alpha_2 = \alpha_a < s >$, hence:

$$\frac{\Delta n}{n} = \frac{(n^2 + 2)^2}{n^2}\frac{\alpha_a\ N_0^2 \mu_0 \eta_a}{54 V_M \alpha(T - T^*)} H^2 \tag{8.60}$$

With a constant magnetic field, Δn should thus vary as $(T - T^*)^{-1}$; this property has been verified experimentally.

d) Static Distortions in a Nematic Single Crystal

In the ideal case of a nematic single crystal, all molecules will tend to align in the same direction defined by directional vector $\pm\boldsymbol{n}$. The medium is uniaxial. In practice, the situation is more complicated, since this ideal configuration is generally incompatible with the stresses exerted on the interfaces constraining the nematic medium on one hand, and with the applied external fields that act on the molecules on the other hand (magnetic field, for example). In this situation, deformation of the alignment is observed and the direction vector is no longer constant but varies continuously within the material; in this case, the directional vector will be defined by $\pm\boldsymbol{n}$, where $\boldsymbol{r}(x, y, z)$ defines a point in the material.

In most situations, $a\nabla\boldsymbol{n} \ll 1$, where a is the length of the molecule. As the change of $\boldsymbol{n}(\boldsymbol{r})$ is imperceptible on the molecular scale, it is assumed that the degree of alignment of the molecules along $\boldsymbol{n}(\boldsymbol{r})$ is not altered by the deformation and the linear deformation regime persists.

The three principal deformations of the directional vector capable of appearing in a nematic are shown in Fig. 8.16; they successively correspond to: fan-shaped opening; curvature and twisting.

In treating the general case, J. L. Eriksen (1966) showed that the free deformation energy F_d (called the Frank free energy) can be written as:

$$F_d(\boldsymbol{r}) = \tfrac{1}{2}K_1(\operatorname{div}\boldsymbol{n}(\boldsymbol{r}))^2 + \tfrac{1}{2}K_2(\boldsymbol{n}(\boldsymbol{r}).\operatorname{rot}\boldsymbol{n}(\boldsymbol{r}))^2 + \frac{1}{2}K_3(\boldsymbol{n}(\boldsymbol{r}) \times \operatorname{rot}\boldsymbol{n}(\boldsymbol{r}))^2 \tag{8.61}$$

where the three terms respectively correspond to the three basic types of deformation represented by:

- $1/2\,K_1(\operatorname{div}\boldsymbol{n})^2$ for fan-shaped opening of $\boldsymbol{n}$(splay);
- $1/2\,K_3(\boldsymbol{n} \times \operatorname{rot}\boldsymbol{n})^2$ for bending of $\boldsymbol{n}$.
- $1/2\,K_2(\boldsymbol{n}.\operatorname{rot}\boldsymbol{n})^2$, for twisting of $\boldsymbol{n}$;

The three elastic constants K_i $(i = 1, 2, 3)$ must be positive for the minimum of F_d to correspond to an undeformed nematic (div $\boldsymbol{n} = 0$ and rot $\boldsymbol{n} = 0$). If F_d is expressed in Joule per m^3, the constants, K_i, are expressed in Joule per meter, that is, in Newton. The order of magnitude of these constants can be obtained with purely dimensional arguments. In effect, we could predict that $K_i = U/a$ where U is the interaction energy between neighboring molecules and a is the characteristic length of the rigid part of

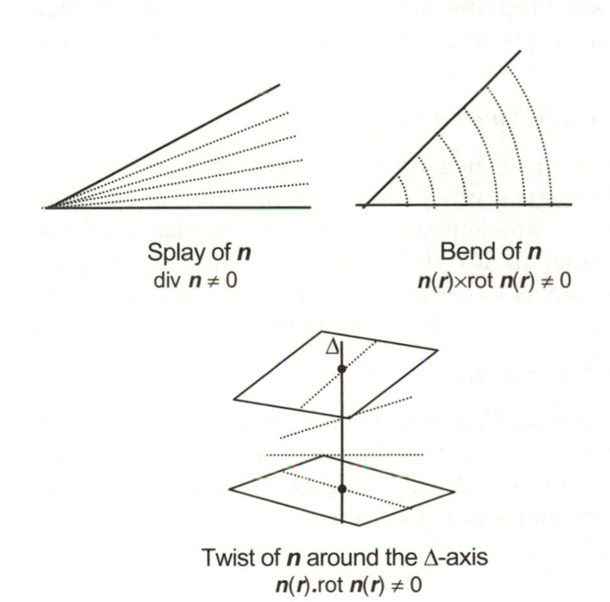

Fig. 8.16. The three principal types of deformation in a nematic monodomain.

the molecules. As the interaction energy is of the order of 2 kcal per mole, U is typically of the order of 1.4×10^{-20} J per molecule with $a \approx 2$ nm. The values obtained for K_i hence are typically of the order of 0.7×10^{-11} J/m (or N). This result is in good agreement with the values measured in nematics; for para-azoxyanisole (PAA) at 120°C, the measured elastic constants are: $K_1 = 0.7 \times 10^{-11}$N; $K_2 = 0.43 \times 10^{-11}$ N; $K_3 = 1.7 \times 10^{-11}$ N.

In general the elastic constants decrease when the temperature increases.

In the immediate vicinity of an interface, nematic molecules tend to be oriented along a preferred direction which is a function of the nature of this interface. In most cases, the surface forces are very strong and impose a well-defined orientation of directional vector $\boldsymbol{n}$ to the surface; we say that the molecules are firmly "anchored."

◇ *Orientation on an Amorphous Solid Wall*

In the vicinity of an untreated amorphous wall, the preferred orientation direction is generally parallel to the wall; all directions parallel to the wall must be considered equivalent. On the other hand, if this amorphous wall has been finely scratched in a particular direction by rubbing with paper, orientation will be along the direction of rubbing: the fine scratches produced by rubbing are sufficient to impose the preferred orientation direction of the molecules on the wall. This method of anchoring the molecules to the wall was used by Chatelain in 1948. Finally, if a small amount of certain detergents capable of fixing their polar head to the amorphous wall is incorporated in the nematic phase, they impose an orientation on the molecules perpendicular to the surface. It is easy to obtain a nematic single crystal between two parallel walls treated in this way with a preferred axis of orientation (optical axis) perpendicular to the two walls.

◇ *Orientation on the Surface of a Single Crystal*

When the surface in contact with the nematic corresponds to a well-defined crystallographic plane of the single crystal, a series of preferred orientation directions parallel to the surface are generally found. This is what happens when *p*-azoxyanisole is deposited on face (001) of sodium chloride, two preferred orientation directions parallel to the plane of the surface appear; these directions correspond to crystallographic directions (110) and $(1\bar{1}0)$.

◇ *Orientation at the Air–Nematic interface*

In general, the preferred orientation direction of the molecules form an angle ψ with the free surface so that the orientation directions are on a cone whose axis is perpendicular to the interface and whose angle is $\pi - 2\psi$. The angle ψ is a function of the temperature and is probably very sensitive to the presence of contaminants.

e) Effects of a Strong Magnetic Field in the Nematic Phase

Most organic molecules are diamagnetic and their diamagnetism is particularly strong when the molecules have aromatic rings. Consider an aromatic ring exposed to a magnetic field $\boldsymbol{H}$; when its plane is perpendicular to the magnetic field, the current that appears in the ring tends to reduce the flux passing through it and the field lines in the vicinity of the ring are strongly perturbed: they tend to deform the ring and the field-ring interaction energy is maximum. On the contrary, if the magnetic field is in the plane of the ring, no current is induced and the field lines are only very weakly perturbed; in this case, the interaction energy is minimum: an aromatic ring tends to be oriented so that $\boldsymbol{H}$ is in the plane of the ring.

Molecules that form nematic phases such as MBBA or PAA contain two aromatic rings in their rigid part, and their axes are approximately perpendicular to the axis of the rigid part of the molecules (Fig. 8.1). If a magnetic field $\boldsymbol{H}$ is applied along $\boldsymbol{n}$, this field will be in the plane of the rings; the interaction energy of the rings with the field will be minimal and the configuration will be stable. On the contrary, if the magnetic field $\boldsymbol{H}$ is applied perpendicular to $\boldsymbol{n}$, many of the molecules will have the axis of their rings with an orientation close to the direction of the field and the overall interaction energy will be high; this configuration is not stable, and $\boldsymbol{n}$ will be reoriented in the direction of $\boldsymbol{H}$.

In practice, the nematic phase is limited by the walls, which impose constraining conditions incompatible with complete reorientation of $\boldsymbol{n}$ in the entire nematic medium. There will be competition between walls-effects which cause anchoring of the molecules in a certain direction, $\boldsymbol{n}_s$ and the magnetic field effect due to $\boldsymbol{H}$ which favors the alignment of $\boldsymbol{n}$ along $\boldsymbol{H}$. Hence deformed configurations will be obtained.

To analyze these effects more precisely and to see their advantage in measuring the elastic constants of a nematic, a quantitative analysis must be performed.

When a magnetic field $\boldsymbol{H}$ is applied to a deformed nematic, characterized by vector $\boldsymbol{n}(\boldsymbol{r})$, the magnetization induced in $\boldsymbol{n}$, $\boldsymbol{M}(\boldsymbol{r})$, is written:

$$\boldsymbol{M}(\boldsymbol{r}) = \chi_\perp \boldsymbol{H} + (\chi_\parallel - \chi_\perp)[\boldsymbol{H}.\boldsymbol{n}(\boldsymbol{r})]\boldsymbol{n}(\boldsymbol{r}) \tag{8.62}$$

where the principal susceptibilities, $\chi_\parallel$ and $\chi_\perp$, are both negative (diamagnetism), with $\chi_a = \chi_\parallel - \chi_\perp > 0$ in most nematics.

The moment per unit of volume acting on the magnetization is equal to:

$$\Gamma_M(\boldsymbol{r}) = \mu_0 \boldsymbol{M}(\boldsymbol{r}) \times \boldsymbol{H} = \mu_0 \chi_a [\boldsymbol{H}.\boldsymbol{n}(\boldsymbol{r})]\boldsymbol{n}(\boldsymbol{r}) \times \boldsymbol{H} \tag{8.63}$$

with $\mu_0 = 4\pi\, 10^{-7}$ H/m.

The free energy per unit of volume, $F(\boldsymbol{r})$, is then written:

$$F(\boldsymbol{r}) = F_d(\boldsymbol{r}) + F_M(\boldsymbol{r}) \tag{8.64}$$

with:

$$F_M(\boldsymbol{r}) = -\mu_0 \int_0^H \boldsymbol{M}(\boldsymbol{r})\mathrm{d}\boldsymbol{H} = -\frac{1}{2}\mu_0\chi_\perp H^2 - \frac{1}{2}\mu_0\chi_a[\boldsymbol{n}(\boldsymbol{r}).\boldsymbol{H}]^2 \quad (8.65)$$

Only the last term is a function of $\boldsymbol{n}(\boldsymbol{r})$; it is minimal when $\boldsymbol{n}(\boldsymbol{r})$ is parallel to $\boldsymbol{H}$.

Two equivalent methods can be used to establish the equilibrium conditions in the presence of an applied magnetic field:

- either the free energy $f = \int_V F(\boldsymbol{r})\mathrm{d}\boldsymbol{r}$ is minimized
- or we write that in any elementary volume of the nematic phase, the moment of magnetic origin acting in the volume is counterbalanced by the moments exercised on its surface by neighboring regions.

Consider (Fig. 8.17) a nematic phase bounded by a flat wall (plane Oxz), where the nematic fills the space corresponding to $y > 0$. The wall was treated to impose anchoring of the molecules along axis $\pm\mathbf{O}\boldsymbol{z}$. In the absence of other stresses, the directional vector $\boldsymbol{n}$ is at equilibrium directed along $\pm\mathbf{O}\boldsymbol{z}$. At any point in the nematic, we will thus have: $n_x = 0$, $n_y = 0$, $n_z = 1$. Now assume that a magnetic field $\boldsymbol{H}$ is applied along $\mathbf{O}\boldsymbol{x}$ axis. On the one hand, far enough away from the wall, the directional vector $\boldsymbol{n}$ will be oriented along $\boldsymbol{H}$, that is, along $\mathbf{O}\boldsymbol{x}$; on the other hand, near the wall, the directional vector will form an angle θ, a function of distance y to the wall, since the wall imposes stress.

The moment created per unit of volume will be, according to (8.63):

$$\Gamma_M(y) = \mu_0\chi_a[\boldsymbol{H}.\boldsymbol{n}(y)]\boldsymbol{n}(y) \times \boldsymbol{H} = \mu_0\chi_a H^2 \sin[\theta(y)]\cos[\theta(y)] \quad (8.66)$$

If we consider a thin layer $[y; y + \mathrm{d}y]$ of surface unit, the moment of magnetic origin acting on this volume will be $\mu_0\chi_a H^2 \sin[\theta(y)]\mathrm{d}y$, while the moments of mechanical origin acting on surfaces y and $y+\mathrm{d}y$ will respectively be $-K_2(\mathrm{d}\theta/\mathrm{d}y)$ in y and $K_2(\mathrm{d}\theta/\mathrm{d}y)$ in $y + \mathrm{d}y$; that is, globally:

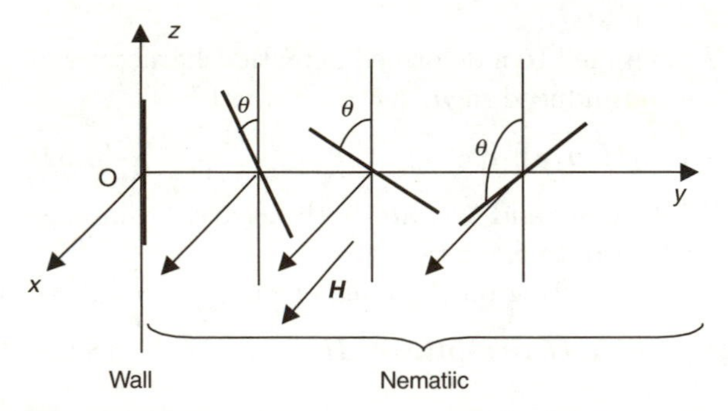

Fig. 8.17. Effect of a strong magnetic field on the nematic phase. Competition between the effect of anchoring the molecules to the wall in direction Oz and the effect of the magnetic field $\boldsymbol{H}$ which tends to align the molecules of the nematic phase in direction Ox.

$$K_2 \frac{\mathrm{d}\theta}{\mathrm{d}y}\bigg|_{y+\mathrm{d}y} - K_2 \frac{\mathrm{d}\theta}{\mathrm{d}y}\bigg|_{y} = K_2 \frac{\mathrm{d}^2\theta}{\mathrm{d}y^2}\mathrm{d}y \tag{8.67}$$

Equilibrium is realized when the sum of the moments is zero, that is:

$$K_2 \frac{\mathrm{d}^2\theta}{\mathrm{d}y^2} + \mu_0 \chi_a H^2 \sin[\theta(y)] \cos[\theta(y)] = 0 \tag{8.68}$$

Let us introduce a length $\xi_2(H)$, defined by:

$$\xi_2(H) = \left(\frac{K_2}{\mu_o \chi_a}\right)^{1/2} \frac{1}{H} \tag{8.69}$$

Equation (8.68) becomes:

$$\xi_2^2(H) \frac{\mathrm{d}^2\theta}{\mathrm{d}y^2} + \sin[\theta(y)] \cos[\theta(y)] = 0 \tag{8.70}$$

that is, by integration:

$$\xi_2^2 \left(\frac{\mathrm{d}\theta}{\mathrm{d}y}\right)^2 = \cos^2\theta + C^{te} \tag{8.71}$$

Far from the wall ($y \to \infty$), we predict that θ will be equal to $\pi/2$ and $(\mathrm{d}\theta/\mathrm{d}y)$ will be equal to 0; the integration constant is thus zero so that:

$$\xi_2 \frac{\mathrm{d}\theta}{\mathrm{d}y} = +\cos\theta \quad \text{or} \quad \xi_2 \frac{\mathrm{d}\theta}{\mathrm{d}y} = -\cos\theta$$

These two equations correspond to opposite twisting in the vicinity of the wall; these two solutions are allowed and equally probable. Selecting the first case and setting $u = \pi/2 - \theta$, the equation is written $\mathrm{d}y = -\xi_2 \mathrm{d}u/\sin u$.

Setting $t = \tan(u/2)$, we obtain $\mathrm{d}y = -\xi_2 \mathrm{d}t/t$, that is by integration:

$$\log t = -\frac{y}{\xi_2} + C^{te} \tag{8.72}$$

Selecting the constant so that $\theta = 0$ for $y = 0$ (which implies that $t = \tan(\pi/4) = 1$ for $y = 0$), we finally obtain:

$$\tan\left(\frac{\pi/2 - \theta}{2}\right) = \exp\left(-\frac{y}{\xi_2}\right) \tag{8.73}$$

This result shows that ξ_2 determines the thickness of the transition zone.

For magnetic induction of 1 T in a vacuum $\boldsymbol{B}_0$ corresponding to a magnetic field $\boldsymbol{H}$ of approximately 8×10^5 A/m, with typical values of K_2 and χ_a ($K_2 = 10^{-11} N$ and $\chi_a = 1.25 \times 10^{-6}$), one finds $\xi_2 = 3 \times 10^{-6}$ m $= 3\mu$.

Other configurations resulting from the competition between the wall-anchoring effect and the applied field effect are naturally possible. In general, in this type of interaction, the alignment of the molecules from the wall obeys an exponential law; the thickness ξ of the transition zone is always a weighted average of the three fundamental coherence lengths, $\xi_i(H) = (K_i/\mu_0 \chi_a)^{1/2}(1/H)$, $(i = 1, 2, 3)$, where the "weight" of each characteristic length varies from one configuration to another.

◇ *Frederiks Transition*

Now consider a thin layer of a nematic bordered by two solid walls. The thickness d of the layer is of the order of 20 μ. It is assumed that there is ideal anchoring of the molecules in direction $\boldsymbol{n}_0$ to the surface. At any point of the layer delimited by two solid walls, the directional vector is initially uniform: $\boldsymbol{n}(\boldsymbol{r}) = \boldsymbol{n}_0$. A magnetic field $\boldsymbol{H}$ is then applied perpendicular to $\boldsymbol{n}_0$. The three fundamental cases are shown on Fig. 8.18.

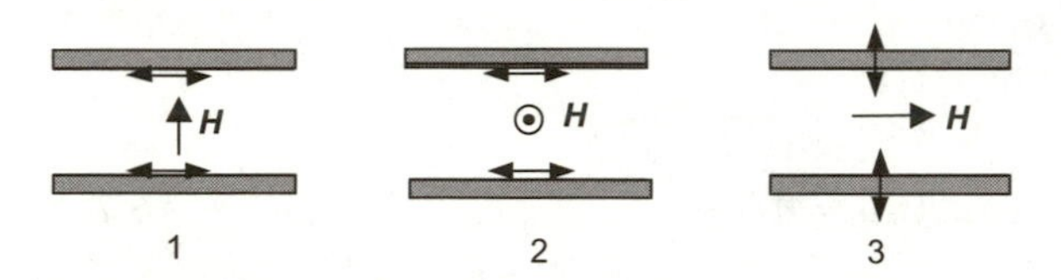

Fig. 8.18. Frederiks transitions in three principal configurations. In the three cases, the magnetic field $\boldsymbol{H}$ is applied perpendicular to the direction of anchoring to the walls. For a weak magnetic field, the molecules in the layer retain a preferred orientation parallel to the direction of anchoring to the walls. However, when the field $\boldsymbol{H}$ exceeds a critical field $\boldsymbol{H}_0$, the molecules in the central zone of the layer rotate and tend to be oriented along $\boldsymbol{H}$.

In the layer, the magnetic moment Γ_M per unit of volume is given by (8.63). In the initial configuration, which corresponds at all points to $\boldsymbol{H}.\boldsymbol{n}(\boldsymbol{r}) = 0$, we can state that local equilibrium is realized. However, for the moment we do not know the conditions in which this equilibrium is stable or unstable. To answer this question, we must study how susceptible weak perturbation of this equilibrium state is to being amplified or cancelled out.

First consider the case corresponding to scheme 2 in Fig. 8.18. The molecules are parallelly anchored to the walls in direction $\mathbf{O}\boldsymbol{x}$, while the magnetic field $\boldsymbol{H}$ parallel to the walls is applied along axis $\mathbf{O}\boldsymbol{y}$ perpendicular to $\mathbf{O}\boldsymbol{x}$. We will designate the axis perpendicular to the walls by $\mathbf{O}\boldsymbol{z}$.

In the absence of a magnetic field, the directional vector is oriented along $\pm\mathbf{O}\boldsymbol{x}$ at any point in the sample and this configuration is stable.

To study the stability of the directional vector in the presence of the magnetic field $\boldsymbol{H}$, consider a slight deviation of the directional vector in the direction of the field:

$$\boldsymbol{H}(\boldsymbol{r}) = \boldsymbol{n}_0 + \delta\boldsymbol{n}(\boldsymbol{r}) \tag{8.74}$$

with $|\delta\boldsymbol{n}(\boldsymbol{r})| \ll 1$ and $|\boldsymbol{n}(\boldsymbol{r})| = |\boldsymbol{n}_0| = 1$. The moment per unit of volume is then written:

$$\Gamma_M = \mu_0\chi_a[\boldsymbol{H}.\boldsymbol{n}(\boldsymbol{r})]\boldsymbol{n}(\boldsymbol{r}) \times \boldsymbol{H} = \mu_0\chi_a[\boldsymbol{H}.\delta\boldsymbol{n}(\boldsymbol{r})]\boldsymbol{n}_0 \times \boldsymbol{H} \tag{8.75}$$

In effect, for a slight deviation of the directional vector toward the direction of the field, $\delta\boldsymbol{n}(\boldsymbol{r})$, directed along $\boldsymbol{H}$, is almost perpendicular to $\boldsymbol{n}_0$ and the moment is proportional to the deviation $\delta\boldsymbol{n}(\boldsymbol{r})$. Moreover, it is reasonable to assume that the distortion is only a function of z.

The distortion energy F_d and the magnetic energy F_M can be written as

$$F_d = \frac{1}{2}K_2(\boldsymbol{n}.\mathrm{rot}\ \boldsymbol{n})^2 = \frac{1}{2}K_2\Big(\frac{\mathrm{d}\delta n}{\mathrm{d}z}\Big)^2 \tag{8.76}$$

$$F_M = -\frac{1}{2}\mu_0\chi_\perp H^2 - \frac{1}{2}\mu_0\chi_a[\boldsymbol{n}(\boldsymbol{r}).\boldsymbol{H}]^2 = C^{te} - \frac{1}{2}\mu_0\chi_a\delta n(z)^2H^2 \tag{8.77}$$

Assuming ideal anchoring of the molecules on the walls, $\delta n(z)$ should be zero on the walls located at $z = 0$ and $z = d$; it is thus necessary for $\delta n(0) = \delta n(d) = 0$. Performing a Fourier expansion on $\delta n(z)$, we obtain:

$$\delta n(z) = \sum_{k=1}^{\infty} \delta n_k \sin kz \tag{8.78}$$

with $k = n\pi/d$ and n being a whole number. $\mathcal{F}$ is the free energy per unit of area of the layer is equal to:

$$\mathcal{F} = \int_0^d (F_M + F_d)\mathrm{d}z = \frac{d}{4}\sum_{k=1}^{\infty} \delta n_k^2(K_2k^2 - \mu_0\chi_a H^2) \tag{8.79}$$

The unperturbed state corresponding to $\boldsymbol{n}(\boldsymbol{r}) = \boldsymbol{n}_0$ will be stable if $\mathcal{F}$ increases regardless of the changes in δn_k^2; whatever k may be, it is then necessary that:

$$K_2k^2 > \mu_0\chi_a\ H^2 \tag{8.80}$$

The smallest value of k is π/d; it corresponds to distortion whose half-wavelength is equal to d. The value of the magnetic field H_{C2} after which distortion appears is equal to:

$$H_{C2} = \frac{\pi}{d}\left(\frac{K_2}{\mu_0\chi_a}\right)^{1/2} \tag{8.81}$$

The transition from a state where $\boldsymbol{n}$ is spatially uniform to a situation where this vector is aligned with the field is called the **Frederiks transition.** For each of the three configurations shown in Fig. 8.18 (case of $i = 1, 2$, and 3), the threshold fields are respectively:

$$H_{Ci} = \frac{\pi}{d}\left(\frac{K_i}{\mu_0\chi_a}\right)^{1/2}, \quad i = 1, 2 \text{ and } 3 \tag{8.82}$$

In the nematic phase, $K_i \approx 10^{-11}$ N and $\chi_a \approx 1.25 \cdot 10^{-6}$. For $d = 20\ \mu\mathrm{m}$, the critical magnetic induction $B_{Ci} = \mu_0$, H_{Ci} is of the order of 0.5 T.

f) Effects of an Electric Field in the Nematic Phase – Frederiks Transitions

We will limit our analysis to ideally isolating nematics. In this case, the effect of an electric field $\boldsymbol{E}$ is similar to the effect of a magnetic field.

When an electric field $\boldsymbol{E}$ is applied to a deformed nematic characterized by directional vector at $\boldsymbol{n}(\boldsymbol{r})$, the polarization in $\boldsymbol{n}$, designated $\boldsymbol{P}_e(\boldsymbol{r})$, is written:

$$\boldsymbol{P}_e(\boldsymbol{r}) = \varepsilon_0[\chi_{e\perp}\boldsymbol{E} + (\chi_{e\parallel} - \chi_{e\perp})[\boldsymbol{E}.\boldsymbol{n}(\boldsymbol{r})]\boldsymbol{n}(\boldsymbol{r})] \tag{8.83}$$

where the principal dielectric susceptibilities, $\chi_{e\parallel}$ and $\chi_{e\perp}$ are both positive (paraelectricity); the difference $\chi_a = \chi_{e\parallel} - \chi_{e\perp}$ can be positive or negative as a function of the chemical structure of the molecules considered, and the dielectric constant of the vacuum ε_0 is equal to $1/36\pi \times 10^9$.

Taking into account the relation between the displacement vector $\boldsymbol{D}$, field $\boldsymbol{E}$, and induced polarization $\boldsymbol{P}_e$, $\boldsymbol{D} = \varepsilon_0\boldsymbol{E} + \boldsymbol{P}_e$, we have:

$$\boldsymbol{D}(\boldsymbol{r}) = \varepsilon_0[\boldsymbol{E} + \chi_{e\perp}\boldsymbol{E} + (\chi_{e\parallel} - \chi_{e\perp})[\boldsymbol{E}.\boldsymbol{n}(\boldsymbol{r})]\boldsymbol{n}(\boldsymbol{r})] \tag{8.84}$$

Introducing the principal relative dielectric constants

$$\varepsilon_\perp = 1 + \chi_{e\perp}; \quad \varepsilon_{e\parallel} = 1 + \chi_{e\parallel} \quad \text{and} \quad \varepsilon_a = \varepsilon_\parallel - \varepsilon_\perp = \chi_{e\parallel} - \chi_{e\perp}$$

we find:

$$\boldsymbol{D}(\boldsymbol{r}) = \varepsilon_0[\varepsilon_\perp\boldsymbol{E} + (\varepsilon_\parallel - \varepsilon_\perp)[\boldsymbol{E}.\boldsymbol{n}(\boldsymbol{r})]\boldsymbol{n}(\boldsymbol{r})]$$

The moment acting on the induced polarization per unit of volume is equal to:

$$\Gamma_E(\boldsymbol{r}) = \boldsymbol{P}(\boldsymbol{r}) \times \boldsymbol{E} = \varepsilon_0\varepsilon_a[\boldsymbol{E}.\boldsymbol{n}(\boldsymbol{r})]\boldsymbol{n}(\boldsymbol{r}) \times \boldsymbol{E} \tag{8.85}$$

The free energy per unit of volume, $F(\boldsymbol{r})$, is written:

$$F(\boldsymbol{r}) = F_d(\boldsymbol{r}) + F_E(\boldsymbol{r}) \tag{8.86}$$

with:

$$F_E(\boldsymbol{r}) = -\int_0^E \boldsymbol{D}(\boldsymbol{r})\mathrm{d}\boldsymbol{E} = -\frac{1}{2}\varepsilon_0\varepsilon_\perp E^2 - \frac{1}{2}\varepsilon_0\varepsilon_a[\boldsymbol{n}(\boldsymbol{r}).\boldsymbol{E}]^2 \tag{8.87}$$

The minimum of $F_E(\boldsymbol{r})$ corresponds to the maximum of $\varepsilon_a[\boldsymbol{n}(\boldsymbol{r}).\boldsymbol{E}]^2$.

In the nematic phase, if the molecules have a permanent dipole moment approximately parallel to their axis, application of an electric field $\boldsymbol{E}$ along the preferred orientation direction, $\boldsymbol{n}$, this will lead to a preferred orientation of the dipoles along $\boldsymbol{n}$ with no appreciable change in the order parameter. Alternatively, an electric field $\boldsymbol{E}$ perpendicular to $\boldsymbol{n}$ will have little effect on the preferred orientation of the dipoles. When the molecules of a nematic have an axial dipole moment, $\chi_{e\parallel} - \chi_{e\perp}$ should be positive (this is the case of molecules having a –C≡N group attached to their rigid part). If the molecules have a permanent dipole moment approximately perpendicular to their axis, the situation is reversed and $\chi_{e\parallel} - \chi_{e\perp}$ is negative (this is the case of PAA).

By applying a strong enough electric field $\boldsymbol{E}$, it is possible, in principle, to alter the direction of the alignment of a nematic. Equilibrium conditions can be obtained in the presence of an electric field either by minimizing the free energy of the system or by writing that in any elementary volume of the nematic phase, the electric moment acting in the volume, Γ_E, is balanced by the moments exercised on its surface by neighboring regions.

Consider once more a thin layer of nematic with a thickness d of 20 μ limited by two solid walls. We will discuss a special case where:

- on the surface of each wall, there is an ideal anchoring of the molecules in the direction $\boldsymbol{n}_0$ parallel to the wall;
- at any point of the layer, the directional vector is initially uniform: $\boldsymbol{n}(\boldsymbol{r}) = \boldsymbol{n}_0$; we will designate the axis perpendicular to the walls by $\mathbf{O}\boldsymbol{z}$ and the axis parallel to $\boldsymbol{n}_0$ by $\mathbf{O}\boldsymbol{x}$;
- The difference $\chi_{e\|} - \chi_{e\perp}$ is positive.

If an electric field $\boldsymbol{E}$ is applied perpendicular to the wall and thus to $\boldsymbol{n}_0$, the electric moment Γ_E per unit of volume is equal to:

$$\Gamma_E(\boldsymbol{r}) = \varepsilon_0\varepsilon_a[\boldsymbol{E}.\boldsymbol{n}(\boldsymbol{r})]\boldsymbol{n}(\boldsymbol{r}) \times \boldsymbol{E} \tag{8.88}$$

In the initial configuration corresponding at all points to $\boldsymbol{E}.\boldsymbol{n}(\boldsymbol{r}) = 0$, local mechanical equilibrium is realized. Let us investigate the stability of this equilibrium. In the absence of an electric field, the directional vector is oriented along $\pm\mathbf{O}\boldsymbol{x}$ at any point of the sample: $\boldsymbol{n}(\boldsymbol{r}) = \boldsymbol{n}_0$, and this configuration is stable.

In the presence of the electric field $\boldsymbol{E}$, consider a small deviation of the directional vector in the direction of the field:

$$\boldsymbol{n}(\boldsymbol{r}) = \boldsymbol{n}_0 + \delta\boldsymbol{n}(\boldsymbol{r}) \tag{8.89}$$

with $|\delta\boldsymbol{n}(\boldsymbol{r})| \ll 1$ and $|\boldsymbol{n}(\boldsymbol{r})| = |\boldsymbol{n}_0| = 1$. The moment per unit of volume is then written:

$$\Gamma_E(\boldsymbol{r}) \approx \varepsilon_0\varepsilon_a[\boldsymbol{E}.\delta\boldsymbol{n}(\boldsymbol{r})]\ \boldsymbol{n}_0 \times \boldsymbol{E} \tag{8.90}$$

Moreover, it is reasonable to assume that the distortion will be a function of z only. Per unit of volume, the distortion energy F_d and the electric energy F_M are then written:

$$F_d = \frac{1}{2}K_1(\text{div}\ \boldsymbol{n})^2 = \frac{1}{2}K_1\left(\frac{\mathrm{d}\delta n}{\mathrm{d}z}\right)^2 \tag{8.91}$$

$$F_E(\boldsymbol{r}) = -\frac{1}{2}\varepsilon_0\varepsilon_\perp E^2 - \frac{1}{2}\varepsilon_0\varepsilon_a[\boldsymbol{n}(\boldsymbol{r}).\boldsymbol{E}]^2 = C^{te} - \frac{1}{2}\varepsilon_0\varepsilon_a\delta n(z)^2E^2 \tag{8.92}$$

Assuming ideal anchoring of the molecules on the walls, $\delta n(0) = \delta n(d) = 0$. By analogy with the situation for the presence of a magnetic field, performing Fourier expansion of $\delta n(z)$, we obtain:

$$\mathcal{F} = \int_0^d (F_E + F_d)\mathrm{d}z = \frac{d}{4}\sum_{k=1}^{\infty} \delta n_k^2 (K_1 k^2 - \varepsilon_0 \varepsilon_a E^2) \tag{8.93}$$

The unperturbed state corresponding to $\boldsymbol{n}(\boldsymbol{r}) = \boldsymbol{n}_0$ will be stable if for all changes in δn_k^2:

$$K_1 k^2 > \varepsilon_0 \varepsilon_a E^2 \tag{8.94}$$

The smallest value of k is π/d; it corresponds to distortion whose half-wavelength is equal to d. The value of the electric field E_{C1} from which distortion appears is equal to:

$$E_{C1} = \frac{\pi}{d}\left(\frac{K_1}{\varepsilon_0 \varepsilon_a}\right)^{1/2} \tag{8.95}$$

This is a Frederiks transition.

In the nematic phase, $K_i \approx 10^{-11}$ N and with $\varepsilon_a \approx 5$. For $d = 20\,\mu$, the critical electric field E_C is of the order of 50 kV/m, corresponding to voltage of 1 V between walls, $V_C = E_C d$. The relatively low value of the voltages that must be applied to induce rotation of the molecules makes possible the simple applications of this transition on which liquid-crystal displays are based.

More complicated effects appear when the nematic is sufficiently conducting to be a charge transfer site. In this case, when the field is applied perpendicular to the initial $\boldsymbol{n}_0$ with plane electrodes, periodic deviations of the directional vector appear with respect to the initial direction; these deformations are associated with periodic distribution of positive and negative charges moving in the direction of the field; this charge segregation is at the origin of the appearance of spontaneous dielectric polarization perpendicular to the field. Finally, when the difference in potential between the electrodes becomes sufficient, this periodic distribution becomes unstable and gives rise to convective instability.

g) Nematic–Smectic A Transition

The nematic phase is characterized by the existence of preferred orientation of the molecules in a direction designated $\mathbf{O}\boldsymbol{z}$. In this phase, the centers of gravity of the molecules are randomly distributed; if we consider a two-dimensional layer of thickness δe comparable to the size of the molecules, the average density of the molecules in this layer will be independent of the choice of layer.

The nematic–smectic A transition corresponds to the appearance of density modulation ρ along axis $\mathbf{O}\boldsymbol{z}$, which corresponds to the appearance of layers. This modulation can be represented as follows:

$$\rho(z) = \rho_0 + \rho_1 \cos(q_s z - \phi_1) + \rho_2 \cos(2q_s z - \phi_2) + \dots \tag{8.96}$$

where $\rho_1 \cos(q_z - \phi_1)$ corresponds to the first harmonic of the density modulation. ϕ_1 is an arbitrary phase that is a function of the reference system selected.

It is now believed that this transition takes place continuously; in other words, the density modulation appears without discontinuity (we note that in the nematic phase, all $\rho_i (i = 1, 2, 3, \ldots)$ are zero).

In the smectic A phase in the vicinity of the nematic–smectic A transition, it is reasonable to assume that the first harmonic dominates; in other words:

$$\rho(z) = \rho_0 + \rho_1 \cos(q_s z - \phi_1) \tag{8.97}$$

ρ_1 is then a parameter characteristic of smectic order.

In the vicinity of the S_A–N transition, the free energy per unit of volume, F_S, corresponding to the appearance of smectic order can be expanded in powers of ρ_1 (Landau model):

$$F_S(\rho_1) = \frac{1}{2} r \rho_1^2 + \frac{1}{4} u_0 \rho_1^4 + \frac{1}{6} u_1 \rho_1^6 + \mathcal{O}(\rho_1^8) \tag{8.98}$$

Only the terms for even powers of ρ_1 are retained, since replacing ρ_1 by $-\rho_1$ does not correspond to two different physical states but to simple translation along $\mathbf{O}\boldsymbol{z}$. Setting $r \cong \alpha(T - T_0)$ and assuming that $\alpha > 0$ and $u_0 > 0$, the anticipated S_A–N transition is second order and occurs at the temperature $T_{AN} = T_0$, in the smectic phase with

$$< \rho_1 >= \sqrt{\frac{\alpha(T_0 - T)}{u_0}} \tag{8.99}$$

In fact, it is necessary to take into account the coupling between the smectic order, defined by ρ_1, and the nematic order parameter $< s >$: the appearance of smectic order due to the formation of a periodic layered structure strengthens the interactions between molecules in the same layer. The value of s at equilibrium no longer coincides with the value of $< s_0 >$ obtained in the absence of smectic order. Setting $\delta s = s- < s_0 >$, we can write the coupling term F_1 per unit of volume:

$$F_1(\rho_1, \delta s) = -C \rho_1^2 \delta s \tag{8.100}$$

since only even powers of ρ_1 are allowed. C is a positive constant.

In the absence of smectic order, the change of the free energy F_N per unit of volume in the vicinity of the equilibrium value $< s_0 >$ can be written, up to second-order in δs, as:

$$F_N(s) = F_N(< s_0 >) + \frac{1}{2\chi(T)} \delta s^2 \tag{8.101}$$

where $1/\chi(T) = \frac{\partial^2 F_N}{\partial^2 < s >}$ for $< s_0 >$. $\chi(T)$ can be obtained from the Maier–Saupe or the Landau–de Gennes model. This second-order expansion is better justified the farther we are from the isotropic–nematic transition T_{NI}, in the vicinity of which χ increases suddenly while remaining finite.

Given (8.100), we can calculate the total free energy F near the smectic A–nematic transition. Writing the equilibrium conditions (Problem 8.2), we then find that in the transition:

$$< \rho_{1AN} >^2 = \frac{3C^2}{2u_1} \chi(T_{AN}) \tag{8.102}$$

$$T_{AN} = T_0 + \frac{3C^4}{a\alpha u_1} \chi(T_{AN})^2 \tag{8.103}$$

8.1.5 The Many Applications of Liquid Crystals

a) Technology and Instrumentation

In 1930, physicists suggested that nematics could be used to make display devices that consumed much less power than cathode-ray tubes. However, the first attempts to produce such systems failed because the compounds available at the time were insufficiently stable when exposed to heat or light. It was only in 1960 that a family of stable nematics, the alkylcyanobiphenyls, were synthesized (Gray). A typical example of a display device is composed of a thin layer of nematic, single domain, held between two crossed polarizers. The surfaces of the polarizers on the nematic side are first coated with a very fine transparent and conducting layer of indium and lead oxide, then a layer of polyimide – a transparent material which, when rubbed in one direction, induces forces that align the nematic molecules in this direction. By rubbing the polyimide layers in perpendicular directions, the nematic is forced to twist by 90° in the layer (Fig. 8.19a). For this reason, the devices of this type are called "twisted nematic device." The light entering the nematic layer is polarized; the 90° twisting of the nematic crystal imposes rotation of 90° from the polarization direction, so that the light passes through the second polarizer with no attenuation. An electric field applied perpendicular to the

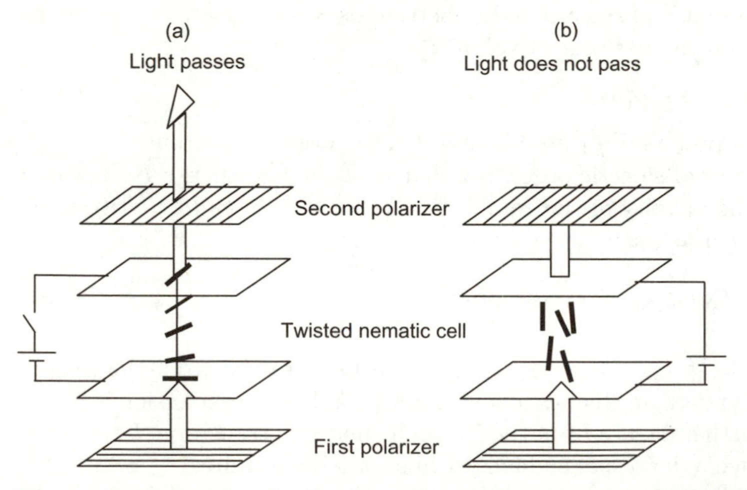

Fig. 8.19. Principle of a display cell utilizing a nematic liquid crystal. There is a liquid crystal in the center of the device which undergoes orientation (twist) rotation.

layer (corresponding to a difference in potential of several volts) is sufficient to align the directional vector perpendicular to the layers in most of it; it allows overcoming the effects resulting from anchoring of the molecules to the surface of the polyimide layers (Fig. 8.19b). In this configuration, the polarization of the light is no longer rotated in crossing the nematic layer and is thus no longer transmitted by the second polarizer. It has been proposed to replace this orientation technique by an ionic irradiation of an inorganic coating of the polarizers. By varying the ions energy one is able to shift the orientation of the nematic molecules.

Devices of this type are extremely well suited for making small displays such as those in digital watches. High-resolution, large television screens could be manufactured with this principle. Some types of smectic liquid crystals, which work differently, offer more interesting possibilities than nematics.

Cholesteric liquid crystals have the property of selectively reflecting light as a function of the temperature. If a monodomain cholesteric liquid crystal is illuminated along the axis of its helix with white light, reflection is selective at a wavelength corresponding to the helical pitch characteristic of their structure. As the pitch of the helix varies as a function of the temperature, the reflected wavelength follows this change. A certain color is thus obtained for a certain temperature by reflection. This phenomenon is totally reversible. Cholesteric liquid crystals, called thermochromic, reflect all colors in the visible spectrum, from red to purple, as a function of the temperature.

These materials have numerous applications:

- display and measurement of temperature (digital thermometers, heat indicators) and medical thermography (detection of breast cancer, diagnosis of vascular diseases, pharmacological tests, skin grafts, etc.);
- nondestructive tests (detection of surface and subsurface defects in metals and defects in electronic components and circuits), determination of turbulent flow in aerodynamic models, heat transfer studies, etc.;
- radiation detectors (thermography, holography of infrared, microwave, and ultrasound radiation ...);
- esthetic use (advertising, decoration, wall coverings, jewelry, fabrics, etc.).

The possibility of creating a ferroelectric fluid from chiral molecules is obviously of interest, and when the possibility of obtaining a fast, bistable electro-optical switch from this type of fluid was demonstrated, the great technological interest of these materials for display systems was immediately perceived. Ferroelectric liquid crystals became the class offering the most interesting applications. Hundreds of liquid crystals of the smectic C* type were then synthesized, and flat screens of excellent definition were made with these materials. Screens utilizing these materials are now available commercially.

As explained, polar order (ferroelectricity) in a liquid crystal phase is only possible if the molecules are chiral. Several attempts have nevertheless been made to obtain polar order with achiral molecules: by fabricating bowl-shaped molecules which intrinsically pack in polar columns; by segregation of distinct chemical subspecies; with more complex systems, such as polymers with branching side chains organized in fish-bone geometry.

The interest in obtaining polar order in fluids in the absence of chirality was stimulated again (Nori, 1996) by demonstration of ferroelectric switching using a smectic phase composed of **curved rod-shaped achiral molecules** (banana-shaped).

In this type of smectic liquid crystal (Fig. 8.20), the particular shape of the molecules, rigid and curved rods (banana-shaped), imposes a particular molecular organization. As in a classic smectic C phase, the axis of alignment of the molecules (direction vector $\boldsymbol{n}$) is the same in all layers and forms angle θ with $\mathbf{O}z$, the normal to the layers, but the structure of the molecular packing differs in each layer. At equilibrium, this structure corresponds to the maximum density in each layer; it is realized when the molecules aligned along $\boldsymbol{n}$ point their "arcs" along direction $\mathbf{O}y$ (or $-\mathbf{O}y$) perpendicular to $\mathbf{O}z$ and $\boldsymbol{n}$, creating spontaneous polarization parallel to the plane of the layers. This molecular arrangement implies that for a given direction of $\boldsymbol{n}$, two equivalent layer structures exist with antiparallel spontaneous polarizations, and these two structures are mirror images of each other in the plane defined by the normal to the layers, $\mathbf{O}z$, and directional vector $\boldsymbol{n}$. As the molecules are achiral, these two structures are equally probable. In practice, it does not seem to be difficult to obtain very wide domains corresponding to the same molecular arrangement and thus the same polarization.

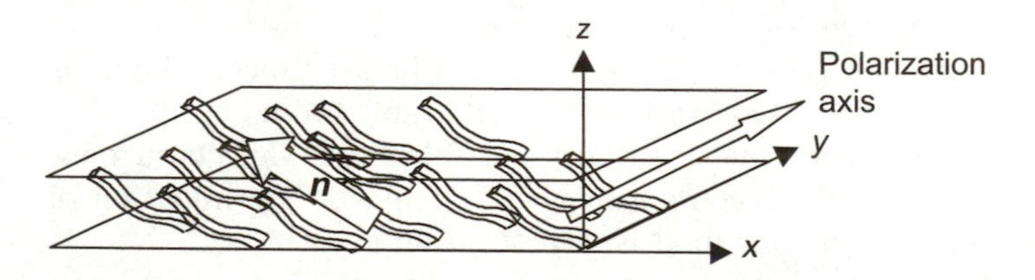

Fig. 8.20. Representation of a chiral layer in a smectic phase constituted of curved, rod-shaped achiral molecules. Axis n of preferred orientation of the molecules forms angle θ with respect to Oz. The molecules aligned along n point their "arcs" along direction Oy (or $-$Oy) perpendicular to Oz and n; the appearance of polarization parallel to the plane of the layers along Oy is observed.

Plane Ozx along which the molecules are tilted is in principle arbitrary in the absence of an applied electric field. However, since there is polarization perpendicular to plane Ozx, this plane can be oriented by application of an electric field $\boldsymbol{E}$ parallel to the plane of the layers; it will then become perpendicular to $\boldsymbol{E}$. By utilizing the ferroelectric property of the material, an electric field $\boldsymbol{E}$ can be used to control the orientation of directional vector $\boldsymbol{n}$ around axis $\mathbf{O}z$. The switching times obtained are very short, approximately 100 times shorter than those obtained with mixtures of conventional nematic liquid crystals. These materials thus have the most interest prospects for application.

Finally it is necessary to note that liquid crystals have been used as structural materials. Fibers have been fabricated from **lyotropic polymers** which yield mesomorphic phases in concentrated solution. This is the case of Kevlar fibers fabricated from concentrated solutions of poly-(p-phenylene terephthalamide); they have a higher breaking strength than steel. Xydar is another aromatic copolymer with a liquid crystal phase ($T_f \approx 420°\mathrm{C}$) which has good corrosion resistance. It is used in aeronautics and in the automobile industry.

8.1.6 Mesomorphic Phases in Biology

We cannot end this presentation of mesomorphic phases without mentioning the important place they have in living organisms. In Sect. 8.1, we mentioned the liquid crystal phase of myelin, a substance that surrounds the axons of nerves and the white matter of the brain, but many other biological materials also have mesomorphic phases. A cell membrane is in the form of a closed surface (with no free edge). It is composed of a liquid film which is immersed in another immiscible liquid and is essentially constituted of a phospholipid bilayer with cholesterol, proteins,and polysaccharides incorporated inside (Fig. 8.21). Phospholipids are amphiphilic molecules composed of a hydrophilic polar *head* illustrated by a circle and a hydrophobic part formed by a double chain, schematized as a hairpin. The hydrophobic paraffin chains are directed toward the inside of the film while the hydrophilic poles cover the two outer faces of the film. Phospholipids thus tend to orient their chains perpendicular to the film. They form a bidimensional ordered liquid. The proteins contained in this bilayer have more or less hydrophilic or hydrophobic parts as a function of the nature of the amino acids in their polypeptide chain. The hydrophobic parts are generally buried in the bilayer while the hydrophilic parts emerge from the film. Amphiphilic cholesterol molecules are intercalated here and there between the phospholipids. The membrane is a highly dynamic structure; the different molecules composing it move with respect to each other while remaining in their monolayer.

It has been shown that slightly twisted nematics form emulsions in various normal or pathological tissues. Arteriosclerosis often involves deposition of a *fatty slurry* in the walls of the arteries; examination with the polarizing microscope shows that it is an emulsion of cholesteric droplets. These anomalies are especially manifested in the elderly.

Mesomorphic emulsions have been observed in other organs undergoing different experimental treatments.

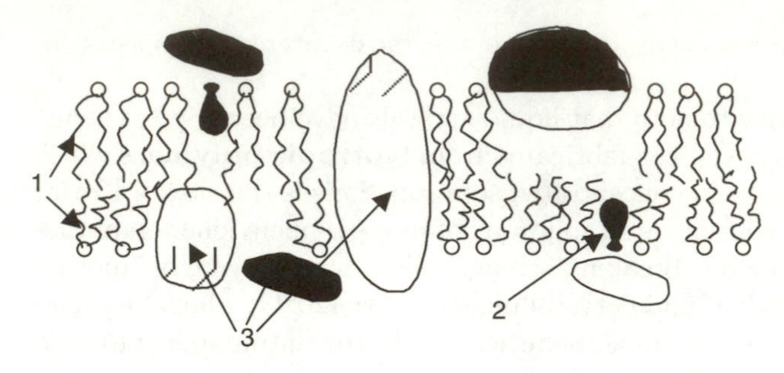

Fig. 8.21. Simple model of a cell membrane. (**1**) Phospholipid molecules with their two paraffin chains (zigzags) and their polar head (circle); these molecules, arranged in a bilayer, form the membrane; they move with respect to each other while remaining confined in their monolayer except when they reverse their orientation (flip-flop) by passing into the opposite layer; (**2**) cholesterol molecules; (**3**) proteins with "grayish" polar regions and "white" nonpolar part (Singer and Nicholson, *Science*, No. 75 (1972), copyright Amer. Assoc. Adv. Sc.).

8.2 Superfluidity of Helium

There are two distinct isotopes of helium, helium 3 and helium 4; the natural abundance of helium 3 is 1.3×10^{-4} %, hence very low.

The term "superfluid" was introduced by P. Kapitza (1937) to describe the very surprising hydrodynamic behavior of liquid helium 4 at very low temperatures. In effect, below 2.17 K, liquid helium 4 can flow through microscopic pores with no pressure loss and thus with no apparent viscosity. Due to more recent discoveries, we know now that the term superfluid (superfluidity) can also be applied to liquid helium 3 at very low temperatures and by analogy to superconductors.

In 1972, it was discovered that the atoms in helium 3 exhibited several superfluid phases below 2.17 mK. The interpretation of this phenomenon is based on the association of ^{3}He atoms into pairs similar to the association of electrons into pairs in superconductors. We have also seen (Chapter 7) that the cancellation of electrical resistance observed in superconductors is explained by considering that some of the electrons are associated into pairs and that these pairs can move through the lattice without loss of energy. As for the neutrons in "neutron stars," they are considered superfluids. Certainly, the temperatures are very high, but the neutron density is sufficient for the quantum effects to be important.

8.2.1 Helium 4

a) The Discovery of Superfluidity

The discovery of superfluidity was the culmination of research conducted between 1908 and 1937. It all started when the liquefaction of helium was obtained by M. Kamerlingh Onnes in 1908 in Leyden: he was the first to liquefy helium 4 at 4.2 K and atmospheric pressure; the path to discovery of superfluidity was open; it was demonstrated several years later. In 1924, M. Kamerlingh Onnes and J. D. A. Boks observed a surprising phenomenon: a density maximum of liquid helium at 2.3 K, and in 1927 W. H. Keesom and M. Wolfke suggested the existence of a phase transition between two distinct liquid phases, helium I present between 4.2 and approximately 2.3 K, and helium II at temperatures below approximately 2.3 K.

In 1930, W. H. Keesom and K. Clusius, who investigated the behavior of the specific heat as a function of the temperature, showed that the specific heat exhibited a peak at 2.19 K (2.172 K in the current temperature scale). The shape of this peak resembled the letter λ, so they decided to call this transition the "λ transition" (Fig. 8.26). In 1935, E. F. Burton showed that the viscosity of helium II is much lower than the viscosity of helium I, then in 1936, W. H. and A. P. Keesom observed that the thermal conductivity of helium at 1.5 K is very high, 200 times higher than the thermal conductivity of copper at ambient temperature, for example.

Finally, in 1937, P. Kapitza, who attempted to measure the viscosity of helium II by studying its flow between two polished disks pushed together, observed that helium II flowed with no apparent friction; he stated that the viscosity could not be measured, in each case at least 1000 times lower than the viscosity of helium I, and he very naturally suggested calling helium II "superfluid." A little later (1938), J. F. Allen and H. Jones discovered the "fountain effect" (Fig. 8.25), while J. G. Daunt and K. Mendelssohn observed a somewhat unexpected phenomenon: superfluid helium is capable of flowing outside a container through the thin film of superfluid helium that covers the walls of the container (Fig. 8.22); although the film covering the walls is very thin (several tens of nanometers), the lack of appreciable viscosity allows establishing very rapid flow of the superfluid in the film, and this flow tends to equalize the inner and outer levels of superfluid helium.

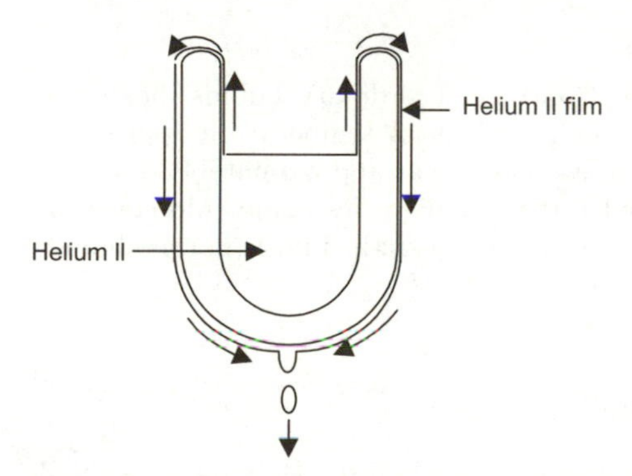

Fig. 8.22. Daunt–Mendelssohn effect. A Dewar filled with helium II empties by itself due to the flow established in the very thin helium film formed on its wall.

In 1938, F. London hypothesized that the appearance of superfluidity resulted from of helium atoms. This idea was taken up and developed by Tisza, who established the bases for the "two-fluid model," mixture of "condensed atoms" (in the sense of Bose–Einstein condensation), and "excited (uncondensed) atoms." The foundations of the superfluidity of helium 4 were established.

b) Pressure–Temperature Diagram of Helium 4

Figure 8.23 shows the phase diagram of helium 4. Note that contrary to most liquids, helium 4 does not solidify as long as the pressure remains below 25 atm. Helium 4 has two liquid phases separated by a transition line λ corresponding to $T_\lambda \approx 2$ K. The λ transition is second-order (with no latent heat or density discontinuity). When going through the transition, the high-temperature liquid phase ($T > T_\lambda$), called helium I, is transformed into a superfluid liquid phase called helium II. There are two triple points: the helium I, helium II, vapor coexistence point (called λ point); the helium I, helium II, solid coexistence point.

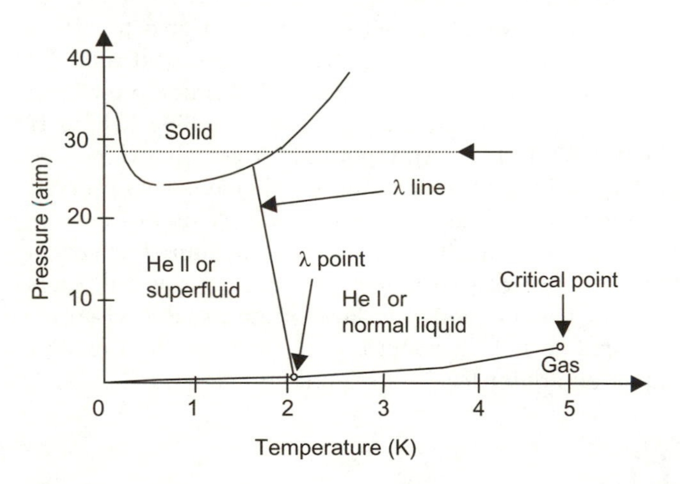

Fig. 8.23. Phase diagram of helium 4. Contrary to ordinary liquids, helium 4, at a pressure below 25 atm, does not solidify when its temperature approaches absolute zero; if cooled while maintaining its pressure at approximately 30 atm, it begins by solidifying, then if it is cooled further, it liquefies again. Moreover, at a pressure below 25 atm, it has two distinct liquid phases: helium I (normal liquid) and helium II (superfluid).

c) The Properties of Superfluid Helium 4

In many situations, the condensed component of helium II behaves as if it does not exist; it can flow with no viscosity as long as the atoms constituting it remain in the ground state. In this case, the viscosity of helium II (condensed component + excited component) is zero if the condensed component alone is set in motion. This only occurs if the liquid flows through a medium of reduced size (porous or capillary medium, for example) to prevent any movement of the uncondensed component, blocked by its viscosity, on one hand, and the velocity of the helium is less than the critical velocity mentioned above on the other hand. Beyond this critical velocity, superfluidity disappears; transitions from the condensed state to an excited state can be induced by interaction between flow of the liquid and irregularities of the walls.

The existence of permanent currents is the first consequence of this property: if flow of the condensed component in a porous ring with velocities lower than the critical velocity begins, at constant temperature, no slowing can be observed in time. A second consequence was revealed with the Andronikashvili experiment (Fig. 8.24): the oscillation frequency of a set of disks in helium II was measured and it was observed that movement of the disks causes movement of the excited component with no effect on the superfluid component.

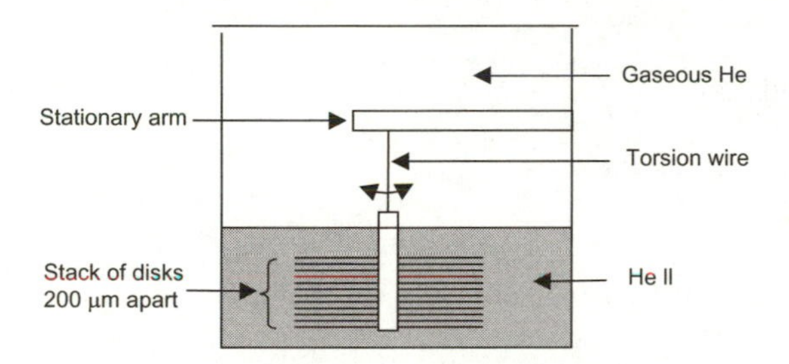

Fig. 8.24. Andronikashvili Experiment.

On the contrary, in an experiment in which both components were set in motion, a resulting viscosity can be detected. This is the case when an object is moved in helium II; the superfluid component does not exercise more friction force to oppose movement, but the normal component is now stressed (more or less strongly as a function of the shape of the object considered) by movement of the object. A nonzero viscosity which is close to the viscosity of helium I is then measured.

Assume that two containers of He II, indicated as A and B, are connected by a porous medium with very small open pores. Container A is initially full and container B is half-full. A superfluid helium current is established between A and B due to the difference in levels in the two containers and the resulting pressure difference between the extreme sections of the porous medium.

Since the superfluid has no entropy, the entropy per unit of mass in container A will increase while the entropy per unit of mass in container B will decrease. The temperature in container A will increase while the temperature in container B will decrease. This is the **mechanocaloric effect**.

Note that if the porous filter is not fine enough, normal fluid flow which is not negligible will take place in the opposite direction of the superfluid current and will be responsible for heat transfer which will limit the mechanocaloric effect.

Conversely, if one of the containers is heated, a superfluid current appears, going from the cold container to the hot container, and a pressure difference is established. With sufficient heating, J. F. Allen and H. Jones succeeded in creating a true jet of superfluid helium in 1938 (Fig. 8.25). This is why the creation of a pressure difference by heating one of the containers is called the **fountain effect**.

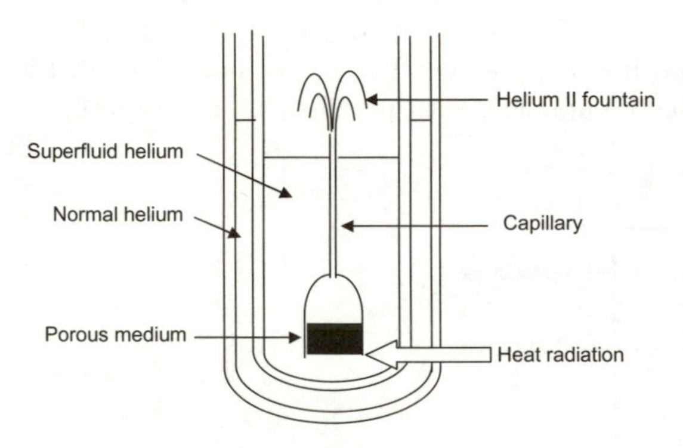

Fig. 8.25. Fountain effect. Local heating of helium at the base of the porous medium causes spurting of helium II.

d) Bose–Einstein Condensation and Superfluidity

Bose–Einstein condensation is not included in classic mechanics. It is a quantum phenomenon.

Let us briefly recall that in quantum mechanics, particles with no mutual interaction are classified in two categories as a function of their spin value. These two categories obey different statistical distribution laws. Free particles of half-integral spin, called fermions, obey Fermi–Dirac statistics, while free particles of integral spin, or bosons, obey Bose–Einstein statistics.

A quantum–mechanical treatment is needed when the de Broglie thermal wavelength, designated Λ, becomes less than d, the average distance between particles. Knowing that $(h^2/2\pi mkT)$, where m is the mass of the particle and h is the Planck constant, quantum treatment is required when the condition $(h^2/2\pi mkTd^2) \ll 1$ stops being observed. In liquid helium 4, $d \approx 3.76 \times 10^{-10}$ m, $m = 0.665 \times 10^{-26}$ kg; for $T = 2\text{K}$, $(h^2/2\pi mkTd^2)$ is of the order of 2.68; quantum treatment is thus required.

Helium 4 of zero spin is a boson. If the interactions between particles can be neglected, it should satisfy the Bose–Einstein distribution function; if the ground level energy of a particle is arbitrarily set at 0, the number of particles $n(\varepsilon, T)$ in energy state ε at temperature T is then given by:

$$n(\varepsilon, T) = \frac{1}{\mathrm{e}^{(\varepsilon-\mu)/kT} - 1} \tag{8.104}$$

where μ is the chemical potential.

It would seem unacceptable to neglect a priori the interactions between atoms in a liquid medium, but in analyzing the situation of liquid helium 4 in detail, we find that liquid helium is far from having the structure of a classic liquid. In effect, the molar volume is very high, corresponding to an abnormally high average distance l between atoms, for a liquid, $l_p \approx 3.76\,\text{Å}$; in this liquid with an expanded structure, the atoms can move freely over significant distances. The transport properties are not very different from those of a classic gas. Finally, the critical temperature is very low, $T_c = 5.2\,\text{K}$; the binding energy between neighboring atoms, which is proportional to T_c, is thus very low.

The density of energy states ε of a free particle, $D(\varepsilon)$, is equal to:

$$D(\varepsilon) = \frac{V}{4\pi^2}\left(\frac{2m}{\hbar^2}\right)^{2/3}\varepsilon^{1/2} \tag{8.105}$$

where V is the accessible volume of the particle (volume of the gas). At temperature T, the number of helium 4 atoms, $N_0(T)$, in the ground state and the number $N_e(T)$ in excited states must be such that $N = N_0(T) + N_e(T)$, where N is the total number of atoms in volume V.

$N_e(T)$, the number of helium 4 atoms in excited states, is given by:

$$N_e(T) \cong \int_0^\infty D(\varepsilon)n(\varepsilon, T)\mathrm{d}\varepsilon \tag{8.106}$$

We can verify that this integral only gives the number of particles in excited states since $D(\varepsilon) = 0$ for the ground state ($\varepsilon = 0$). The atoms in the ground state, whose number is $N_0(T)$, must be counted separately.

Bringing the expressions for $D(\varepsilon)$ (8.105) and $n(\varepsilon, T)$ (8.104) into (8.106):

$$N_e(T) \cong \frac{1.306V}{4}\left(\frac{2mkT}{\pi\hbar^2}\right)^{3/2} = \frac{2.612V}{V_Q} \tag{8.107}$$

where V_Q, the quantum volume, is defined by: $V_Q = \left(2\pi\hbar^2/mkT\right)^{3/2}$. The fraction of excited atoms is then found:

$$\frac{N_e(T)}{N} \cong 2.612\frac{V}{NV_Q} = \frac{2.612}{nV_Q} \tag{8.108}$$

where $n = (N/V)$ is the numerical density of the atoms.

The Bose–Einstein condensation temperature T_0 is the temperature below which $(N_e(T)/N)$ becomes less than 1. The condensation temperature T_0 is thus the temperature for which $N_e(T) = N$ in (8.108); we then find:

$$T_0 = \frac{2\pi\hbar^2}{km}\Big(\frac{N}{2.612V}\Big)^{2/3} \tag{8.109}$$

For $T \leq T_0$, (8.109) is then written:

$$\frac{N_e(T)}{N} \cong \Big(\frac{T}{T_0}\Big)^{3/2} \tag{8.110}$$

The number of atoms in the ground state is then:

$$N_0(T) = N - N_e(T) = N\Big[1 - \Big(\frac{T}{T_0}\Big)^{3/2}\Big] \tag{8.111}$$

The specific heat at constant volume is then equal to:

$$C_v = 1.92Nk\left(\frac{T}{T_0}\right)^{3/2} \tag{8.112}$$

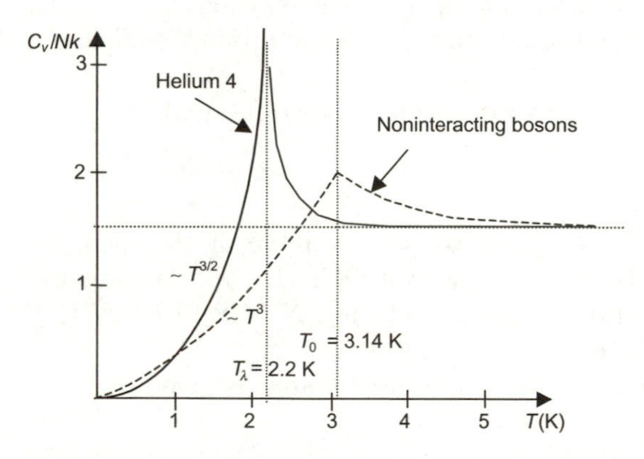

Fig. 8.26. Change of the heat capacity C_V/Nk as a function of the temperature in the vicinity of the lambda transition. The solid curve corresponds to the measured values. The dashed curve corresponds to the theoretical changes predicted for the bosons; the He I–He II transition (lambda transition) is second order (with no latent heat of transformation and with no density discontinuity); for $T \gg T_\lambda$, the specific heat approaches the classic value ($C_V/Nk = 3/2$) and when T approaches zero, it approaches zero, in accordance with the Nernst principle; however, helium differs from a boson gas with no interactions; in the vicinity of $T = T_\lambda$, C_v diverges; in the vicinity of $T = 0$, C_v does not vary as $T^{3/2}$ but as T^3 like the phonons in a solid.

According to (8.109), the Bose–Einstein condensation temperature of liquid helium 4 with no interaction between atoms will be $T_0 = 3.14$ K; this temperature is very close to the "helium I–helium II" transition temperature observed experimentally in helium 4 (λ transition), $T_\lambda = 2.17$ K. It is then tempting to assume that the λ transition is a Bose–Einstein condensation modified by the presence of molecular interactions; helium II corresponds to $N_0(T) > 0$ obtained for $T < T_0$ and helium I corresponds to $N_0(T) = 0$

obtained for $T > T_0$. The interactions nevertheless have four important consequences:

- the transition temperature is 0.97 K lower and decreases slightly with the pressure (Fig. 8.23);
- only 10% of the atoms are condensed at zero temperature instead of the 100% predicted by Bose–Einstein condensation (8.110);
- the divergence of C_v, specific heat at constant volume, in the vicinity of $T = T_\lambda$ (Fig. 8.26);
- the fact that C_v does not vary as $T^{3/2}$ but as T^3 in the vicinity of $T = 0$, like the phonons in a solid (Fig. 8.26);
- elementary excitations are collective modes whose group velocity exhibits a minimum. Since the work by L. D. Landau (1941), N. Bogolyubov (1947), and R. P. Feynman (1954), it has been known that the existence of these collective modes is at the origin of superfluidity. If an object moves in helium at velocity v, it can only be slowed by creating elementary excitation in the fluid; such a process is forbidden if v is less than the velocity of this excitation. Numerous experiments have demonstrated the existence of such a critical velocity; it is generally less than the velocity calculated by L. D. Landau, $v_{cr} = 50$ m/sec.

e) Tisza Two-Fluid Model

This model allows interpreting the properties of helium II (^{4}He below the λ point). Tisza was able to interpret most of the experimental results by assuming that phase II is constituted of two components called "normal fluid" and "superfluid." It is assumed that these two components can be attributed density ρ_n for the "normal fluid" and density ρ_s for the "superfluid." These two fluids, although they cannot be separated, do not interact and can move independently. When the liquid flows at velocity $\boldsymbol{v}$, it is assumed that the two components are moving at two different velocities, $\boldsymbol{v}_n$ for the "normal liquid" and $\boldsymbol{v}_s$ for the "superfluid." Moreover, it is assumed that the density and quantity of movement of the two-component fluid, respectively designated ρ and $\rho\boldsymbol{v}$, are such that:

$$\rho = \rho_n + \rho_s \quad \text{and} \quad \rho\ \boldsymbol{v} = \rho_n\ \boldsymbol{v}_n + \rho_s\ \boldsymbol{v}_s \tag{8.113}$$

The "normal liquid" is assumed to behave like a normal liquid: its viscosity is nonzero.

The "superfluid" has unusual properties: its entropy is zero and it flows without resistance through channels of very small diameter.

Based on these hypotheses, it is possible to understand the strange properties of helium II. We should nevertheless note that although the existence of two components of this type were assumed in the Bose–Einstein model, the quantity ρ_s/ρ is only a concept which helps to describe superfluid helium 4 and should not be confused with the fraction of actually condensed atoms in the ground state.

Within the framework of the two-fluid model, E. L. Andronikashvili proposed a method for determination of the ρ_n ratio in helium 4. The device used for this measurement is shown in Fig. 8.24. The container with helium II is kept at a uniform temperature $T \leq T_\lambda$. A stack of disks several tenths of a millimeter apart is part of a shaft suspended on a torsion wire. The oscillation frequency of the disks is a function of the mass of fluid entrained. Knowing that movement of the disks drags the "normal fluid" but does not drag the "superfluid," it is possible to measure the density of the "normal fluid" as a function of the temperature.

The theories of Landau (1941) and Feynman (1954) attempted to establish a theoretical basis for the two-fluid model in the vicinity of absolute zero.

Propagation of sound in the superfluid liquid is also unusual. From the two-fluid model, two types of acoustic wave can be anticipated in helium II:

- we can imagine a sine wave where ρ_n and ρ_s oscillate in phase; in this case, the total density ρ varies sinusoidally and the classic scheme of acoustic wave propagation is observed; this is called "**first sound**;"
- we can also consider a new oscillation mode where ρ_n and ρ_s oscillate in opposite phase with the same amplitude so that the total density, ρ, remains constant. In this mode, the mass entropy of the normal fluid oscillates while that of the superfluid remains zero. This is not a sound wave, since ρ is constant, but a wave corresponding to oscillation of the entropy per unit of mass at constant density ρ. This new oscillation mode could perhaps be excited by producing a local thermal shock in He II. The locally produced perturbation should not be propagated by diffusion, but should be propagated like a wave with a characteristic velocity. This phenomenon is called **second sound**; it was observed for the first time by V. P. Peshkov in 1946;
- when helium is confined in a porous medium whose pores are open but very small, the second sound is modified since the movements of the normal fluid are blocked by the effect of its viscosity. This is the **fourth sound**;
- when we attempt to propagate a capillary wave along a thin superfluid film, it oscillates parallel to the substrate, while the normal fluid is almost frozen due to its viscosity. At the crest of the wave, the superfluid concentration is higher than at equilibrium, while at the trough, the situation is reversed: the temperature is higher in the trough than in the crest of the wave. However, these temperature fluctuations are strongly attenuated by an evaporation–condensation mechanism, where the vapor produced on the surface of the trough is condensed on neighboring crests. This is the **third sound**. The second, third, and fourth sounds are critical modes; their velocities approach zero at T_λ.

f) Quantum Vortices

Quantum mechanics describes the ensemble of the superfluid component with a wave function $\psi(\boldsymbol{r}) = \psi_0 \mathrm{e}^{\mathrm{i}\varphi(\boldsymbol{r})}$, the solution of the Schrödinger equation. The momentum $\boldsymbol{p}$ of the superfluid component is such that:

$$\boldsymbol{p}\Psi = -\mathrm{i}\hbar\nabla\Psi \tag{8.114}$$

so that:

$$\boldsymbol{p} = \hbar\nabla\varphi \tag{8.115}$$

The velocity of the superfluid component $\boldsymbol{v}_s$ is thus proportional to the phase gradient $\varphi(\boldsymbol{r})$, that is:

$$\boldsymbol{v}_s = \frac{\hbar}{m_4}\nabla\varphi \tag{8.116}$$

where m_4 is the mass of helium 4; it is thus irrotational (rot $\boldsymbol{v}_s = 0$).

Now consider a superfluid component occupying an annular region contained between two coaxial cylinders. Flow along a circle centered on the axis of the cylinders and located in the annular region is equal to:

$$\kappa = \oint \boldsymbol{v}_s \mathrm{d}\boldsymbol{l} = \frac{\hbar}{m_4}\oint \nabla\varphi \mathrm{d}\boldsymbol{l} = \frac{\hbar}{M_4}(\Delta\varphi)_{\text{contour}} \tag{8.117}$$

Since the wave function has a unique value, one turn in a closed circuit must leave it unchanged; this implies that $(\Delta\varphi)_{\text{contour}}$ should be equal to a whole number of times 2π; as a consequence: $\kappa = n(h/m_4)$ with integer n.

The annular region just considered is a special example. We note that quantification of the flow of $\nabla\varphi$ is only possible in the presence of a central cylindrical hole which, in the case described, is physically realized by an inner cylinder.

In practice, cylinder-shaped holes can either be imposed by the surface of a cylindrical solid (as treated here) or appear spontaneously in the superfluid; stable currents of the superfluid component are then inscribed in circles around the cylindrical hole. These **vortices** are very easily formed in the superfluid component and are very important for understanding many properties.

If a container of superfluid helium is rotated, the superfluid component remains immobile at slow speed, then when the rotation rate increase, a system of quantified vortex lines of increasing density appears, finally giving the fluid a normal appearance. These vortices can interact with elementary excitations, which gives rise to exceptional coupling between the movement of the two superfluid and normal components and explains the observation of a critical velocity less than the velocity calculated by Landau.

8.2.2 Superfluidity in Helium 3

The discovery of superfluidity in helium 3 is much more recent. It was discovered in 1972 by D. D. Osheroff, R. C. Richardson, and D. M. Lee. For temperatures below 2.7×10^{-3} K, helium 3 is in several distinct superfluid phases. These phases are anisotropic and magnetic.

The superfluidity of helium 3 is a priori unexpected since the nuclei of helium 3 have 1/2 spin and are thus fermions, but weak attraction of magnetic origin can appear in certain conditions and allows them to form pairs. For these pairs of helium 3 atoms, Bose–Einstein condensation can thus be envisaged. The theory of the superfluidity of helium 3 is much more complicated than for helium 4. Two superfluid phases have been detected. One only appears if pressure greater than 20 atm is applied, called superfluid A, and the other, called superfluid B, appears at a lower temperature at atmospheric pressure, but can also be obtained at higher pressure, up to 20–30 atm (Fig. 8.27). The "superfluid A–normal fluid" and "superfluid B–normal fluid" transitions are first order. The "superfluid A–superfluid B" transition is second order.

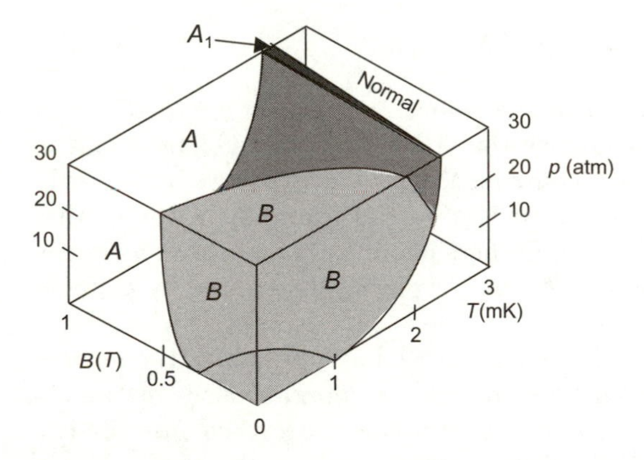

Fig. 8.27. Phase diagram of helium 3 at very low temperature in "pressure–temperature–magnetic field" space. Phases A, B, and A_1 are superfluids; the solid phase is present beyond 34 atm.

Application of a magnetic field has a profound influence on the $p - T$ phase diagram: it reveals a new phase called A_1. Phase A and B have complex and somewhat surprising properties, since they are anisotropic and magnetic. Nevertheless, a model that accounted for the phenomena observed was rapidly developped, inspired by the theory of superconductivity.

We note that the conducting electrons in a metal which have equal and negative electric charges move in a lattice composed of positive ions. If we disregard the lattice effect, they will thus be exposed to important electrostatic forces which will force them away from each other. However, the lattice of positive ions constitutes an effective electrostatic screen and the electrons no longer mutually repel each other (Chapter 7). At low temperatures, in certain metals or alloys, the residual interactions between conducting electrons can become slightly attractive, which allows the electrons to couple in pairs, the Cooper pairs; this is the microscopic mechanism at the origin of superconductivity. In a superconducting Cooper pair, the 1/2 spins of the two quasiparticles are oriented head-to-tail – the result is a zero spin – and the motion of the pairs that corresponds to a zero orbital moment (orbital state *s*). In superconductors, Cooper pairs are all in a single and identical ground state of zero spin; these pairs are thus in a state similar to the helium 4 nuclei that participate in the condensation (zero spin ground state).

A similar approach can be applied to helium 3 nuclei, representing them by weakly interacting quasiparticles; the residual attractive interaction capable of stabilizing pairs of quasiparticles is of magnetic origin. It becomes sufficient to allow the existence of a new phase, around 2mK, in which the quasiparticles form pairs (Cooper pairs).

However, in condensation of helium 3 atoms, the Cooper pairs that appear have much more complicated behavior than the Cooper pairs in the case of electrons. There are many arguments, particularly the fact that in the condensed state, these phases are anisotropic and magnetic, for confirming that the pair formation is of the "*p* wave, spin triplet " type. This means that on one hand, the space wave function of the Cooper pair should be that of an atomic orbital *p*, that is, antisymmetric, and, that the spin wave function, on the other hand should be symmetric, corresponding to a total spin $S = 1$.

P. W. Anderson and P. Morel proposed in 1961 a first model with pair formation with parallel spins, state $\mid\uparrow\uparrow>$ or $\mid\downarrow\downarrow>$. Then R. Ballian and N. R. Werthamer showed in 1963 in a complete treatment that pair formation also included the combination $\mid\downarrow\uparrow> + \mid\uparrow\downarrow>$. Taking into account simple interactions between particles, the state $\mid\downarrow\uparrow> + \mid\uparrow\downarrow>$ would seem to have lower energy than the state $\mid\uparrow\uparrow>$ or $\mid\downarrow\downarrow>$.

On the other hand, the experimental results reveal two phases, A and B, and the NMR studies of these two phases strongly suggest that phase A should be identified with state $\mid\uparrow\uparrow>$ or $\mid\downarrow\downarrow>$ and phase B should be identified with $\mid\downarrow\uparrow> + \mid\uparrow\downarrow>$.

Finally, A. J. Legett, considering the magnetic interactions and based on an analysis of the magnetic properties, showed in 1973 that the two phases predicted indeed correspond to phases A and B as observed by D. D. Osheroff, R. C. Richardson, and D. M. Lee.

These pairs are probably not molecules; the distance between quasiparticles in a pair is of the order of several tens of an Angström, thus very large in comparison to the interatomic distances of a diatomic molecule. Given the density of the molecules in liquid helium, the different pairs are nested in each other and it is not possible to determine which pair is part of a given atom. Moreover, given the bulk of the pairs, it is not possible in principle to obtain the superfluidity of this fluid in pores smaller than the size of the pairs.

The very fact of obtaining nonzero total spin and orbital moment for the pairs implies magnetic anisotropy of the pairs whose effects are particularly marked in phase A, where structures similar to those observed in liquid crystals have been found.

Helium 3, in the milliKelvin region, is simultaneously superfluid, magnetic, and anisotropic; for this reason, it has a very large variety of physical properties. Its superfluidity was verified in experiments similar to those described for helium 4: measurement of the viscosity and critical velocity of the superfluid, identification of the different sounds, etc.

Problems

8.1. Latent Heat of the Nematic Transition

The "nematic liquid crystal–isotropic liquid" transition is described with the Maier–Saupe model.

1. Calculate the latent heat associated with the transition;
2. What conclusion can be drawn?

8.2. "Nematic – Smectic A" Transition

Order parameter ρ_1 is determined at the "nematic liquid crystal – smectic A" transition by calculating the total free energy F taking into consideration the nematic s and smectic ρ_1 orders. A Landau model is used to describe the material and $\delta s = s - < s_0 >$ is introduced, where $< s_0 >$ is the nematic order parameter at equilibrium in the absence of smectic order.

1. Calculate F;
2. Determine $< \delta s >$ at equilibrium from it;
3. Write a new form of F taking into account $< \delta s >$;
4. Determine the value of $< \rho_1 >$ at the transition and the transition temperature. Study the case with the fourth order term in ρ_1 in F being zero.

8.3. Light Scattering in a Liquid Crystal

Consider a nematic liquid crystal whose axis of preferred orientation is $\boldsymbol{n}_0$. Apply a magnetic field $\boldsymbol{H}$ directed along $\boldsymbol{n}_0$ and designate the fluctuations of the directional vector by $\delta\boldsymbol{n}(\boldsymbol{r})$. For weak fluctuations, $\delta\boldsymbol{n}(\boldsymbol{r}) \perp \boldsymbol{n}_0$. These fluctuations will be represented by their Fourier components $\delta n_t(\boldsymbol{q})$ for the component parallel to $\boldsymbol{q}$ and $\delta n_\perp(\boldsymbol{q})$ for the other perpendicular component.

1. Calculate the total free energy associated with the fluctuations as a function of $\delta\boldsymbol{n}(\boldsymbol{q})$;
2. Knowing that the scattered light intensity $I(\boldsymbol{q})$ is proportional to $<| \delta\boldsymbol{n}(\boldsymbol{q}) |^2>$, calculate $I(\boldsymbol{q})$;
3. What conclusion do you draw from this?

9 Microstructures, Nanostructures and Phase Transitions

9.1 The Importance of the Microscopic Approach

With a few exceptions, polyphasic fluids in particular, we have always assumed up to now that materials were constituted of a **single phase**, that is, a domain, solid or liquid, whose properties were uniform and independent of the structure. This macroscopic vision of the structure of materials does not hold up on more detailed observation of the systems; it is only an approximation which does not allow accounting for their properties.

A solid, considered a priori as monophasic, is most often a polycrystalline structure formed from a liquid or vapor phase by a nucleation mechanism. The simple observation of snow flakes or ice crystals deposited on frost-covered glass, with their characteristic arborescences, suffices to demonstrate the polycrystalline nature of ice. Microcrystallites develop in the liquid phase in the liquid/solid transition by nucleation and are at the origin of the presence of **microstructures** within the solid phase whose size is in particular very strongly dependent on the heat treatment the material has undergone. In the case of solids, metals, for example, the existence of these microstructures modifies many of the properties (mechanical, magnetic, electronic, etc.).

It is thus essential to know the mechanisms of formation of microstructures and to understand their influence on the macroscopic properties of a solid material, whether it is metal, glass, or polymer.

Moreover, certain phases are known to be intrinsically heterogeneous; this is especially the case of colloidal suspensions and emulsions. In the case of colloids, we have a system constituted of microscopic particles suspended in a liquid, a solid, or a gas; this is in fact a biphasic material. It is easy to see that microstructures play an essential role in the properties of the special materials which are colloidal systems.

Although we are used to consider microstructures in a solid whose size is of the order of the micron or tenth of a micron, for some years now we have had techniques that allow fabricating solid phases with microstructures of the order of the nanometer (thousandth of a micron) in size. As we will see with solid materials of this nature, **nanomaterials**, we are entering a totally new area with very important technological prospects.

9.2 Microstructures in Solids

9.2.1 Solidification and Formation of Microstructures

The formation of a new solid phase within a liquid is a dynamic phenomenon which usually occurs discontinuously. We dedicated Chapter 2 of this book to all of these questions of the dynamics of phase transitions. With specific attention to the formation of a crystalline solid, we showed that nuclei progressively grow within the solid phase to form **grains** which occupy the entire available volume at the end of the transition (Fig. 2.2). These grains are microcrystals whose growth rate is particularly a function of heat treatment. In most cases, for metals, for example, a polycrystalline structure is obtained (formation of a single crystal is obviously possible). The morphology of these microstructures is a function of the type of materials, as we have seen: in the case of metals, **dendrites** are formed, while they are generally **spherulites** in polymers. **Fractal** forms can also be observed.

Recall that homogeneous nucleation of a solid phase implies relatively important supercooling of the liquid phase, however the presence of impurities which are nucleation sites in the liquid generally facilitates crystallization.

After formation of a first solid nucleus in the mother phase, the new solid phase grows by addition of atoms at the interface between the solid nucleus, grain, and mother phase. The growth of the crystalline phase will thus be strongly dependent on the nature of this interface. A very uneven interface on the atomic scale and consequently its implied roughness thus favors attachment of additional atoms (Fig. 9.1).

In the first case, the crystal growth rate is rapid and obeys an expression of the type $v = K\Delta T$, where ΔT is the supercooling of the liquid. In particular, this is the situation encountered in metals. In the second case, the crystallization rate is slow and we have $v = K' \exp -(\Delta E/k\Delta T)$, where ΔE is the equivalent of activation energy characteristic of the surface (v is only large if supercooling ΔT is important). This growth is characteristic of polymers.

The grains formed within the mother phase no longer remain without contact during the growth phase and end by touching and adjusting with respect to each other.

The region or contact area between two grains is called the **grain boundary**. This is a region in which the crystal structure is locally perturbed but the atomic bonds established between grains are numerous enough so that this perturbation does not globally weaken the solid structure (Figs. 2.5 and 9.2). Grain boundaries very significantly modify the macroscopic properties of a material and especially its mechanical properties. The interfacial energy on grain boundaries plays an important role in determining the equilibrium structure of the polycrystalline solid.

We can use the analogy with a soap film that forms an ensemble of bubbles. We find that the film, to minimize its total energy, tends to form bubbles

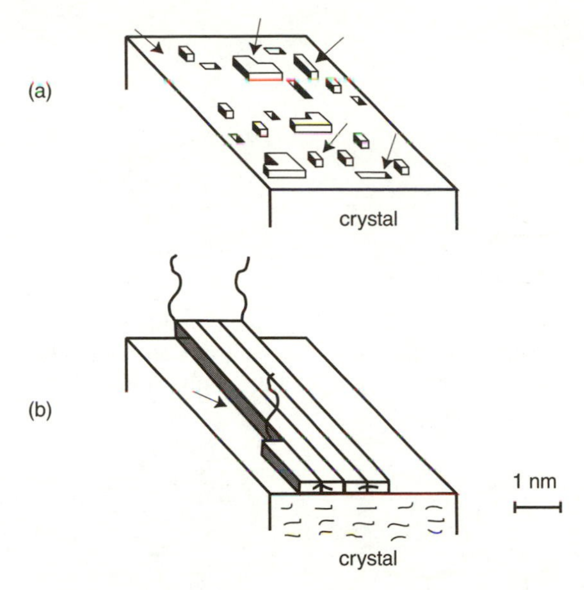

Fig. 9.1. Crystallization and interfacial structures. (**a**) The rough interface has numerous uneven spots that are sites of attachment of atoms to the crystal. (**b**) A smooth interface with no defects requires growth of a complete new layer to form the crystal; this is particularly the case with polymers.

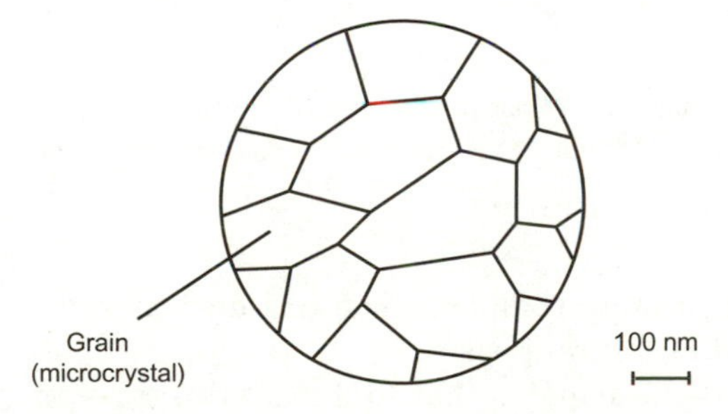

Fig. 9.2. Illustration of grain boundaries. This view of a polycrystalline material is obtained with an optical microscope. The grain boundaries indicated in solid lines are at the interface between two grains (or microcrystals).

such that when three films meet, they form angles of 120°, which allows equalizing the surface tensions. In the case of a polycrystal, which is a three-dimensional structure, if this principle is transposed, the grain boundaries will then tend to constitute planes that intersect along 120° dihedrons and thus minimize the surface energy. Ideally, the space occupied by the material could be filled with grains composed of polyhedrons with 14 faces, called tetrakaidecahedrons to satisfy this condition of minimization of the surface energy (Fig. 9.3).

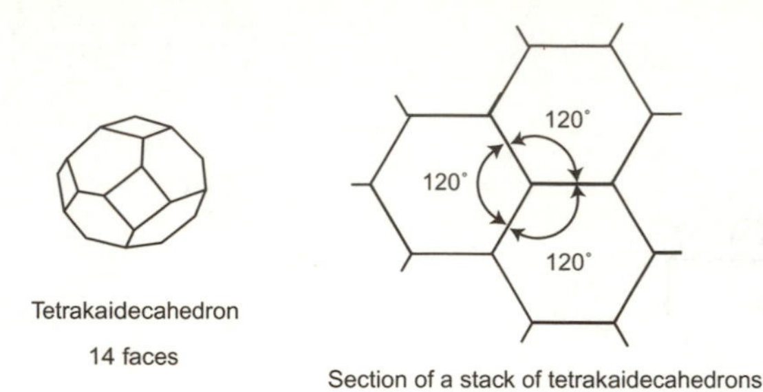

Fig. 9.3. Three-dimensional polycrystal. The surface energy is minimum when the grain boundaries are planar. Equilibrium of surface tensions requires that the angles on the edges of the intersections of the planes be equal to 120°. Polycrystal grains tend to form tetrakaidecahedrons with 14 faces.

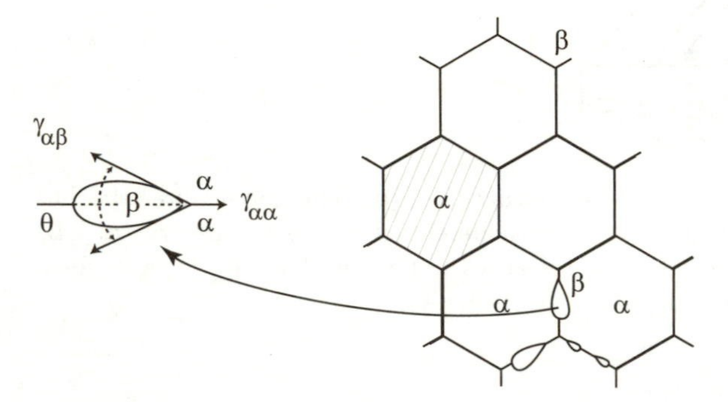

Fig. 9.4. Nucleation of a metal having two phases. The β-phase appears upon nucleation at the grain boundary between two crystals of phase α; $\gamma_{\alpha\beta}$ is the surface tension between the two phases.

The physical situations are often more complex when certain materials, metals, for example, are composed of several phases.

If a polycrystalline metal is two-phase and is thus formed of two types of grains, there are two cases of figures:

- the simplest corresponds to nucleation of a grain of one of the phases within the grain of another phase;
- the second corresponds to nucleation of one of the phases at the grain boundary of the other phase (Fig. 9.4).

If we assume that the interfacial energy is isotropic, the nuclei of one of the phases will tend to assume a spherical or lenticular shape in order to minimize the interfacial energy. Designating by θ the contact angle between the spherical grain cap of the grain in phase β formed at the grain boundary of phase α and introducing the surface tensions corresponding to interfaces (αα) and (αβ) by $\gamma_{\alpha\alpha}$ and $\gamma_{\alpha\beta}$, we can write the equilibrium condition as:

$$2\gamma_{\alpha\beta}\cos(\theta/2) = \gamma_{\alpha\alpha} \tag{9.1}$$

We immediately see in this relation that if $\gamma_{\alpha\beta} < \gamma_{\alpha\alpha}/2$, no angle θ can satisfy (9.1); the only possible equilibrium solution corresponds to $\theta = 0$, that is, covering of the grain boundary by phase β. This situation is not without risks for the material, because if this phase is fragile or if it contains defects, it can favor propagation of cracks. We note in passing, alternatively, that if a ceramic is to be enameled, it is desirable for the liquid forming the enamel layer to completely wet the solid and thus the grains on the surface.

If the interfacial energy is anisotropic, and consequently the surface tension is $\gamma_{\alpha\beta}$, the grains in phase β will tend to form platelets within phase α.

The morphology of polymer crystals is very specific, as we have seen. Observation with the optical and electron microscope reveals the existence of **spherulites** (Fig. 3.14) composed of crystallized lamellae emanating from a central nucleus whose thickness is of the order of magnitude of approximately 10 nm and whose length is of the order of several hundred nm. These lamellae are connected by polymer chains that stabilize the structure. The polymer in the solid state is polycrystalline; an amorphous phase persists between spherulites. The degree of crystallinity of solid polymers generally varies from 20 to 60%; it is a function of their chemical nature and the heat treatment of the liquid phase (in particular, the cooling rate).

9.2.2 A Typical Example: The Martensitic Transformation

The phase diagram of steel which is an iron–carbon alloy is very complex (Fig. 2.17). This diagram is in fact the diagram of a Fe–Fe_3C system since we have an intermetallic compound, cementite Fe_3C, for a 6.7 wt. % carbon content. In addition to the liquid phase, there are four stable solid phases in all for this system: cementite Fe_3C, austenite or γ iron, δ iron, ferrite or α iron. Low-carbon steels (α and δ phases) are soft steels, while those with a high carbon content (less than 2 wt.%) correspond to hard steels. Cast irons are alloys with a carbon content greater than 2 %.

The nature and morphology of the microstructures formed are strongly dependent on the concentration of carbon in the steel. A structure called ledeburite is obtained when the liquid phase solidifies below 1130°C with a 4.3 % carbon concentration corresponding to the eutectic points in the upper part of the diagram in Fig. 2.17. This structure is in fact a composite formed of α iron platelets alternating with cementite platelets. If the γ austenite phase corresponding to the eutectoid perlite (0.8 % C below 723°C), a composite structure formed of α-phase platelets alternating with Fe_3C platelets, is obtained. Perlite is formed at the grain boundaries of the γ phase (Fig. 9.5).

If the eutectoid solution is rapidly quenched (at a cooling rate of the order of 200°C/sec), martensite, a metastable solid solution of iron and carbon, is formed. The martensite grains are formed and cross at the initial phase boundaries of austenite (face-centered cubic symmetry, Fig. 9.6).

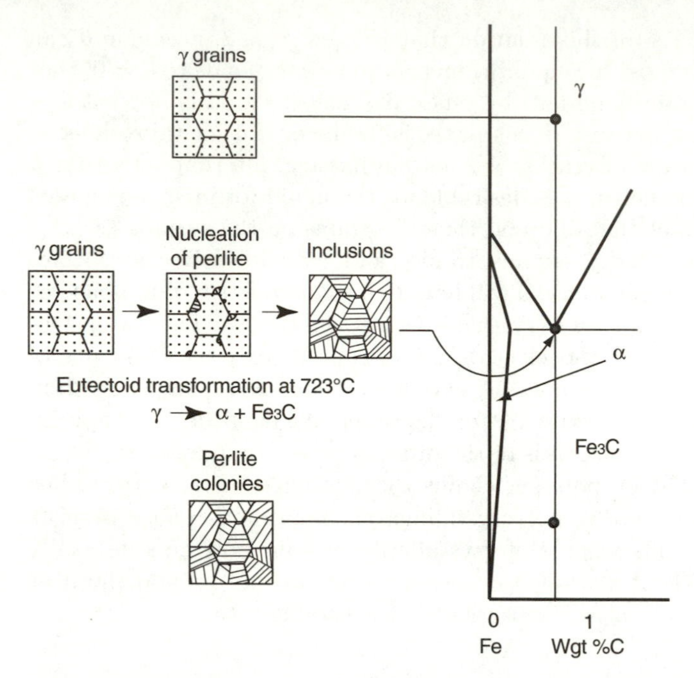

Fig. 9.5. Microstructures in steel. In slow cooling of eutectoid steel (corresponding to the eutectic composition of 0.8% carbon) below 723°C, perlite nucleates at the grain boundaries of the γ phase (austenite). Perlite is a composite structure formed from α phase platelets alternating with cementite Fe_3C platelets (M. F. Ashby and D. R. H. Jones, Materials, Dunod (1991), p. 107).

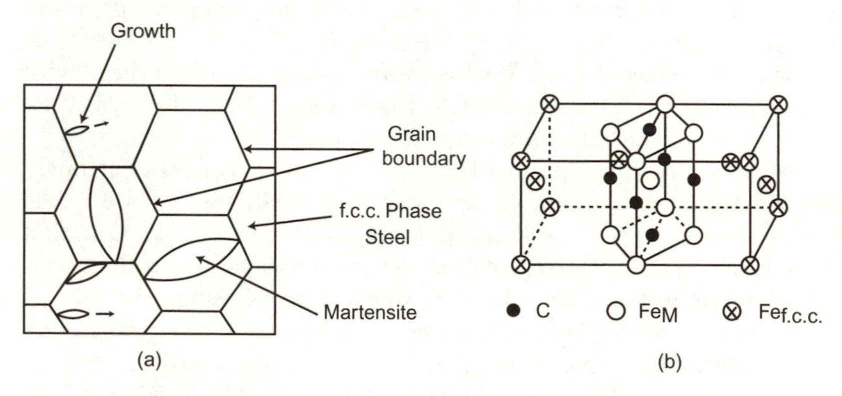

Fig. 9.6. Formation of martensite in steel. (**a**) Martensite grains are formed and grow to the grain boundaries of the initial f.c.c. austenite phase. (**b**) During quenching, carbon atoms will take an interstitial position in the initial lattice, and the martensite is supersaturated with carbon. The unit cell is locally distorted and of centered tetragonal symmetry. All interstitial sites are not occupied by carbon atoms. Fe_M corresponds to the iron atoms in martensite.

Martensite globally has a centered tetragonal structure like ferrite but its lattice is locally deformed by an excess of carbon and has centered tetragonal symmetry (Sect. 2.4.2).

Martensite obtained from eutectoid steel is five times harder than perlite, which is the stable form of the alloy at ambient temperature. The carbon atoms are in effect 40% smaller than the iron atoms, and when iron is quenched to make martensite, the iron atoms move, but the carbon atoms remain frozen in their initial positions. The face centered cubic phase corresponding to perlite is only stable with a carbon content of 0.035% at the maximum, and martensite will thus be supersaturated in carbon. These carbon atoms will move to interstitial positions in the initial lattice, distorting the unit cells (one direction of the cube expands while the other contracts) and form a centered tetragonal local structure (Fig. 9.6). When there are dislocations in the lattice, they have great difficulty in propagating in a distorted and thus stressed structure, which *ipso facto* increases the hardness of martensite. Remember that a phase transition of this type, where there is collective movement of atoms which migrate within the lattice by inducing local rearrangements, is called a **displacive transition**.

Transformations of this type where a metastable phase such as martensite is formed are not limited to iron and steel. By generalization, displacive (or diffusionless) transformations in which the displacements caused by lattice distortion are important enough to dominate the kinetics of the transition, are called **martensitic transformations**. These transformations are encountered in metal alloys, but also in ceramics such as zirconium, ZrO_2, for example. They are often associated with a change in the shape of the material with the temperature. This property is particularly utilized in alloys with **shape memory**. These transformations are generally reversible.

9.2.3 Singular Phases: The Quasicrystals

The discovery of alloys with icosahedral symmetry in 1984 (by Schechtman, Blech, Gratias, and Cahn) opened the way to new crystalline phases named **quasicrystals**.

Initially discovered in an Al–Mn alloy, these phases were subsequently identified in materials with other symmetries (dodecahedral in NiCr, octahedral in VNiSi). The originality of these new phases is that they violate the principle of three-dimensional periodicity of the crystal. The non-periodicity of the crystal structure was demonstrated by the existence of very intense spots in a diffraction diagram corresponding to rotational symmetry of order 5 while the only symmetries compatible with long-range crystal order are of order 2, 3, 4, and 6. Nonperiodic structures of this type had been envisaged previously, particularly by the mathematician Penrose. Symmetry of order 5 has also been observed in liquid phases by X-rays diffraction; it seems to be characteristics of the liquid state.

Another characteristic of these quasicrystals is that they correspond to systems in a metastable state. They are prepared by rapid quenching of a liquid alloy or from an amorphous, and thus nonequilibrium state. The rapid quenching methods allow avoiding nucleation of phases in thermodynamic equilibrium. A method of implantation of Mn ions deposited on thin aluminum films has also been used to produce AlMn quasicrystals.

The morphology of the microstructures formed in quasicrystals is greatly dependent on the method used for synthesis of the material. Dendritic forms such as those in Fig. 9.7 are often obtained.

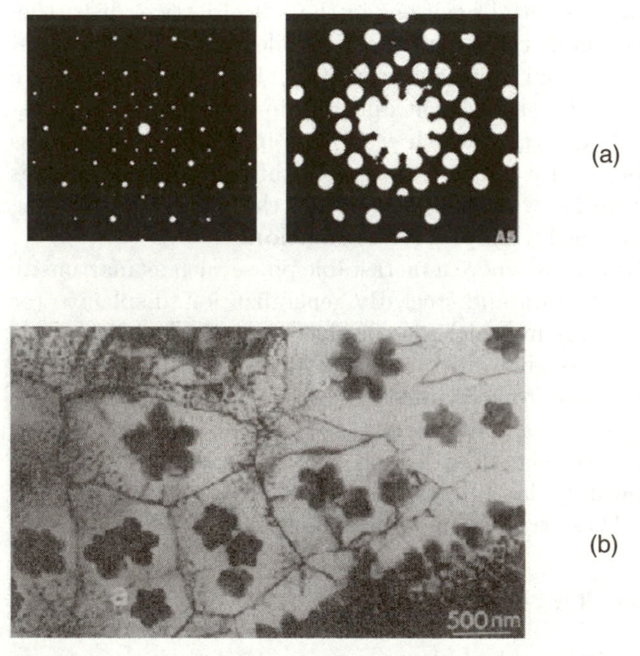

Fig. 9.7. Quasicrystals.(**a**) Electron diffraction pattern of AlMn alloy in fast solidification. The electron beam is parallel to an axis of symmetry of order 5; the spots correspond to this symmetry. (**b**) Electron micrography of dendrites of an AlMn phase (6.8% Mn) solidified by fast quenching. There is growth along axes of symmetry of order 3. These solid phases (quasicrystals) "violate" the principle of three-dimensional crystal periodicity (P. Guyot, P. Kramer, and M. de Boissieu, *Rep. Prog. Phys.*, 1373–1425 (1991)).

9.2.4 The Special Case of Sintering in Ceramics

Ceramics are polycrystalline materials based on oxides (Al_2O_3, MgO, ZrO_2, etc.) or chemical compounds such as SiC and WC. They are materials with a high melting point produced by a method called sintering. It consists of heating powders at high temperature, perhaps under pressure, and their coalescence leads to the formation of a macroscopically homogeneous solid.

Sintering can be defined as the consolidation of a material initially made of a disperse phase (a powder) by heating without attaining melting. The polycrystalline solid is generally formed by a process that involves diffusion of atoms between particles in the solid phase.

In some cases, the solid particles of powder undergo surface melting and a liquid film is formed at the interface between grains and thus favors their "welding;" we then have **liquid phase** sintering. This is the case of porcelain, where the constituting aluminum silicate crystals ($3\,Al_2O_3 \cdot 2\,SiO_2$) are welded together by a silica glass. This is also the method used for manufacturing cobalt and tungsten carbide cutting tools.

In most cases, the chemical composition of the material after the sintering operation is identical to the composition of the initial powder; in a few exceptional cases, heat treatment results in a new chemical compound via solid–solid or solid–gas reaction (it is used for synthesis of silicon nitride ceramics, for example). The ceramic formed is always denser than the initial powder; there is thus volume shrinkage within the material. This shrinkage is induced by a reduction in the size of the pores between the grains or the disappearance of some of them.

The sintering temperature is always high (above 1000°C) but it is a function of the chemical nature of the powders and thus the materials used. The size of the microstructures is a function of the heat treatment the material has undergone, generally approximately 10 μ, but it can be much less for nanomaterials, as we will see.

Sintering is a process that redistributes the atoms between powder grains by establishing bonds (Fig. 3.19). This redistribution tends to minimize the surface energy of the divided medium, since a powder (a metal oxide, for example) has excess free enthalpy with respect to the ideal single crystal corresponding to the surface energy of the grains. There are two ways for the grains to reduce their interfaces:

- enlargement which reduces their number;
- welding by sintering, which eliminates the pores present in the powder and increases the density (Fig. 9.8).

In the case of sintering, the solid–vapor interfaces are replaced by new interfaces, grain boundaries formed by welding of the grains. If γ_{SS} designates the surface tension of the grain boundaries and γ_{SV} is the surface tension associated with the solid–vapor interface, the change in the free enthalpy $\mathrm{d}G$ in this process is written as:

$$\mathrm{d}G = \gamma_{SS}\mathrm{d}A_{SS} + \gamma_{SV}\mathrm{d}A_{SV} \tag{9.2}$$

where $\mathrm{d}A_{SS}$ is the change in area associated with the creation of interfaces between the grains, and we have $\mathrm{d}A_{SS} > 0$. $\mathrm{d}A_{SV}$ is the change in area corresponding to the decrease in the solid–vapor interfaces and $\mathrm{d}A_{SV} < 0$.

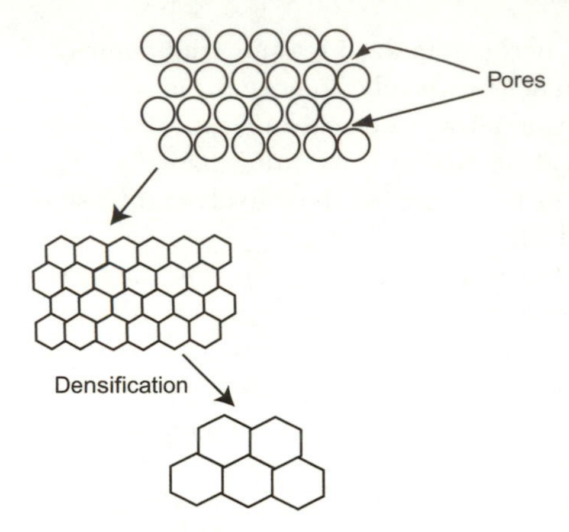

Fig. 9.8. Sintering process. Sintering is the basic process for fabrication of a ceramic. The ceramic formed is denser than the initial powder, and densification (enlarged figure) is obtained by reducing the pore size or by the disappearance of some pores.

Since the energies of the grain boundaries are much lower than the energies of the solid–vapor interface, $\mathrm{d}G < 0$ and the system attains new equilibrium due to formation of grain boundaries.

As for the development of two spherical grains in contact, the connection angle defined by the tangents at the contact point is initially zero (Fig. 9.9). As the spheres get closer, this angle increases to attain equilibrium value θ so that:

$$2\gamma_{SV} \cos(\theta/2) = \gamma_{SS} \tag{9.3}$$

This relation is obtained by requiring that the resultant of the surface tension vectors in the contact zone is zero (they are projected, for example, on a line in the contact plane). For the value of the connection angle corresponding to (9.3), we have $\mathrm{d}G = 0$ and the grain system will be in equilibrium.

Since the connection angle of two grains increases as sintering progresses, the sintering possibilities will be better if θ is larger, which requires high values of γ_{SV} and low values of γ_{SS}. Sintering will thus be strongly controlled by the surface energies.

Diffusion of matter during sintering takes place under the influence of chemical potential gradients corresponding to the stress gradients in the interfaces caused by very important curvatures: for example, in the gas phase, the pressure in the convex part of the surface of the grains is greater than the pressure far below the connection zone; in the solid phase, in the center of the grain boundary, the state of the material corresponds to matter under a plane surface and thus free of stresses. Several diffusion mechanisms are possible under the effect of these gradients: matter can diffuse to the grain boundary by the gas route from the surface of the grains; it can diffuse through the

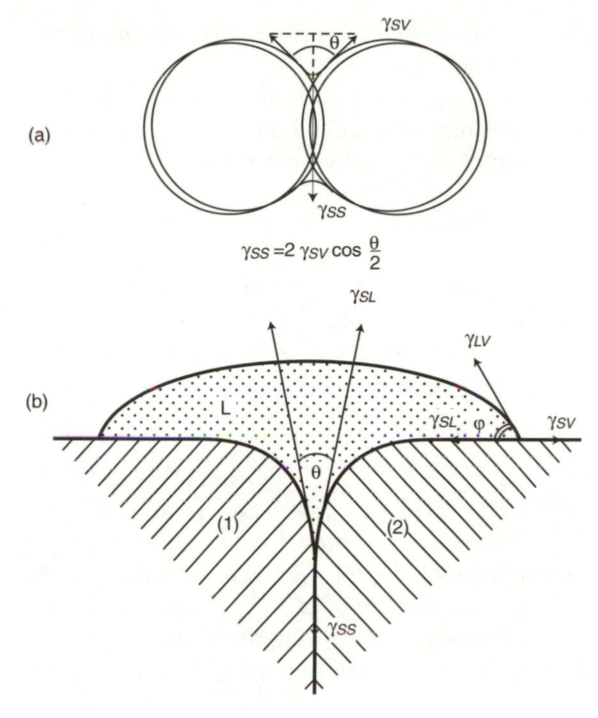

Fig. 9.9. Connection and wettability. (**a**) Evolution of the connection angle θ during the approach of two spherical grains. γ_{SS} is the equilibrium value. (**b**) Wetting of two solid grains by liquid L. Angles θ and φ at equilibrium are defined by (9.3) and (9.4).

surface or through the inside of the grains; it can migrate from the center of a grain to the grain boundary or from dislocations. Each of these mechanisms affects the evolution and kinetics of the microstructures.

Liquid-phase sintering is a method of manufacturing ceramics from powders composed of at least two solid constituents utilizing a phase transition. One of these solid constituents can melt at the sintering temperature, ensuring the existence of a liquid phase that can form a bridge between solid particles. This liquid interface between grains favors the formation of grain boundaries and an increase in the density of the material due to the reduction of the pore size.

In particular, liquid-phase sintering is used for manufacturing some metal alloys from powders (with mixtures such as tungsten–nickel, tungsten–nickel–iron, copper–tin, etc.), amalgams such as silver–tin–mercury, and oxide-based ceramics.

The liquid bridges formed between the solid particles create forces of capillary origin between them. They can be repulsive or attractive as a function of the wettability of the solid by the liquid and thus the relative values of the surface tensions of the liquid, solid, and vapor.

Wetting of two solid grains by the liquid is shown in Fig. 9.9b. The wetting angle φ characterizes the equilibrium of a solid–liquid–vapor line and is classically defined by the Young–Dupré relation by projection on the surface of the solid:

$$\gamma_{LV} \cos\varphi + \gamma_{SL} = \gamma_{SV} \tag{9.4}$$

If $\gamma_{SV} > \gamma_{LV} + \gamma_{SL}$, the only equilibrium situation corresponds to total spreading of the liquid over the surface of the solid (ideal wetting).

The angle θ corresponding to the solid (1)–liquid–solid (2) equilibrium is defined by an equation similar to (9.3):

$$2\gamma_{SL} \cos(\theta/2) = \gamma_{SS} \tag{9.5}$$

If $2\gamma_{SL} > \gamma_{SS}$, we have $0 < \theta < \pi$; if $\gamma_{SL} = \gamma_{SS}$, no value of θ can satisfy equilibrium relation (9.5), the solid (1)/liquid and solid (2)/liquid system containing two solid/liquid interfaces is more stable than the system with only one solid/liquid interface, and the liquid penetrates along the grain boundary between the two solid grains. Finally, if $\varphi > \pi/2$, the liquid does not wet the solid and no sintering is possible with a liquid phase.

The mechanisms involved in liquid-phase sintering are still far from completely elucidated.

Finally, we note that certain ceramics are constituted of microcrystalline phases dispersed in a glassy phase; this is the case, for example, of porcelain and materials called glass ceramics.

9.2.5 Microstructures in Ferromagnetic, Ferroelectric, and Superconducting Phases

At a temperature below the Curie temperature, the individual electronic magnetic moments of a ferromagnetic material are all aligned in the same direction if the material is examined on the microscopic scale. However, the value of the magnetic moment for the macroscopic sample is generally less than the moment corresponding to saturation; it is necessary to apply an external magnetic field to obtain this saturation.

In fact, the magnetic material is not strictly homogeneous; it is composed of a very large number of small magnetized regions called **domains** as shown on (Fig. 9.10). Each domain has a magnetization corresponding to saturation, but the direction of the magnetization varies from domain to domain and the resultant of all these magnetized zones corresponds to a total magnetic moment which is much smaller than the saturation value and which can even be zero for a ferromagnetic material.

The physicist P. Weiss, one of the founders of the modern theory of magnetism along with the Curies, was responsible for this notion of domain, but ferromagnetic domains were first directly observed by F. Bitter in 1931. He

Fig. 9.10. Magnetic domains. A magnetic material is composed of a large number of magnetized regions called domains. The domains in a thin film (6 microns) of YGdTm are shown here. (**a**) Non-magnetic phase; (**b, c, d**) magnetization in an increasing magnetic field; magnetic bubbles are formed (A. H. Bobeck and E. Della Torre, in: *Magnetic Bubbles*, North Holland, 1975).

deposited a drop of a colloidal ferromagnetic suspension (a ferrofluid) on the surface of a ferromagnetic solid, and the pattern obtained reproduced the domain structure of the material.

Regions structured in domains are also found in ferroelectrics and superconductors. The analogy with the situation encountered in a polycrystalline material is immediately obvious, but of all of these materials, magnetic, ferroelectric domains or domains in a superconducting state correspond to spatial dependence of the order parameter: each one is associated with a different orientation of the order parameter. The existence of domains affects the properties of a material; this is an important problem to which we will return.

The formation of a domain is governed by the free energy of the system, which must be minimized. As in nucleation within a liquid, gas, or solid, the formation of magnetic or ferroelectric domains is associated with the creation of an interface that costs energy. Their size is an optimum corresponding to a minimum of the free energy of the material. If we consider a ferromagnetic or an antiferromagnetic, we do not abruptly pass from one domain to an-

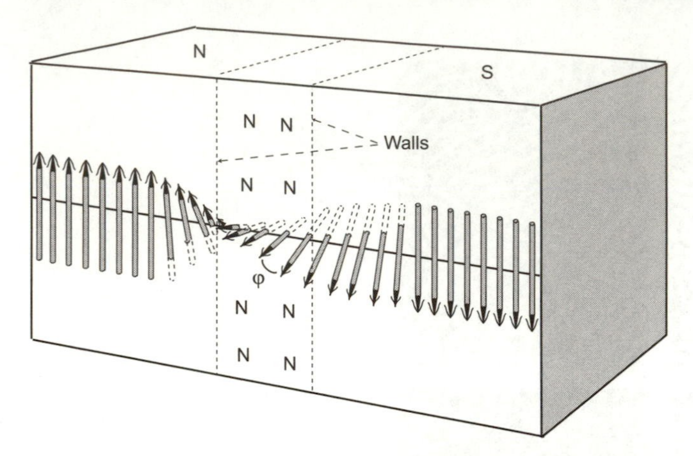

Fig. 9.11. Bloch wall. Bloch walls separate magnetic domains and have a finite thickness; φ is the angle between neighboring spins.

other. F. Bloch was the first to recognize that this transition was progressive. Two contiguous domains are in fact separated by a wall of finite thickness, called a **Bloch wall** in a ferromagnetic material and a **Néel wall** in an antiferromagnetic material. The magnetic orientation changes direction within the wall (Fig. 9.11).

The energy of the wall can be simply calculated. The interaction or exchange energy between neighboring spins (i) and (j) is given by a Heisenberg Hamiltonian and can be written as:

$$U = -2JS_iS_j \tag{9.6}$$

We can show (Problem 9.3) that for a line of $N + 1$ spins, the total exchange energy is:

$$NW_e = JS^2\pi^2/N \tag{9.7}$$

The larger N is, the lower this energy is and the thickness of the Bloch wall will tend to increase to minimize this exchange energy. The phenomenon is in fact limited by the existence of **anisotropy** or **magnetocrystalline** energy. The origin of this energy is as follows: a ferromagnetic crystal has crystallographic axes which are preferred directions for magnetization and are naturally a function of the crystal structure, so that any deviation of spins with respect to these preferred orientations is associated with anisotropy energy.

Consider a Bloch wall parallel to the face of the cube of a crystal of cubic symmetry containing N atomic planes; the energy per unit of surface area of the wall U_p will be the sum of the exchange energy U_e and the anisotropy energy U_a. The exchange energy for each line of atoms perpendicular to the plane of the wall is given by (9.7) and the anisotropy energy U_a can be considered proportional to the wall thickness, that is, $U_a = KNa$, where K

is the characteristic anisotropy constant of the material and a is the lattice constant. As there are $1/a^2$ lines per unit of surface area, $U_e = JS^2\pi^2/Na^2$, where S is the value of the spin, and we thus have:

$$U_p = U_e + U_a = JS^2\pi^2/Na^2 + KNa \quad (9.8)$$

This energy is minimum for $(\partial U_p/\partial N) = 0$, that is:

$$-JS^2\pi^2/N^2a^2 + Ka = 0, \quad \text{and} \quad N = (JS^2\pi^2/Ka^3)^{1/2} \quad (9.9)$$

In the case of iron, $N \cong 300$, which corresponds to a wall thickness of the order of 100 nm.

Now assume that a magnetic field is applied with the magnet parallel to a preferred direction of easy magnetization. The domains oriented in the direction of the field will tend to increase to the detriment of those with differently oriented magnetization, and the Bloch walls will move to favor this increase. Impurities and defects in the material will "pin" the Bloch or Néel walls, and are in a way the equivalent of nucleation sites for them. The ease with which these movements take place will determine the **coercive force** of the material, that is, the magnetic field that must be applied to cancel out induction. This is a very important characteristic property for magnetic materials. This magnetic field can vary from 0.03 A/m (0.0004 Oe) for permalloy to 0.72×10^6 A/m (9000 Oe) for Co_5Sm alloys. The value of the coercive field determines the shape of the hysteresis curve of a magnet. This coercive field is itself highly dependent on the size of the microstructures in the material. Magnetic materials formed of powders with very fine grains have high coercivity.

If the intensity of the magnetic field applied to a material is increased, in some cases the magnetic domains will evolve by assuming a cylindrical shape (Fig. 9.10c, d) and end in forming **bubbles**. The diameter and stability of the bubbles are a function of the intensity of the magnetic field. They can be moved in the material by applying a magnetic field gradient, and these microstructures can be used to store bits of information in a magnetic memory.

Situations comparable to those in magnetic materials in ferroelectric crystals are encountered. In the absence of an electric field, a ferroelectric crystal is not uniformly polarized in one direction, but a very large number of domains which each have electric polarization with its own orientation, where the polarization varies from domain to domain is observed. As in magnetic systems, ferroelectric domains are separated by walls which are regions within the crystal where a change in polarization orientation is observed. In some ferroelectrics such as TGS (triglycine sulfate), there are only two possible antiparallel orientations of the polarization, so that the walls separate domains with antiparallel polarizations. In other ferroelectrics such as $BaTiO_3$, the structures are more complex, with several possible orientations for the domains (Fig. 9.12).

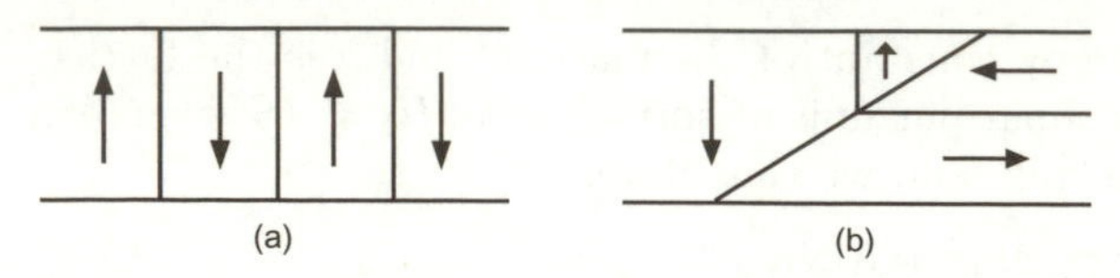

Fig. 9.12. Domains in a ferroelectric. (**a**) Wall at 180°. (**b**) Wall at 90°. They delimit the domains at equilibrium.

The domains in a ferroelectric can be shown by an optical technique such as birefringence; they are induced by spontaneous polarization of the material. As the polarization varies from one domain to another, the birefringence is different in two adjacent domains, and this difference appears by observation of the crystal in polarized light microscopy.

The application of an electric field to a ferroelectric alters the number and size of the domains: when an electric field is applied to a ferroelectric crystal, domains are formed on the surface (this is the equivalent of nucleation), then they grow inside the material and end in coalescing to form a single domain corresponding to the maximum polarization.

With respect to the long-range order associated with Cooper pairs in a **superconductor**, there are analogies with the situation characterizing a magnetic material. In a ferromagnetic, in effect, the interaction between spins, which is cooperative, induces alignment of all spins in the same direction below the Curie temperature, while in a superconductor, the long-range order is associated with the complex wave function ψ of the Cooper pairs (written as $\psi = |\psi| e^{i\varphi}$, where $|\psi|$ is its amplitude and φ is its phase), and the direction of ψ (φ phase) plays the role of spin orientation. The Cooper pairs must conserve long-range phase coherence in the material. By analogy with solid mechanics, a "rigidity modulus" of the phase of the superconducting state ρ_S which becomes zero at T_C is defined. However, in two dimensional superconductors (of the copper oxide type), there can be Cooper pairs without long-range phase coherence. Moreover, one of the characteristics of a superconductor is to "expel" an applied magnetic field: this is the **Meissner effect**.

Magnetic fields can penetrate **type I** superconductors by only destroying the superconductivity. There is an intermediate situation in **type II** superconductors: weak magnetic fields are expelled, while a field of intensity greater than $\boldsymbol{H}c_1$ penetrates the material unevenly, forming **vortices**; when the magnetic field intensity exceeds a critical value $\boldsymbol{H}c_2$, superconductivity disappears. The more the magnetic field intensity is increased above $\boldsymbol{H}c_1$, the greater the number of vortices formed in the material; this is the equivalent of a nucleation phenomenon. The critical field values $\boldsymbol{H}c_1$ and $\boldsymbol{H}c_2$ are dependent from the superconductor material. In the new high-temperature superconductors (copper oxide family), $\mu_0 \boldsymbol{H}c_2$ attains 15 T at 77 K but can increase to 50 T at 4 K.

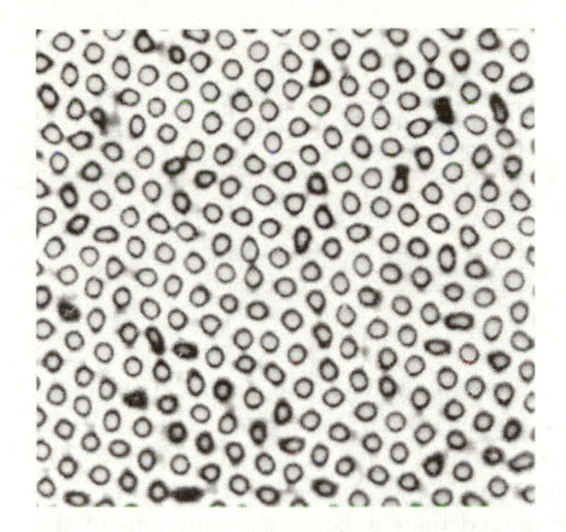

Fig. 9.13. Vortices in a superconductor. The vortices are the magnetic flux lines that penetrate a superconducting material like tubes. This image represents the flux lines in a crystal of the BiSrCaCuO type; they are demonstrated by the magnetic decoration technique. The vortices form a quasi-lattice, and the separation of the slots is of the order of 1μ (K. Harada et al., *Phys. Rev. Lett.*, **71**, 3371, 1993, copyright Am. Inst. Phys.).

The vortices are indeed magnetic flux lines that penetrate the solid like tubes (Fig. 9.13). They can be revealed by the magnetic decoration technique: the grains of a ferromagnetic powder placed on the surface of the material will be attracted by the flux lines and will localize the vortices in this way. These grains can be formed by inducing their vapor phase nucleation, for example, and they can be identified by transmission electron microscopy. Note also that the vortices are positioned on the surface of the superconductor, forming a **triangular lattice**.

The magnetic flux is quantized, and each flux line carries n flux quanta ϕ_0, where the individual quantum is $\phi_0 = h/2e$. If we consider an individual vortex, we find that the diameter of a flux line is the characteristic length λ which is the penetration length of the magnetic field in the superconductor. The region where the amplitude of the wave function ψ is zero or almost zero has a diameter ξ which is the coherence length of the superconductor (as shown on Fig. 9.14).

In superconductors of the copper oxide family, ξ is of the order of 1–2 nm at low temperature, while λ is of the order of 100 nm.

The vortices are "pinned down" by structural defects in the material (they can result from precipitation of one phase into another) or by impurities. Movement of the vortex lines within the superconducting material can be considered a nuisance because it is an energy dissipation factor and causes a drop in the electric potential (movement of the current lines can be compared with the velocity field around the eye of a cyclone, with all proportions

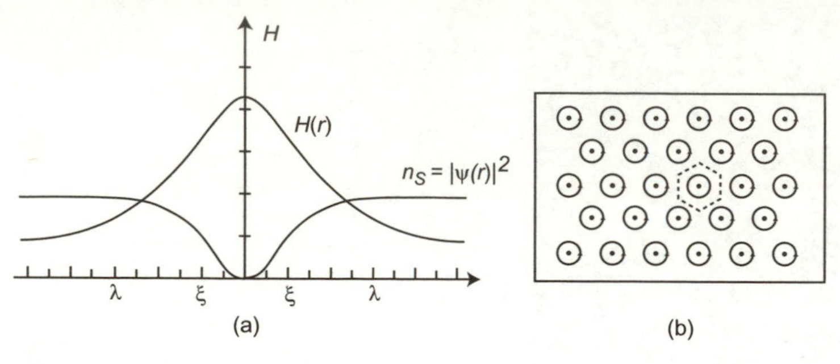

Fig. 9.14. Structure of a vortex in a superconductor. (**a**) The change of the magnetic field $H(r)$ within a vortex (r is the distance with respect to the center) and the square of the amplitude of the wave function of a Cooper pair inside a vortex in a type II superconductor is shown. (**b**) Vortex lattice: the central nucleus is surrounded by a circular current; the dotted lines indicate the unit cells of the lattice.

retained). In practice, to expand the applications of superconductors, the vortices should thus be "pinned" to prevent movements of the flux lines; point defects are thus a stabilizing factor for the superconductor.

In the new high temperature superconductors, vortex pinning is related to the planar structure of the materials; two copper oxide planes, Cu_2O, where most of the electrons in the superconducting state are localized, define the configuration of the vortices; they are separated by insulating layers or layers with metal conductivity, which oppose circulation of electrons between superconducting planes by a tunnel effect. In superconductors, it is important to stabilize the vortices by strengthening the coupling between adjacent Cu_2O planes, for example, by reducing the distance between planes and by metallizing the intermediate layers by chemical substitution.

The vortex lattice in the superconducting phase can be considered similar to a classic three-dimensional crystal lattice (the vortices are the periodic lattice points of the crystal), and it also undergoes the equivalent of a solid/liquid phase transition when the temperature is increased and the critical temperature T_C is approached. In the "pseudoliquid state" associated with the vortices, the type II superconductor has nonzero resistance which ends in destroying the superconducting state. This situation can be described by a solid/liquid phase diagram (Fig. 9.15): when the solid vortex/liquid vortex transition line is crossed, the vortex lattice "melts" and the superconducting state tends to disappear (the critical current that it can withstand approaches zero). By analogy with a real solid or a real liquid, the system can be described by the interactions between vortices and they determine the transition. Latent transition heat associated with this solid-liquid transition was demonstrated in a superconducting compound of the YBaCuO type with a differential calorimetry technique, and this transition is thus first order. Understanding phenomena of this type is important because

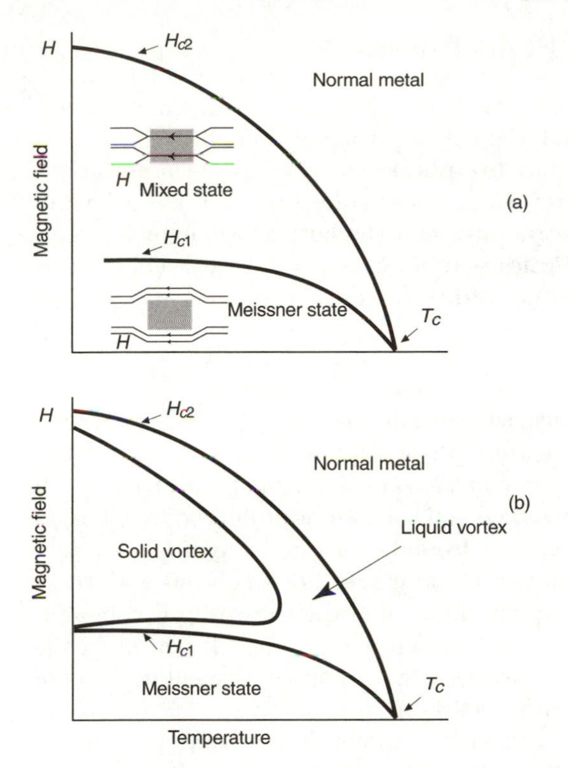

Fig. 9.15. Phase diagram of a superconductor. (**a**) Type II superconductor with penetration by vortex lines in the mixed state. (**b**) Type II high-temperature superconductor with a so-called liquid vortex state; in this region, the material has nonzero resistance that limits its practical interest. In the Meissner state, the field lines are expelled outside the material (D. Bishop, *Nature*, **382**, 360 (1996), copyright Macmillan Magazines Ltd.).

the structure of the vortices and their interactions determine the properties of type II superconductors and their capacity for withstanding high critical currents, which is obviously determining for their technological applications, particularly in electrical engineering.

The long range phase coherence is destroyed by the presence of vortices, which are strongly correlated when they exist at low temperature. In two-dimensional systems, the **Kosterlitz–Thouless transition** precisely involves destruction of phase coherence and destructuring of the vortex lattice, a phenomenon found in bidimensional superconductors, superfluid helium films, and liquid crystal monolayers. For a superconductor, the corresponding transition temperature is $T_{KT} = 8\rho_S/\pi$.

9.3 Microstructures in Fluid Phases

Milk, water colors, oil and vinegar dressing, or fog have a common characteristic: they are phases constituted of microstructures in equilibrium in a fluid (liquids and a gas). They are in fact two-phase systems that are in precarious equilibrium which is easily perturbed: a phase transition causes breaking of equilibrium between the microstructures and the host phase, inducing their precipitation The surface properties play an essential role in keeping these phases, which are **emulsions** or **colloidal systems** in equilibrium.

9.3.1 Microemulsions

Emulsions, and thus microemulsions, are systems in which the dispersing medium and the dispersed phase are both immiscible liquids. The most frequent and best-known case is microemulsions of water in oil, or inversely, oil in water. The mixture that forms the emulsion can be stabilized by addition of surface-active molecules having a hydrophilic part and a hydrophobic part and which are absorbed on the water–oil interfaces (for emulsions with these two constituents). These mixtures are fluids, and their viscosity is comparable to the viscosity of water. In fact, the molecules are not dispersed on the molecular scale, and it is a two-phase system, composed of small droplets of water (or oil), generally of the order of 10 nm in diameter.

To create an emulsion from a water–oil system, it is necessary to supply mechanical energy (by stirring, for example) at least equal to the interfacial energy that is to be created. The system is not stable in the long term and returns to its initial state, the two liquid phases, after a more or less long time. The role of the surfactants added to the mixture of the two liquids is to stabilize the emulsions, on one hand by strengthening the mechanical stability of the droplets, and on the other, by creating barriers that oppose their coalescence (Fig. 9.16).

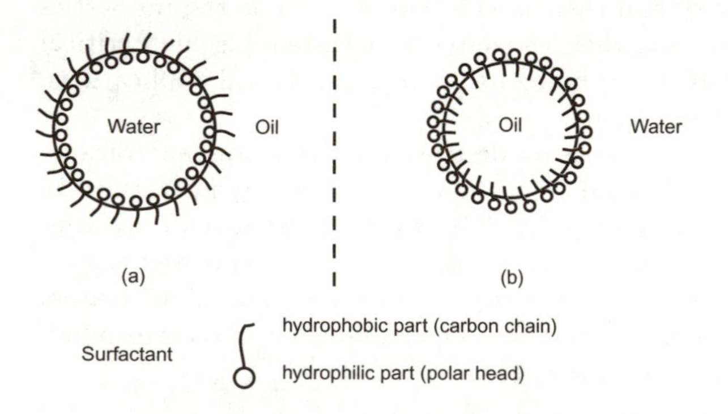

Fig. 9.16. Microemulsions. Droplets of microemulsions are shown: (**a**) water in oil; (**b**) oil in water. The surfactant molecules are attached to the surface of the droplets whose mechanical stability they strengthen.

Surfactants are molecules that tend to be absorbed on free surfaces or interfaces between two liquids and between a liquid and a solid. These molecules are composed of two parts which have very different affinities (for example, a hydrophobic chain and a hydrophilic head because it can be ionized), as in the case of sodium dodecyl sulfate, $CH_3(CH_2)_{11}SO_4^-Na^+$.

In general, molecules having two different parts are called amphiphiles. They are surfactants if they modify the surface energies when absorbed on free surfaces.

The surface tension at the interface between two liquids (drops of oil in water, for example) decreases due to absorption of the surfactant on the surface of the droplets which form a bidimensional film in this way. The situation will be more favorable the closer the interfacial tension is to zero. If γ is the oil/water surface tension in the absence of a surfactant and $\pi(A)$ is the surface tension due to the monomolecular layer of surfactant, the surface tension of the drops is $\gamma = \gamma_0 - \pi(A)$.

The entropy of dispersion will increase if the surface tension γ is close to zero and the microemulsion will be stable for a value A such that $\pi(A) = \gamma_0$ (the medium emulsifies). In general, surfactants which are alcohols are used. The radius R of a droplet covered with a surface-active monomolecular film can be calculated. If φ is the volume fraction of the dispersed phase and n_s is the number of molecules of surfactant per unit of volume, we have:

$$R = 3\varphi/n_s A \tag{9.10}$$

Microemulsions are used in many areas of daily life and in numerous industrial sectors: the food and cosmetics industry use a very large number of systems which are emulsions (milk is an emulsion). Microemulsions are also used in the chemical industry. In particular, they are excellent catalysts due to their high degree of division; these catalysts can be stereospecific because the surfactant film at the interface is capable of orienting the substrates of the reaction. Finally, microemulsions are of interest for assisted oil recovery. When water is injected in an oil well to displace the oil trapped in rock pores, some of the oil remains trapped by capillary forces which are inversely proportional to the pore radius. By reducing the water/oil surface tension by injecting a surfactant, these capillary forces and thus the water pressure necessary for driving out the oil are decreased by its very nature. A microemulsion containing the surfactant is in fact injected. This technique, which would theoretically allow recovering 100% of the crude oil, is now only cost-effective in a limited number of fields given the relatively high cost of the surfactants. Fuels (in particular, for diesel engines) which are emulsions of water in gasoil stabilized by a surfactant have also been considered. They would have the advantage of reducing emissions of pollutants (solid particles and NO_x).

9.3.2 Colloids

The Chinese have known since antiquity that a fine aqueous suspension of carbon black no longer decanted by separating from the solution when gum arabic (a natural polymer) was added to it. This mixture, which constitutes Indian ink, no longer underwent a phase transition (demixing in some way) as it was stabilized by an additive; it constitutes a **colloidal system**. The phenomenon is strictly the opposite when a wine maker finds that the wine in a barrel has become cloudy and induces separation of suspended organic particles by adding a polymer (egg white, which is albumin, or gelatin) to the wine. This is **clarification** of the wine, which is associated with phase separation, **flocculation**.

In general, a system formed by a substance dispersed in a solvent and whose molecules are grouped in **micelles** bearing an electric charge of the same sign is called a **colloidal state**. A micelle is a globular aggregate typically 3–5 nm in diameter. The stability of the colloidal system is conditioned by the existence of an electric charge and a solvation shell in the dispersed particles. Colloidal particles absorb a certain species of ions; the system is globally neutral since the charges of the particles are compensated by the charges of the ions of opposite sign present in the solution (Fig. 9.17).

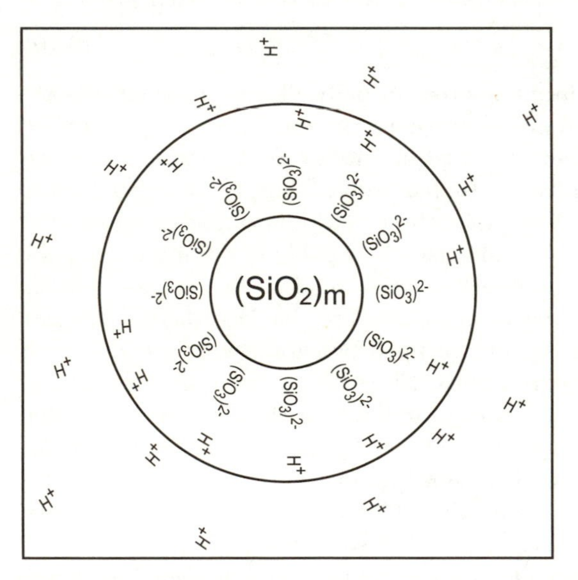

Fig. 9.17. Structure of a micelle. The micelle is composed of a nucleus $(SiO_2)_m$. The surface of this nucleus reacts with water to form silicic acid, H_2SiO_3 which is ionized. The whole micelle is globally electrically neutral, as the SiO_3^{2-} charges are balanced by H^+ ions which together form an electric double layer.

The fact that the particles of a colloidal system are charged clearly plays an essential role in its stability: their charges oppose coagulation of the system. We also find that if an electric field is applied to a colloidal system, all of the suspended particles will move in the direction of one of the electrodes. In colloidal systems composed of iron, aluminum, or chromium hydroxide particles, they have a positive charge, while colloidal gold, silver, silica, starch, and gum arabic particles are negatively charged.

There is a very large number of colloidal systems with particles sizes between 1 and 200 nm. Some are natural, like milk and blood (the solvent is a liquid medium), and agate and opals correspond to a solid system. Others are of industrial origin and are found in the fluid suspensions used in the chemical and oil industries or in metallurgy. Colloidal systems are also encountered in aerosols, mists, or foams, systems in which the dispersing medium is a gas.

The stability of a colloidal system is determined by the interaction forces that exist between the particles in the medium. The theory of Derjaguin, Landau, Verwey and Overbeek accounts for this stability by implicating attractive forces (the van der Waals force) on one hand and repulsive forces of electrostatic origin on the other hand. Suspended colloidal particles are charged, but they attract charges of opposite sign (counterions) on their surface; they form a **double layer** on their surface. This is the origin of the electrostatic repulsive force which decreases exponentially as a function of the distance d between particles in the medium and which is a function of its ionic strength. The van der Waals attractive term is short-range and varies as $-d^{-n}$ (Fig. 9.18).

The situation of thermodynamic equilibrium corresponds to the total interaction energy minimum $E(d)$. Figure 9.18 shows that a series of structures corresponding to different forms of function $E(d)$ can be found. An energy barrier W which corresponds to the maximum of function $E(d)$ and the minimum for the potential energy, the position of stable thermodynamic equilibrium (the corresponding distance d can be of the order of 2–3 nm), can generally be found on the curves.

When $d > d_{max}$, if the particles have no sufficient external energy source (mechanical stirring, for example), they cannot cross the potential barrier W to pass into the attraction zone; they then remain separated from each other in a metastable state.

The situation is similar when there is a second minimum for $E(d)$; in this metastable state, the particles form aggregates which can be dispersed by stirring, for example. W can also be decreased by reducing the surface charges, for example, by addition of salt which increases the ionic strength. When the energy barrier has disappeared, the particles will tend to agglomerate in the position corresponding to the unique minimum of $E(d)$: we say that there is **flocculation**.

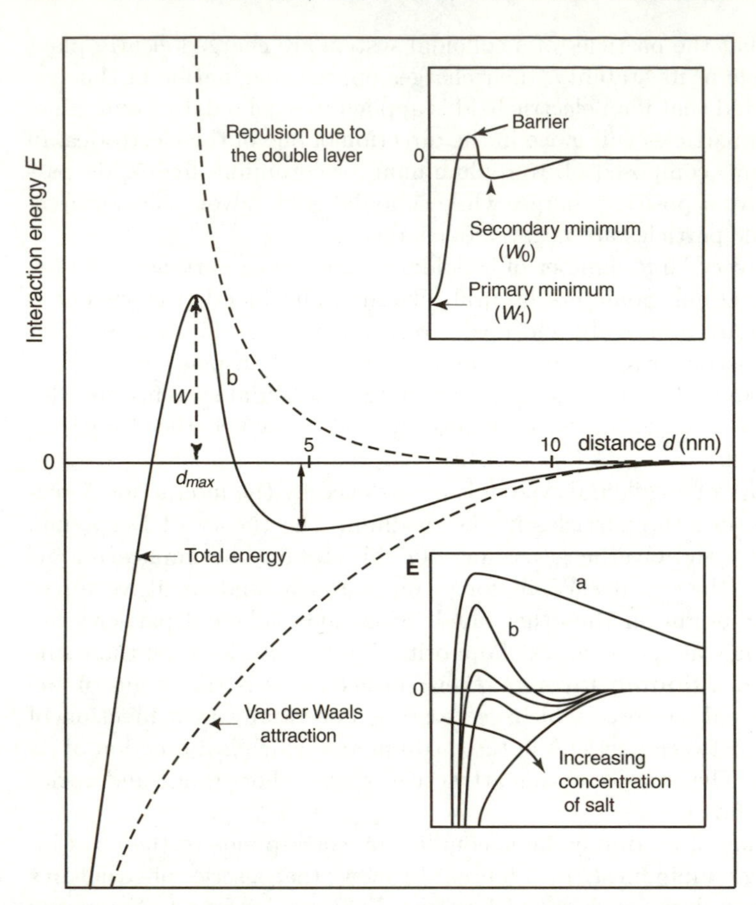

Fig. 9.18. Stability of a colloidal system. The stability of a colloidal system is determined by the interaction forces between particles in the medium. The total energy $E(d)$ is a function of the distance d between particles: it is the sum of the attractive part of the van der Waals potential and the repulsive energy due to the double layer. There is an energy barrier W corresponding to the maximum of $E(d)$ and a minimum Wo. If $d > d_{max}$ (2–3 nm), the particles cannot cross the barrier to pass into the attractive zone and minimize their energy: there is no coalescence. W can be decreased by adding a salt to the solution, which decreases the surface charges. The van der Waals potential has a repulsive part (*dashed curves*) at short distances. When the barrier W has disappeared, there is flocculation (e) (J. Israelachvili, *Intermolecular and Surface Forces*, Academic Press (1992), p. 248).

9.4 Microstructure, Nanostructures, and Their Implications in Materials Technology

Phenomena such as melting and solidification can be influenced by the size of the materials, as we have seen (Chapter 3). Supercooling of metal drops several tens of nanometers in size over several tens of degrees and superheating of small nanometer particles over several tens of degrees are observed.

The formation of microstructures affects very significantly the mechanical properties of a material:

- An amorphous Al–Ni–Fe–Ce alloy fabricated with aluminum nanoparticles 3–10 nm in diameter has a breaking strength a factor of three higher than the breaking strength of the highest performing conventional alloys (Fig. 9.19);
- Cu–Pd nanophases formed with particles 5–7 nm in diameter have hardness and breaking strength five times higher than ordinary alloys (nanoparticles have fewer defects).

In polycrystalline structures, the grain size is an important factor that conditions the mechanical properties: the breaking strength increases when the size decreases, as the preceding examples show. This effect, already known for micrometer structures, is called the **Hall–Petch** effect. It is explained by blocking of dislocations by grains which can no longer propagate in the material, thus delaying rupture. The smaller the grain size, the more effective this blocking is. This effect could disappear for very small grains, as gliding

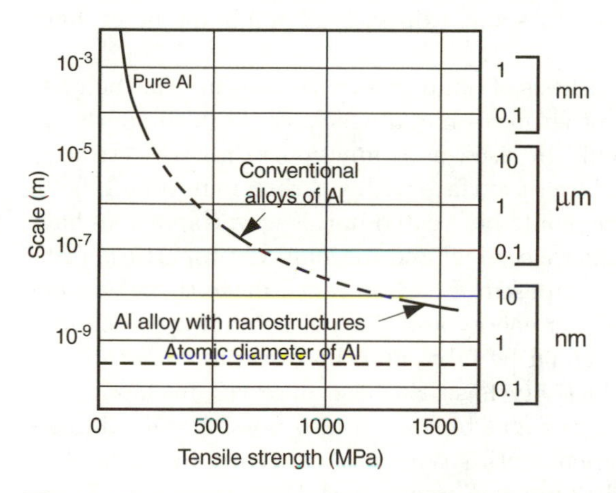

Fig. 9.19. Nanostructure and mechanical strength. The tensile mechanical strength of aluminum alloys is compared here. Alloys with nanostructures formed by rapid quenching have high strength. These materials are composed of crystalline nanaoparticles dispersed at a concentration of 20–25% in an amorphous matrix (A. L. Greer, *Nature*, **368**, 688 (1994), copyright Macmillan Magazines Limited).

would occur at the grain boundaries. This has been found in copper and palladium for grains ≈ 10 nm in size.

In ceramics, microstructures improve their ductility because they glide more easily with respect to each other. Fabrication of nanocomposite ceramic materials is also possible. A composite manufactured from nanoparticles of carbon and silicon nitride (less than 500 nm in diameter) is superplastic at a temperature above 1600°C and can be molded.

It is necessary to emphasize that when nanomaterials composed of 1000–10,000 atoms (less than 10 nm in size) are synthesized, a new physics of matter is possible because the electronic properties of nanomaterials are of another nature. These atomic aggregates have collective properties of quantum nature and they can be modulated by changing their composition.

Nanodiodes that emit in the red with aggregates of aluminum and indium arsenide can be made. By changing the size of semiconducting materials such as CdSe, they can be made to emit in the red or green. Their energy levels are in effect a function of the size of the aggregates, also called quantum dots. These nanomaterials should allow making electronic and optical devices such as micro- or nanolasers.

These techniques for production of electronic components (microlithography utilizing electrons, X-rays, and UV) and the advances in microscopy (scanning tunneling and atomic force microscopy) now allow fabricating nanostructures.

Synthesis of compounds such as fullerenes (spherical groups of 60 carbon atoms) also allows fabricating new chemical architectures and opens up other prospects for nanomaterials. "Nanotubes," three-dimensional assemblies of carbon atoms constituting true tubes approximately 15 nm in diameter, have been synthesized in this way.

It is possible to imagine synthesis of nanotubes with one end that behaves like a semiconductor and the other like a metal which would in some way be a "biphasic" material that could be used as a junction in microelectronics. Moreover, atoms can be moved over a surface with a carbon nanotube 100 nm long. A true molecular nanotool could be created in this way. Nanotubes have interesting prospects due to their electrical and mechanical properties. Their tubular structure, with walls composed of carbon rings, make them particularly suitable for absorbing impact energy and thus withstanding mechanical stresses. Composite materials could be fabricated by inserting nanotubes in polymer matrices. Superconductivity has also been observed in nanotubes. The marriage of colloid chemistry and biology is also possible for fabricating nanomaterials. An experiment of this type was conducted by assembling nanoparticles of colloidal gold 13 nm in diameter with DNA fragments. These nanomaterials can have interesting optical and electronic properties that can be controlled by changing the particle size and length of the DNA sequences.

Problems

9.1. Droplets in Equilibrium with their Vapor

Consider the equilibrium of a droplet of radius r in equilibrium with its saturated vapor at pressure p and temperature T; the liquid/vapor surface tension is designated by γ and the equilibrium pressure with the liquid as a whole is designated by p_∞. We set

$$S = \frac{p}{p_\infty}$$

which is the saturation factor.

Calculate S as a function of r, γ, T. We will assume that the vapor is an ideal gas.

9.2. Creation of Holes

Holes which will modify the structural properties are created in a solid. ε_f is the energy of formation of a hole. N is the total number of sites, and the concentration of holes is $c = n_l/N$, where n_l is the number of holes.

1. Calculate the change in free enthalpy associated with the formation of holes.
2. Determine the concentration of holes at equilibrium.
3. $\varepsilon_f = 1$ eV; determine c at 300 K and at 1000 K.

9.3. Energy of Bloch Walls

The interaction energy between neighboring spins i and j is given by a Heisenberg Hamiltonian $U = -2JS_iS_j$ (Fig. 9.11).

1. Calculate the energy change associated with rotation of these neighboring spins by angle φ.
2. Assuming that total inversion of orientation of magnetization is observed in N spins, calculate the total exchange energy for the N spins of a wall.

9.4. Molecular Film

Surfactant molecules are adsorbed on the surface of a liquid and the pressure π they exercise on a floating mobile barrier separating the surface of the liquid into two compartments, one of them with no surfactants, is measured. The number of moles of surfactant adsorbed on the surface of a liquid of area A is designated by

$$\Gamma = \frac{n^s}{A}$$

If γ is the surface tension, we have $\mathrm{d}\gamma = -\Gamma \mathrm{d}\mu$ (Gibbs relation), where μ is the chemical potential of the adsorbed molecular species. Assume that γ is a decreasing linear function of the concentration x of surfactant.

1. Write the relation between γ, Γ and γ_0, the surface tension of the pure liquid.
2. Write an equation of state. What conclusion can be drawn from this?

10 Transitions in Thin Films

10.1 Monolayers at the Air–Water Interface

10.1.1 The Role of Surfactants

Flow of oil on the surface of water has been observed since the 19th century, but it was only at the beginning of the 20th century that Lord Rayleigh suggested that films are composed of a monolayer of molecules. It was necessary to wait for another 20 years for Langmuir and Harkins to independently demonstrate the formation of films composed of a monolayer of molecules on the surface of water and reveal that the formation of such layers is only possible if the molecules constituting the film are amphiphilic (Sect. 9.3.1):

- insoluble in water due to the presence of a hydrophobic chain (paraffin chains, for example);
- having a hydrophilic polar "head" at one end (such as, a –COOH carboxyl or –OH hydroxyl or ionic termination).

In fatty acid films on water, Langmuir and Harkins were the first to suggest that the stability of a monolayer was due to the fact that the hydrophilic polar "heads" are immersed in the surface of the water while the hydrophobic paraffin chains remain above the surface of the water.

The establishment of this theoretical scheme in large part explains the increasing interest in the study of monolayers. The dual goal was to understand the structure of these elementary systems on one hand, and to determine the degree to which some of their characteristics are shared by other more complex systems such as bilayers and micelles on the other hand.

The subsequent research was strongly dependent on the available methods of analysis. Until very recently, the essential experimental tool was the Langmuir trough (Fig. 10.1) which allows measuring the relation between the surface pressure, π, and the molar surface area A in a monolayer deposited on the surface of water at a given temperature. This technique allows thermodynamic characterization of the monolayer, but it obviously cannot supply accurate information on the structure of the monolayer and in particular, the presence or absence of phase coexistence. We also note that the experiments conducted in Langmuir troughs are extremely difficult to control

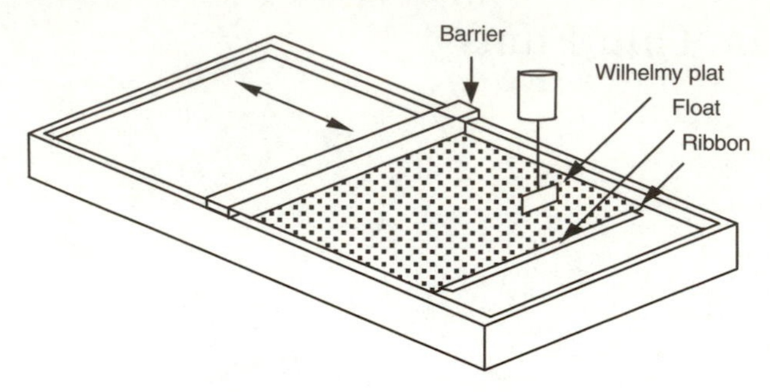

Fig. 10.1. Schematic diagram of a Langmuir trough. The monolayer is deposited on the surface of the water in the region bordered by the barrier and the float. A thin and very flexible tape on the ends of the float stops any movement of the layer behind the float. The float thus separates the surface covered by the layer from the uncovered water. To vary the area accessible to the monolayer, the barrier can be moved while remaining parallel to the float. The surface pressure can be measured either by determining the force exercised by the layer on the float or by measuring the difference between the forces exercised on the Wilhelmy plate when the plate is suspended in pure water and in the water covered by the monolayer.

because of the risks of contamination of the monolayer on one hand, and the interpretation of the results is often ambiguous and subject to argument on the other.

The recent renewed interest in the study of monolayers was stimulated by the development of new techniques: imaging, reflection, and diffraction methods, spectroscopic methods. These methods allowed clarifying many controversial points and obtaining accurate information on the structure of the phases in the monolayers.

10.1.2 Examples of Molecules Forming Monolayers

The simplest monolayers in the most recent studies are composed of:

- fatty acids (carboxyl polar "head");
- alcohols (hydroxyl polar "head");
- phospholipids (phosphatide polar "head").

These three types of amphiphilic molecules (Fig. 10.2) all contain one or two long hydrophobic saturated hydrocarbon chains with $n = 12$ to 20 $-CH_2-$; a minimum of approximately 12 units of $-CH_2-$ is necessary for the molecules to no longer be practically soluble in water.

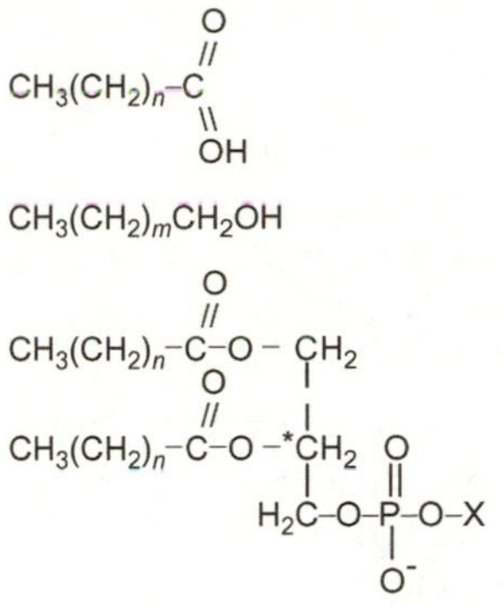

O
//
$CH_3(CH_2)_n$–C with n = 12 to 20
\\
OH

$CH_3(CH_2)_mCH_2OH$ with m = 16 to 19

O
//
$CH_3(CH_2)_n$–C–O – CH_2 with n = 12 to 12
O |
// |
$CH_3(CH_2)_n$–C–O –*CH_2 O X corresponds to $^-CH_2CH_2\overset{+}{N}(CH_3)_3$ ou ^-H
| ||
H_2C–O–P–O–X *C corresponds to an asymmetric carbon
|
O^-

Fig. 10.2. Examples of amphiphilic molecules.

10.1.3 Preparation and Thermodynamics Study of Monolayers

a) Monolayers on the surface of water

A monolayer can be obtained by depositing a solution of amphiphilic molecules (composed of a hydrophilic "head" and a paraffin chain) in a volatile solvent immiscible with water on the surface of pure water; during evaporation of the solvent, the film, called a **Langmuir film**, forms spontaneously.

The surface pressure π is defined as the difference between the surface tension of pure water, γ_0, and the surface tension in the presence of the film, γ:

$$\pi = \gamma_0 - \gamma \tag{10.1}$$

At constant temperature, the relation between π, the surface pressure, and the area accessible to the monolayer is obtained with a Langmuir trough, shown in Fig. 10.1. The film is deposited on a surface closed by a fixed barrier and a float connected to the walls by two flexible tapes; an average area per molecule, designated A, is imposed in this way. It is imperative that the surface of the water behind the float be free of impurities. The surface pressure π can be measured:

- either by determining the force F acting on the float of length L:

$$\pi = \frac{F}{L} \tag{10.2}$$

- or by measuring the surface tensions of the pure water surface γ_0 and the surface of the water covered by the film, γ, with a Wilhelmy plate; the surface pressure, π, is then determined from (10.1).

By performing these operations:

- for different positions of the barrier and thus different values of A;
- for different temperatures, T;

it is possible to obtain the equation of state for the film $\pi = f(A, T)$.

b) Monolayers on the Surface of a Solid

The monolayers obtained by the technique just described can be transferred to a solid substrate. To deposit amphiphilic molecules on a substrate, the molecules are first dissolved in a volatile solvent; this solution is immediately spread on the water–air interface in a Langmuir trough (Fig. 10.3a). The area of the layer can then be modified by moving the barrier to change the surface density and local organization of the amphiphilic molecules. The previously immersed substrate is removed from the water by keeping its plane perpendicular to the surface supporting the monolayer (Fig. 10.3b). A monolayer can be deposited on the substrate in this way. A new layer is deposited each time the plate is transferred through the surface of the water. The films deposited in one or more layers on the surface of a solid are called **Langmuir–Blodgett films** (Fig. 10.3c and d).

The monolayers obtained on a solid substrate can provide information on Langmuir monolayers, but it is necessary to make sure that the surface of the support (plate) does not modify the structure of the monolayer.

10.1.4 Phase Diagram of a Monolayer

A typical (π, A)-isotherm for a fatty acid or a phospholipid monolayer is shown in Fig. 10.4.

In zone G, corresponding to very high values of A ($A > 400$ Å^2/ molecule for the plotted isotherm), the monolayer is a gas G, in two-dimensional space.

When the value of A decreases, a first plateau (zone EL–G) terminated by A_G and A_{EL}, corresponding to condensation in a liquid state called the "expanded liquid" state EL, is found. This plateau thus corresponds to coexistence of the EL and G phases.

Again decreasing the value of A beyond A_{EL}, the surface pressure rises again until a new plateau appears; the monolayer is in the "expanded liquid" state (zone EL) in this zone.

The new plateau (zone DL–EL) delimited by A'_{EL} and A_{DL} corresponding to condensation of the "expanded liquid" state into a "dense liquid" state is designated by DL.

Again decreasing the value of A beyond A_{DL}, in some cases, a break in the slope in the dependence of A as a function of the surface pressure is observed; this break in the slope is interpreted as a transition from the "dense liquid" state to a two-dimensional solid state.

The two-dimensional crystal obtained can eventually be transformed into another more stable crystal form with a higher surface pressure.

Identification of the EL–G plateau as a phase coexistence region is unambiguous: this plateau is horizontal and the EL–G transition is thus first order. On the contrary, in the EL–DL zone, most of the experimental results seem to indicate that the plateau is not perfectly horizontal, so that identifying this zone as the zone of coexistence of EL and DL phases by saying that the

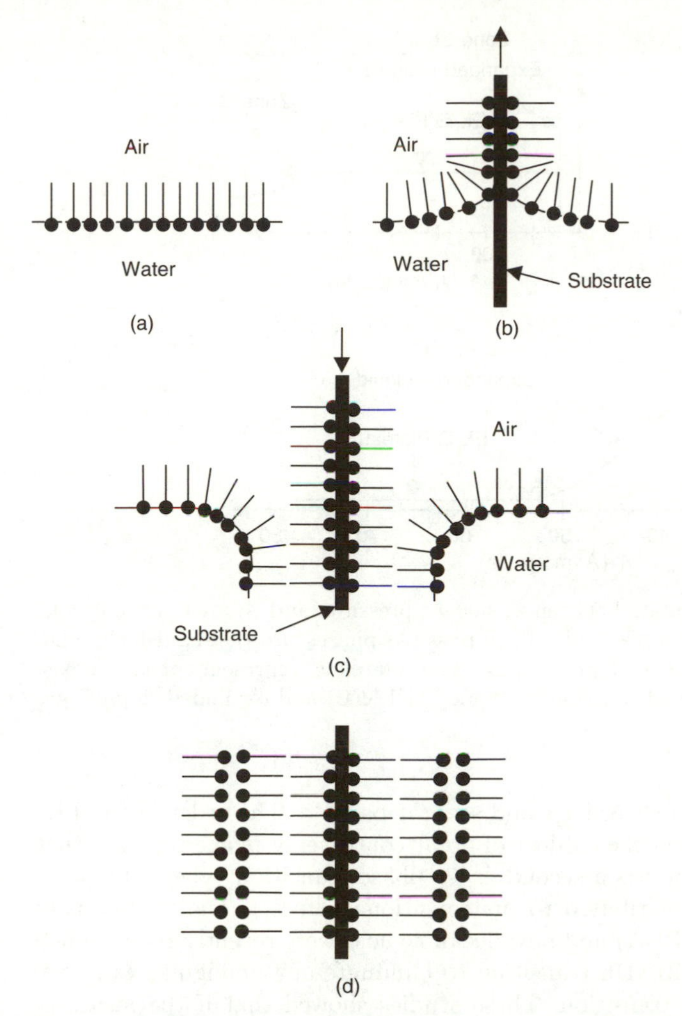

Fig. 10.3. Diagram of the Langmuir–Blodgett technique. The amphiphilic molecules are first dissolved in a volatile solvent; this solution is immediately spread to the water–air interface in a Langmuir trough (**a**). The area of the layer can then be modified by moving the barrier to alter the surface density and local organization of the amphiphilic molecules. To form a Langmuir–Blodgett film, the substrate, which was previously immersed, is vertically removed from the trough, passing through the Langmuir layer (**b**). Each new transfer of the substrate through this layer deposits a new monolayer on the substrate (**c** and **d**).

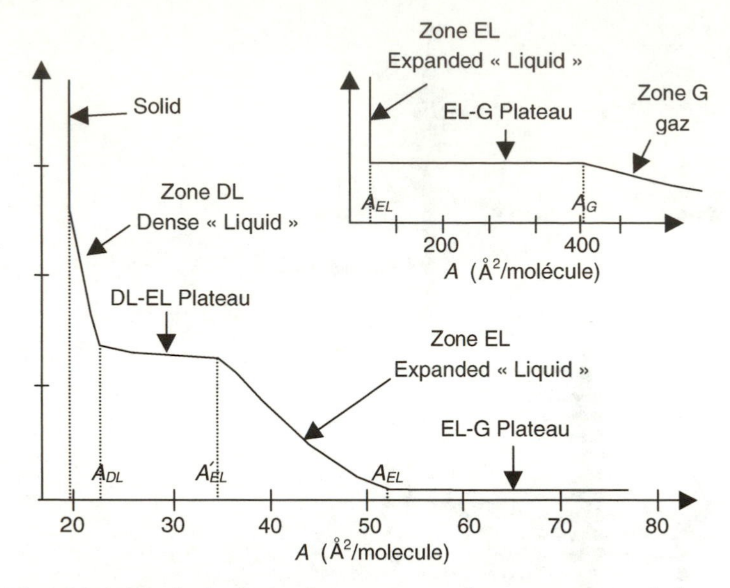

Fig. 10.4. Typical relation between π, surface pressure, and A, area per molecule, for fatty acids and phospholipids. Four possible phases are distinguished: solid, dense "liquid", expanded "liquid," gas. The plateaus represent phase coexistence zones: dense "liquid"–expanded "liquid" (DL–EL) and expanded "liquid"–gas (EL–G).

EL–DL transition is first order could seem debatable. The order of the EL–DL transition has been the subject of great controversy (some believed that the EL–DL transition was a second order phase transition, where the pseudoplateau was then attributed to pretransitional effects). The isotherms of pentadecanoic acid (PDA) and hexadecanoic acid were recently re-examined in the region of the EL–DL transition to eliminate any ambiguity as to the order of the EL–DL transition. These studies showed that if the measurements were carefully performed to avoid any contamination of the surface by extraneous molecules from the environment of the film, the plateaus corresponding to the EL–DL transition are completly horizontal: the EL–DL transition is thus indeed first order.

Until recently, characterization of the phases of monolayers was only based on the analysis of the (π, A)-isotherms, where the different parts of the isotherm were analyzed with the procedure just described. This method, although undeniably interesting, does not allow accurately defining the exact limits of the phase coexistence zones; for this reason, it cannot be used to demonstrate the possible existence of critical points. Other approaches are more suitable, fluorescence microscopy in particular, which can provide information on the microscopic structure of the monolayers in phase coexistence zones.

The amphiphilic molecules constituting the monolayer can be doped with fluorescent molecules. These molecules have to be amphiphilic to remained trapped in the monolayer. To be suitable for the study, their fluorescence intensity must be a function of the characteristics to be investigated: local density or orientation of the molecules in the film.

A diagram of a fluorescence microscopy setup is shown in Fig. 10.5. The incident light comes from a mercury lamp or a laser; after passing into a narrow-band filter and through a dichroic mirror, it is focused with a microscope lens on the monolayer investigated, previously deposited on the surface of the water. The part of the light emitted by the fluorescent molecules in the layer, received by the microscope lens, is reflected by the dichroic mirror. After passing into a narrow-band filter tuned to the fluorescence wave length, it forms an image of the fluorescent zones in the layer which is amplified and recorded by a video camera. This image can be displayed on a television screen and recorded for further processing.

The introduction of fluorescent impurities can obviously modify the phase diagram of the layer and cause demixing of fluorescent impurities; for this reason, it is imperative to evaluate the effects of a progressive increase in the concentration of molecules on the phase diagram to determine the concentration limits not be exceeded. This is thus a difficult method to implement, but if it is used correctly, it allows direct visualization of the zones occupied by coexisting phases.

Figure 10.6a shows an example of a fluorescence image obtained with a microscope. This concerns a monolayer of PDA (pentadecanoic acid) doped with fluorescent molecules of NBD-hexadecylamine in a molar concentration of 1%. The layer was prepared in the EL–G coexistence region. Phase G corresponds to the dark zones, while the light zones correspond to phase EL. Expansion of the monolayer leads to formation of a foam structure (Fig. 10.6b).

The technique just summarily described allows determining the limits of the phase coexistence zones (EL–G zones and CL–EL zone) for different temperatures with relatively good accuracy. This method allowed better establishing the phase diagrams of monolayers. For fatty acids, the fluorescence studies of the isotherms seem to converge toward the complex phase diagram shown in Fig. 10.7. When the surface pressure increases, the EL–G coexistence zone exhibits clear contraction that suggests closing of this zone and the existence of a critical point.

A similar situation is suggested for the DL–EL transition zone, where the phase coexistence zone seems to close again when the surface pressure increases; this should normally suggest the existence of another critical point. However, the problem is more complicated here; this is all a function of the symmetry of the DL phase, which could not be clearly identified for the moment. If the DL and EL phases have the same symmetry, closing of the DL–EL transition zone will correspond to the existence of a critical point.

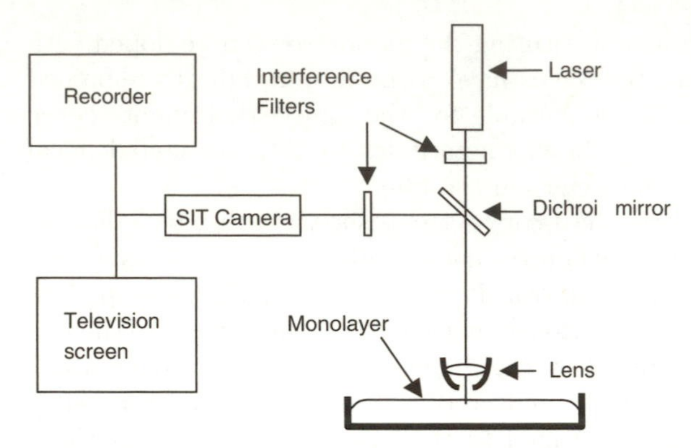

Fig. 10.5. Diagram of the apparatus used for fluorescence microscopy of monolayers. The exciting light is produced by a laser or a xenon arc; after passing through a narrow-band filter, the incident beam passes through the dichroic mirror, then it is focused by a microscope lens on the layer investigated, which contains fluorescent molecules. The light emitted by fluorescence and transmitted by the microscope lens is reflected by the dichroic mirror, then filtered by an interference filter tuned to the fluorescence wavelength; it forms an image of the fluorescent zones in the layer, which is amplified and recorded by a video camera. This image can be displayed on a television screen and recorded for further processing.

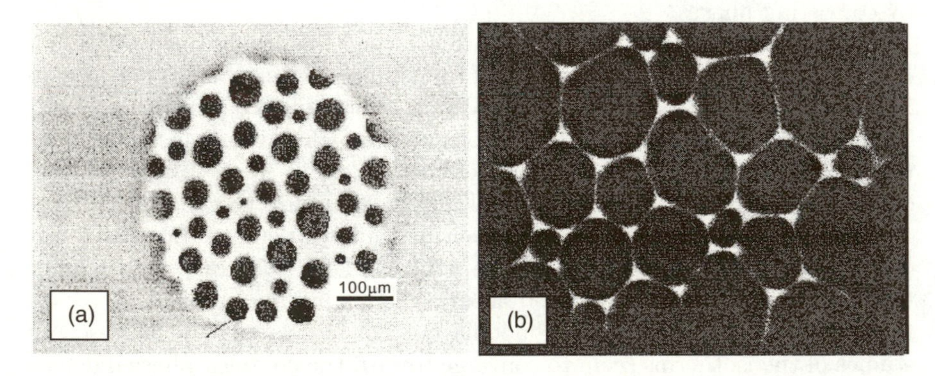

Fig. 10.6. Examples of fluorescence images obtained with a microscope. This concerns a monolayer of PDA in the region of EL–G coexistence. To be observable, the layer was doped with fluorescent molecules of NBD-hexadecylamine in a molar concentration of 1%. Phase G corresponds to the dark zones while the clear zones correspond to phase EL. Figure 10.6a shows the initial structure of the layer prepared at 20°C in the EL–G transition region. Its expansion leads to the formation of a foam structure (see Fig. 10.6b). The bar corresponds to 100 µm (B. G. More, C. M. Knobler, S. Akamatsu, and F. Rondelez, *J. Phys. Chem.*, **94**, 4588–4595 (1990) and C. M. Knobler, *Adv. Chem. Phys.*, **77**, 397–449 (1990)).

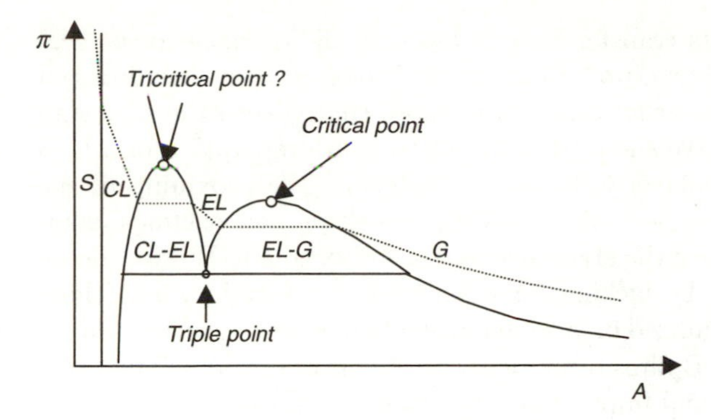

Fig. 10.7. Phase diagram of a fatty acid. The dashed line is an isotherm; there is a triple point corresponding to the coexistence of phases DL, EL, and G; a critical point will culminate the EL–G coexistence zone; a critical or tricritical point will culminate the DL–EL coexistence zone.

On the contrary, if the DL and EL phases do not have the same symmetry, which means that the DL phase is an ordered phase (two-dimensional solid), the transition zone cannot close without the appearance of a second order transition line between the DL and EL phases. Two situations can then be encountered:

- either the second order transition line intercepts the coexistence curve of DL–EL phases below the critical point (in this case, the critical point is an ordinary critical point);
- or the second order transition lines intercepts the DL–EL phase coexistence curve at the top of the coexistence zone of the DL and EL phases, and the critical point is then a tricritical point.

10.2 Monolayer on the Surface of a Solid

In Sect. 10.1.3, we described the method advocated by Langmuir and Blodgett in 1930 for depositing a monolayer of amphiphilic molecules on the surface of a solid. This method, simple in principle, called the L–B technique, was the first technique for preparation of monolayers on a solid substrate. In the conventional description of this technique, organization of the monolayer on the water is assumed to be transferred without any major changes in the substrate. This hypothesis is now strongly questioned. In effect, although the methods of analysis of monolayers at the water–air interface (X-ray scattering, phase diagram, fluorescence microscopy, Brewster-angle optical microscopy) differ from those used to characterize monolayers on substrates (electron diffraction, X-ray diffraction, atomic force microscopy); the new methods of analysis show that the molecular organization of the monolayer generally evolves during its transfer from the water–air interface to the surface of the substrate. The structure of Langmuir–Blodgett layers is influenced by a large number of factors that differ from those that govern the structure in Langmuir monolayers. We should not forget that in preparation of a L–B layer, transfer of the monolayer is followed by drying; the proximity of the solid substrate with its own periodicity and the resulting new chemical environment inevitably influence the structure and stability of a L–B layer. Some basic knowledge seems to be lacking for explaining the way in which these complex molecules are organized on a solid substrate of given structure. Moreover, the elaboration of L–B films often poses problems of reproducibility due to the presence of defects and impurities on the surface of the substrates. All of these problems have delayed the development of practical applications for L–B films.

In addition to their scientific interest, the properties of new organic films and particularly thin polymer films, molecular monolayers and multilayers offer technological opportunities for various components. At present, the use of organic films in microelectronics, either in new applications or to replace existing materials, can be envisaged.

As soon as we know how to induce and control modification of their structure, molecular films should offer interesting opportunities. The size of currently used electronic devices is continuously decreasing and will soon be close to the molecular scale. It is fascinating to consider the possibility of processing and storing information on the molecular scale, in layers.

In the current state of knowledge:

- chemists can synthesize molecules by optimizing the desired property; and there are already numerous possibilities;
- films can be fabricated by the Langmuir–Blodgett method, by related methods, or by other techniques currently under development;
- the structure and properties of the films can be analyzed by numerous techniques for characterization of surfaces.

New films adapted to an application could thus be produced, characterized, and checked to make sure they have the desired properties. Molecular films thus have many interesting prospects.

We already know how to fabricate switchable ferroelectric Langmuir–Blodgett films by fabricating films of approximately 15 monomolecular layers of a ferroelectric polymer (polyvinylidene fluoride) or a ferroelectric copolymer (copolymer of 70% vinylidene fluoride and 30% trifluoroethylene).

11 Phase Transitions under Extreme Conditions and in Large Natural and Technical Systems

In almost all phase transitions that we have studied up to now, extreme conditions (in particular, very high pressures and very high temperatures) in which new states of matter such as plasmas, or phase transitions such as the transition between a conducting metallic state and an insulating state, have been excluded. It is thus useful to consider the behaviors of matter on exposure to the very high pressures or temperatures that can be created in the laboratory today. It is also necessary to emphasize that these "extreme" conditions are found in natural systems, for example, in the earth's core where materials are exposed to high pressures (3.5 Mbar) and temperatures of the order of 4000 K.

As we noted in several chapters, numerous very "classic" phase transitions, solidification and vaporization, for example, occur in natural systems that affect life on earth (meteorological conditions and the evolution of the climate in particular). Phase changes are also found in numerous technical systems, where they are used to produce or store energy, for example. These aspects of the physics of phase transitions are at the limit of science and technology and also merit investigation.

11.1 Phase Transitions under Extreme Conditions

Although physicists have been interested in the behavior of matter at high pressures since the 18th century, the work of Amagat on gases up to 3 Kbar at the end of the 19th century, then of Bridgman at the beginning of the 20th century really opened up the field of experimental studies in extreme conditions (pressures greater than 1 Mbar and temperatures of several thousand kelvin).

11.1.1 Experimental Methods

We now know how to apply static pressures of several Mbar on a material of limited volume using a relatively simple experimental device, the **diamond anvil**. Pressure is produced in the small volume holding the material studied by forces mechanically transmitted to a pair of diamonds (Fig. 3.5). The interest of such a device is not only the hardness of the diamond, which allows transmitting loads through its faces (their area is less than 1 mm^2) but also its transparency, which allows performing measurements both in visible light and with X-rays in samples of the order of 1–20 μ in size.

Pressures slightly above 2.5 Mbar (250 GPa) applied to hydrogen have been obtained (experiments conducted by H. K. Mao).

At present, it is only possible to go beyond pressures of several Mbar under dynamic conditions, that is, by using passage of a shock wave in a material. The use of chemical explosives to send a missile to a target by utilizing detonation energy creates pressures of several Mbar. Devices of the shock tube type (Fig. 11.1) are used. The hydrostatic pressure behind the shock wave can be calculated and is called the **Hugoniot pressure**. Very high temperatures of several thousand K can also be obtained with such systems. By varying the pressure and temperature, we can attempt to establish the equation of state for a material. The gas-gun method was used to study melting of solids at high temperature and pressure (5000–6000 K for iron and pressure of 1–2 Mbar).

Very high temperatures can be attained with the gas gun by using metallic material in the porous state. The initial shock wave reduces the porosity of the metal by compressing the metal particles, which causes very important local heating. In the case of porous tungsten, a temperature of 32,000 K with a pressure of 2 Mbar has been reached.

In all of these shock-compression experiments, the temperature is measured by pyrometric methods, assuming that the radiation emitted by the material is in equilibrium with it.

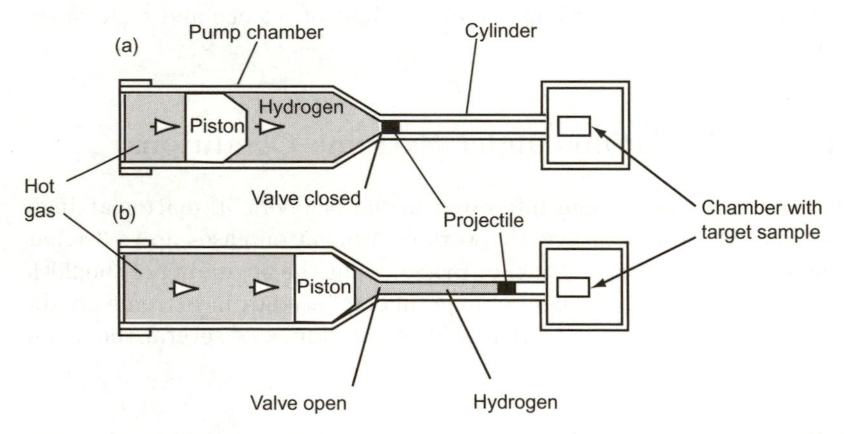

Fig. 11.1. Block diagram of a shock tube. There are two stages: (**a**) the hot gas produced by a detonation moves the piston, compressing the hydrogen, which opens the valve located inside the cylinder. (**b**) The hydrogen propels a projectile to the target material. (W. J. Nellis, *High Pressures Measurement Techniques*, Applied Science Publishers, London (1983), pp. 68–89).

New advances were made in shock-wave research after World War II through programs for the elaboration and development of nuclear weapons, particularly in the United States and the former USSR, but also more recently in France. Very high pressures can be obtained in underground nuclear explosions during testing campaigns (now prohibited by the nuclear test ban treaty). Pressures of several hundred Mbar were reached in experiments with nuclear explosives following the work done by Al'tshuler in the former Soviet Union in particular. Studies have also been conducted on aluminum in the pressure range of 400 to 4000 Mbar.

Another way is to create a shock wave in a solid with a high-energy laser beam pulse. High temperature and pressure conditions are simultaneously created. This method will undoubtedly be developed because the nuclear test ban will lead to construction of large installations with powerful lasers to simulate the conditions of thermonuclear explosions.

11.1.2 Equations of State and Phase Transitions under Extreme Conditions

The theories describing the behavior of matter at ordinary pressure and temperature (from 1 to 10 Kbar and up to a few hundred K) can *a priori* be transposed to very"extreme" conditions (beyond 1 Mbar and several thousand K). However, the theoretical models applicable under ambient conditions may lose their validity when the pressures and temperatures are very high. Moreover, the properties of the matter are strongly dependent on the interactions between atoms and molecules, and application of high pressures will change the interaction potentials by modifying the density and interatomic distances: phase transitions to new structures may appear in this way, and electronic configurations may be modified.

At high pressures and ordinary temperatures, the material is in the solid state and a series of studies was conducted with shock-wave devices to determine the melting curve of solids in a (p, T)-diagram. Under compression by a shock wave, a solid may be made to pass its melting point. Measurements have been performed on compounds such as NaCl, KBr, LiF, and CsI. The melting pressures vary from 250 Kbar for CsI to 2.8 Mbar for LiF. Melting occurs in the temperature range from 3000 K to 6000 K.

Similar experiments have been conducted in metals, where melting is detected by measuring the change in the sound velocity in the material; it is a very sensitive indicator of the disappearance of the shear modulus in the solid and thus of a solid–liquid transition. In iron, melting seems to be observed at 2.5 Mbar for a temperature of 5500–6000 K and in aluminum at 1.3 Mbar for $T \approx 4800$ K.

Hydrogen in the solid phase at very high pressure has been the subject of numerous recent studies conducted in particular with diamond anvil devices. At low pressures, hydrogen crystallizes in the form of an insulating molecular solid, but it has been hypothesized that at very high pressure, it could undergo a phase transition and pass into the metallic state in the molecular or atomic form. Numerous calculations have been performed, particularly with the Monte Carlo type methods , in attempting to predict these phase transitions. The most recent theoretical estimations predict that a transition in a molecular metallic phase will occur at a pressure around 200 GPa (2 Mbar) and in the atomic form at 300 GPa (the hydrogen molecules dissociate). The existence of the superconducting transition has even been predicted at very high pressures (> 1 TPa) and in a temperature region between 145 and 300 K.

Theoretical phase diagrams showing a series of hydrogen phase transitions, particularly structural transitions with solid phases corresponding to different crystal symmetries, have been plotted (Fig. 11.2). The transition temperatures are a function of the form of the hydrogen molecules (*ortho-* and *para-*hydrogen).

Experiments were conducted on solid hydrogen, particularly by Mao with a diamond anvil cell, up to a pressure of 250 GPa. Using optical methods (Raman scattering), it was possible to demonstrate structural transitions in the solid; on the contrary, the transition to a metallic phase has not yet been observed.

The behavior of dense fluids at very high pressure poses another series of questions. It is necessary to modify the classic equations of state to take into account the influence of the pressure over intermolecular distances. At ordinary pressures, the movement of an atom in a liquid is restricted to the

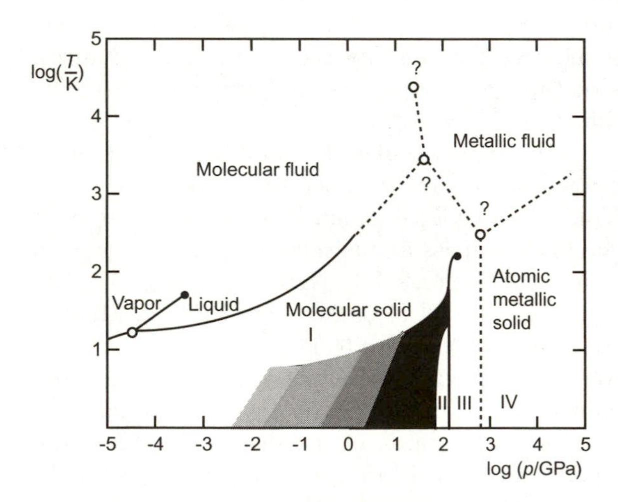

Fig. 11.2. (T, p) phase diagram of hydrogen. Different ordered solid phases of hydrogen appear (I, II, III, IV). The transition temperatures and pressures are a function of the concentrations of *ortho-* and *para-*hydrogen and the concentrations of H, D, T isotopes(various grey and black zones) (W. H. Holzapfel, Physics of solids under strong compression, *Rep. Progr. Phys.*, **89**, 74 (1996)).

interatomic distances in the vicinity of the interaction potential minimum. At very high pressures, the average intermolecular distances are reduced; the repulsive part of the potential then tends to become dominant. The intermolecular potential often selected to represent these situations has the following form:

$$\Phi(r) = \frac{\varepsilon}{\alpha - 6}\left[6 \exp \alpha\left(1 - \frac{r}{r^*}\right) - \alpha\left(\frac{r^*}{r}\right)^6\right] \tag{11.1}$$

where ε and r^* are the amplitude and position of the potential well and α is a constant (for most systems, $\alpha = 13$). The attractive part of the potential r^{-6} represents the contribution of the dipole interaction between molecules, which is long-range. It is sometimes necessary to correct the function (11.1) by introducing terms that take into account the contribution of multipole interactions.

At high pressures, the structure of the liquid phases obtained by melting of solids of the NaCl type tends to change: passing continuously from a denser structure at high pressure where the short-range repulsive forces tend to dominate. The packing of the molecules is more compact, similar to what is found in inert-gas fluids with 12 near neighbors of different charge. This change in the structure of the liquid is accompanied by a decrease in the change of the volume on melting.

Experiments were conducted by compressing liquid nitrogen with a shock wave at up to 0.8 Mbar and 14,000 K. A transition was observed at the pressure of 0.3 Mbar and 7000 K corresponding to breaking of chemical bonds in molecular nitrogen.

It is undoubtedly necessary to note that many of the physical phenomena that intervene at high pressure result from the changes that occur in the electronic properties of solids. By increasing the pressure, most of the elements in the periodic table undergo a series of transitions to more strongly coordinated structures. At very high pressure, we should expect that all materials will become metals with compact atomic packing. To obtain a metallic phase from a solid in the insulating state, it is necessary to either induce delocalization of valence electrons (as in semiconductors with covalent bonding) or to cross an unoccupied conduction band (case of insulators with a filled electron shell). In all cases, electron delocalization is at the origin of a **insulator–metal transition**, called the **Mott transition** (the metal has conductivity). In the case of xenon, the pressure broadens the type $5p$ valence band so that it crosses the $5d$ conduction band, injecting valence electrons into the conduction band. Xenon has been metallized at a pressure of 1.4 Mbar. CsI, which has an electronic structure close to the structure of xenon, is metallized at 1.1 Mbar with a similar process. An insulator–metal transition has been demonstrated in molecular iodine at 170 Kbar; it becomes a monoatomic metal at 210 Kbar.

A Mott transition is observed in many systems: metals inserted in a rare gas matrix, doped semiconductors (Si or Ge with elements from columns III and V in the periodic table), metal oxides, metal–ammonia systems. Several phenomena could be at the origin of this transition, induced by application of high pressure: a density change that allows delocalization of electrons which may become itinerant (when the density is high, electrons produce a screening effect with respect to positive ions, preventing them from being captured); delocalization in a random potential created by impurities.

The reverse transition can also be induced, the transition from the metal phase to the insulator phase, by applying high pressure on a metallic solid. The disorder created by impurities in the lattice or by strong interactions between electrons can cause disappearance of the metallic state. High pressure can also lead to localization of electrons that were itinerant at ordinary pressure and thus loss of the metallic state. A metal–insulator transition was observed in this way in compounds of the $NiS_{2-x}Se_x$ type for a critical pressure p_c around 1.5 Kbar; in this case, the conductivity approaches zero when $T \to 0$ (Fig. 11.3).

In fact there is a clear distinction between metal and insulator only at absolute zero, because insulators can always be the site of circulation of a very weak electric current at nonzero temperature, induced by thermal excitation. A continuous metal–insulator transition (with conductivity reduced to zero) will thus be observed at $T = 0$ only. By analogy with the scales used for describing the critical transitions, the conductivity $\sigma(T)$ can be written as:

$$\sigma(T) = \sigma(0) + AT^{1/2} + BT \tag{11.2}$$

$$\sigma(0) = \sigma_0 \left(\frac{p}{p_c} - 1\right)^{\nu} \tag{11.3}$$

where A and B are constants, p is the pressure applied, and p_c is the critical pressure at which the conductivity is reduced to zero (we then have a second order phase transition). The critical exponent ν is generally close to 1. When the electrons in the metallic phase are very strongly correlated, we will also have a term T^2 in (11.2) for the conductivity (Fig. 11.3).

Finally, it is also necessary to note that the interest stimulated by the new materials in the fullerene family made of spherical C_{60} molecules led to the study of their stability at high pressure. Although these molecules are very stable up to hydrostatic pressures of the order of 20 GPa, a transition was on the contrary observed in a diamond phase in fullerene by subjecting it to nonhydrostatic pressure in the vicinity of 20 GPa at ordinary temperature. We know that diamond can be obtained from carbon in the graphite phase at high pressure but at high temperature ($T > 1000$ K). A transformation from fullerene would thus be more interesting. A graphite phase was also obtained from fullerene by compressing it with a shock wave; the transition was observed at 600°C for a pressure of 17 GPa.

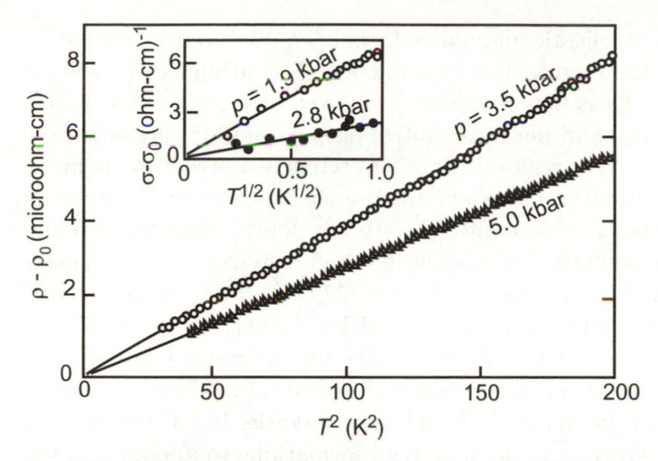

Fig. 11.3. Metal–insulator transition. Graph showing the decrease in resistivity in approaching the transition in a $NiS_{2-x}Se_x$ compound with $x = 0.47$ here. At very low temperatures ($T < 1$ K) and for $p > 1.5$ kbar, the conductivity approaches zero at $T^{1/2}$ when $T \to 0$. Passage into the insulator state (A. Husmann et al., *Science*, **74**, 1975 (1996), copyright Am. Assoc. Adv. Sci.).

11.1.3 Geomaterials

The universe is a natural laboratory where matter is exposed to very high pressures. In the center of the earth, the pressure is 3.5 Mbar and the temperatures are of the order of 5000 K. In the center of the planet Jupiter, it is estimated that the pressure can attain 100 Mbar and the temperature would be 24,000 K. In the center of stars and white dwarves, the pressures would be several thousand Mbar and the temperatures would be tens of millions of Kelvin; the pressures would be counted in thousands of Mbar in neutron stars. In these conditions, the density of the matter would be important and it could have very unusual properties.

Many studies conducted in the laboratory to investigate materials at high pressure have the goal of simulating the phase transitions that geomaterials undergo in the earth's crust; these are particularly the reasons for research using diamond-anvil cells.

The center of the earth is composed of a liquid iron core surrounding a solid central core (together they represent 32% of the mass of the planet). The pressure at the boundary between the liquid core and the surrounding mantle is 139 GPa, and it is 364 GPa in the center of the solid central core (4% of the total core volume). The melting points of iron at the boundary of the solid central core can vary from 4000 to 8000 K.

The boundary between this core and the mantle, the solid part of the earth, is located at a depth of 2900 km below the surface of the earth, and it has a very important influence on the thermal and chemical evolution of the interior of the planet. The mantle is essentially composed of crystalline silicates. Very important seismic anomalies have been observed in this separation region between the mantle and central core (discontinuities in seismic wave propagation velocities, for example). Moreover, the properties of this boundary region influence and perhaps control propagation of magnetic field lines outside of the core. This region is highly heterogeneous due to chemical reactions between the liquid iron in the core and the silicates in the mantle.

The lower part of the earth's mantle (located at depths between 650 and 2900 km) is essentially composed of magnesium and iron oxide phases (magnesiowüstite) and iron and magnesium silicate, (Mg, Fe) SiO_3, of perovskite structure. Perovskites have the general composition ABO_3, which are phases in which a cation A is intercalated between BO_6 octahedrons. These mantle perovskite phases are only stable at pressures above 23 GPa. Their properties control the properties of the materials in the lower mantle. It is thus essential to determine their structures and possible transformations to understand the physical and chemical processes that take place in the mantle.

Numerous structural transitions can thus be observed in this type of perovskite in X-ray diffraction experiments conducted with diamond-anvil cells. A structural transition was demonstrated in $MgSiO_3$ at 600 K and 7.3 GPa: from an orthorhombic structure to another structure.

Iron is the dominant component of the earth's core; this explains the interest of the phase diagram of iron at high pressure and high temperature for reproducing in the laboratory the physical conditions which exists in the core. This phase diagram is determined with a diamond anvil where phase changes are detected by a sudden change in resistivity or optical reflectivity. The melting point of iron, $T_m(p)$, could be determined up to a pressure of approximately 200 GPa in this way (at 3900 K). Complementary X-ray diffraction measurements are also generally performed. The sample of iron is heated by a laser pulse. Solid/ solid phase changes at high pressure were observed, but the stability of the different phases is still debated.

Finally, it is necessary to note that microcrystal inclusions formed at high pressure and high temperature on impact with the ground are found in meteorites. The shock induced melting of minerals and perhaps their vaporization and then recondensation. Small inclusions of diamond and SiC have been found in rocks at the site of the city of Nördlingen in Germany. These crystals (100 nm in size) were undoubtedly formed by vapor-phase deposition from surface minerals composed of carbon and silicon which were vaporized on impact of an asteroid which left a crater 1 km in diameter at the actual site of the city approximately 15 million years ago. This is in a way the trace of a natural phase transition.

11.1.4 The Plasma State

When matter is brought to a very high temperature, it is totally ionized and passes into a new state, the **plasma state**, which is a gas of charged particles (electrons and ions). A plasma with only one constituent is ideally a system of positive charged ions immersed in an electron gas which neutralizes the positive charges.

Plasmas are obtained by bringing matter to temperatures of several hundred million kelvin (2×10^8 K) corresponding to a thermal energy equal to several keV (one electron-volt is equivalent to 1.602×10^{-19}J). These plasmas, initially formed from mixtures of deuterium and tritium atoms, are produced in experimental setups developed to attempt to trigger controlled thermonuclear fusion in laboratory conditions. These temperatures are obtained in plasmas either with the **inertial confinement** technique using a very powerful laser beam that compresses a solid target composed of deuterium and tritium or by confinement in a very high magnetic field, where the plasma is heated by an electric current (**tokamak** machines). The conditions for triggering a self-maintained thermonuclear fusion reaction (called ignition) have not yet been realized experimentally.

The gas phase–plasma transition is thus a totally specific phenomenon which reveals a new state of matter in extreme conditions and which is intrinsically unstable.

11.1.5 Bose–Einstein Condensates at Extremely Low Temperature

Einstein had predicted, in 1925, that in a gas of particles, obeying Bose–Einstein statistics and with a fixed number of particles, there should exist a critical temperature T_0 under which a finite number of them would condensate into the same energy state. The discovery of the superfluidity phenomenon (Sect. 8.2.1) gave support to this hypothesis and superfluidity was interpreted as a Bose–Einstein condensation phenomenon although the particles, namely helium atoms, were interacting rather strongly. It was only in 1995, that Bose–Einstein Condensates (B.E.C) were produced in magnetically-trapped atomic alkali gases. The condition for observing a B.E.C with bosons in a volume V is given by (8.108):

$$(N_e/V)\Lambda^3 = 2.612$$

where $n = N_e/V$ is the numeric density of atoms in excited states and Λ the de Broglie wavelength. In a situation corresponding to a homogeneous system, as liquid helium, in which particles are confined to a large volume (a box potential), the particle fraction N_0/N occupying the ground energy level (which is taken as $\epsilon = 0$ by convention) is given by (8.111) and it varies as $[1 - (T/T_0)^{3/2}]$. For B.E.C obtained with atomic gases, experimental conditions impose a strong reduction of the effective volume to which particles may have access, this tends to increase n. One can thus show that:

$$N_0(T) = N[1 - (T/T_0)^3]$$

For $T < T_0$, an increasing fraction of atoms will condense into the lowest energy state. They go through a quantum phase transition: they constitute a new state of matter, the B.E.C behaving as a coherent atomic cloud in which all atoms are occupying the same energy level. To observe a B.E.C one should avoid a "classic " condensation of the atoms to form a liquid or a solid; the only means to avoid such a situation consists in cooling rapidly the diluted atomic gas. Progress in laser cooling techniques and obtained in the years 1980, by C. Cohen–Tannoudji in France and by S. Chu and W. C. Philips in the USA, rendered the production of B.E.C feasible. The first experiment was realised, in 1995, by E. A. Cornell, W. Ketterle and C. Wieman in the USA, on atoms of rubidium 87 at 170 nanokelvins; the atoms are trapped in a cell by a small and heterogeneous magnetic field and by laser light beams. Absorption of several photons by atoms results in their braking and as their velocity is decreasing, their energy is also decreasing. One proceeds to an evaporative cooling of the atoms from the trap: the most energetic atoms leave the trap, and only the atoms with the lowest energy are remaining, they have condensed into the ground energy level. Observation of the condensate is performed with detectors of a video camera. Atoms in the B.E.C have a weak repulsive interaction which avoids their collapsing. B.E.C have been produced with rubidium, sodium, lithium and hydrogen atoms with 10^5–10^7 atoms and at temperature around 100 nK. In 2001, condensates have been produced with helium atoms in their first excited states (one electron is excited by an electric discharge); they are in a metastable state.

By modulation of the laser light in the trap one can produce the equivalent of a 3 dimensional lattice in which atoms from the B.E.C can freely move and one has thus realised a superfluid phase at 10 nK. One observes a phase transition into a non-superfluid phase by changing the shape of the optic trap: this is a so-called quantum phase transition which is induced by quantum fluctuations.

Application of B.E.C can be envisaged to obtain sources of cold and mono-energetic matter which would behave as the equivalent of lasers in optics. Those sources could be used in atomic clocks and possibly in new lithographic processes.

11.2 The Role of Phase Transitions in the Ocean–Atmosphere System

The earth's climate system can be compared to a thermal machine on planetary scale whose heat source is solar energy which heats the tropical regions of the globe more than the polar regions. The thermal gradient on the surface of the planet creates pressure differences in its fluid mantle – the ocean-atmosphere system – which moves and causes heat transfer from the equator to the poles.

The ocean and the atmosphere are strongly linked, each reacting to the fluctuations of the other, but while the volume mass of water is one thousand times larger than the volume mass of the air in the atmosphere, its inertia to motion is much higher. In addition, as the specific heat of water is four times higher than the specific heat of air and the total mass of the oceans is 300 times the total mass of the atmosphere, the heat capacity of the oceans is much higher than the heat capacity of the atmosphere: the first 2.5 m of the water column of the ocean has the same heat energy as the 40 km of the atmospheric column above it.

Solar radiation is obviously the predominant energy source on the surface of the earth (the geothermal energy from the deep layers of the globe has a marginal contribution). Then 70% of the solar radiation is absorbed by the surface of the globe, while the remaining 30% is reflected or scattered in space: the ratio between the reflected part of solar radiation and the total incident radiation is called the **albedo**. The absorbed solar energy has a maximum value of 300 $\mathrm{Wm^{-2}}$ at low latitudes. An important part of the absorbed solar energy induces phase transitions due to evaporation of the water in the ocean to the atmosphere and by melting of ice and the snow cover in certain regions on earth.

This corresponds to a total latent heat flux of the order of 70 $\mathrm{Wm^{-2}}$. A minor part of solar radiation is converted into the kinetic energy of the winds which is in turn dissipated as heat by friction (the corresponding energy flux is 2 $\mathrm{Wm^{-2}}$).

The atmosphere is the smallest reservoir of water on the planet, since it represents only 0.001% of the total mass of water present; the ocean contains 97% of the water, the Antarctic and Greenland ice caps represent 2.4% of the water mass, and natural fresh water reservoirs contain the remainder. The latent heat flux corresponding to vaporization of water (essentially vaporization of the oceans) contributes to the linkage between ocean and atmosphere in a determining way. When atmospheric water vapor is condensed, it returns very important latent heat: for each kg of condensed water vapor, there is heat release corresponding to 2.4 MJ, which corresponds to average energy flux of the order of 80 $\mathrm{Wm^{-2}}$. This flux is the major source of energy that controls general atmospheric circulation. For example, the trade winds that blow over large maritime areas between the equator and the tropics are moved by latent heat release in intertropical regions. The intertropical zone plays a major role in the exchange of heat and water vapor between the ocean and the atmosphere. Above the maritime expanses, it represents a narrow band on either side of the equator. The latent heat of water vapor condensation can also be released several thousand km from the zone of evaporation of ocean water into the atmosphere. Globally, one has an equilibrium: the water condensation and evaporation rates are equal.

11.2.1 Stability of an Atmosphere Saturated with Water Vapor

The earth's atmosphere is in a gravitational field and its density decreases with the altitude. Its structure can be significantly altered by the presence of water vapor.

The study of the equilibrium of an atmosphere saturated with water vapor is evidently a very important question in meteorology.

If ρ designates the density and p is the pressure at altitude z measured from the surface, we have:

$$\mathrm{d}p = -g\rho \mathrm{d}z \tag{11.4}$$

The mechanical equilibrium of the air mass between altitudes z and $z+\mathrm{d}z$ requires that its weight be equalized by the pressure forces p and $p + \mathrm{d}p$ exercised on it.

The equation of state for an ideal gas of molecular weight M_a at temperature T is written:

$$\rho = \frac{M_a p}{RT} \tag{11.5}$$

First consider an atmosphere without liquid water in which the movement of an air mass at pressure p and temperature T and specific volume $V = 1/\rho_a$ is followed. The driving and gravitational forces are equalized and it is not necessary to introduce them in the expression for the first law of thermodynamics which can be written for a unit mass:

$$\mathrm{d}q = C_v \mathrm{d}T + p\mathrm{d}V \tag{11.6}$$

If no heat flux enters the ascending mass (it is assumed to be transparent to all radiation), the movement can be assumed to be adiabatic and we have $\mathrm{d}q = 0$. By differentiation of (11.5), we have:

$$p\mathrm{d}V + V\mathrm{d}p = \frac{R\mathrm{d}T}{M_a} = (C_p - C_v)\mathrm{d}T \tag{11.7}$$

where M_a is the molecular mass of the air. By substitution of $p\mathrm{d}V$ in (11.6), we determine:

$$C_p \mathrm{d}T - V\mathrm{d}p = \mathrm{d}q = 0 \tag{11.8}$$

that is

$$\frac{\mathrm{d}T}{\mathrm{d}z} = -\frac{g}{C_p} = -\Gamma_d \tag{11.9}$$

Γ_d is called the **dry-adiabatic lapse rate** for a dry atmosphere. Γ_d for the earth's atmosphere is of the order of 10 K km^{-1}. This means that if this atmosphere is heated by contact with the earth's surface and is thus made to move vertically, a temperature gradient of this order of magnitude should appear. We see in Fig. 11.4 that a given temperature gradient $\mathrm{d}T/\mathrm{d}z$ in the atmosphere will represent a stable situation if $-\mathrm{d}T/\mathrm{d}z < \Gamma_d$. In effect an air mass in adiabatic vertical ascent from point X on adiabatic line (1) at X$'$ will be surrounded by an atmosphere at the same altitude corresponding to point X$''$ whose temperature is lower. Point X$'$ thus does not correspond to a stable situation and an air mass hotter than its environment will continue

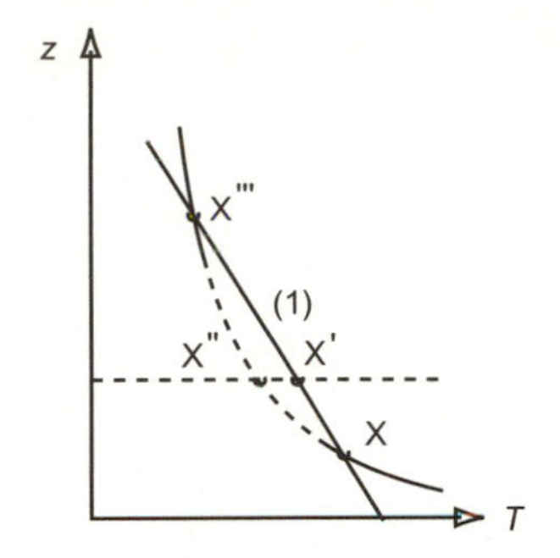

Fig. 11.4. Stability of a temperature profile. At point X′ on adiabatic (1), the ascending air mass is surrounded by an atmosphere at the same altitude corresponding to X″ whose temperature is lower. Point X′ is not stable, and the air mass will continue to ascend to point X‴ which is stable.

to ascend to point X‴ where it is stable, as the temperatures are identical in the ambient atmosphere and the ascending air mass.

If we consider the presence of water vapor in atmospheric air, we are then dealing with vertically ascending humid air. The stability conditions are not altered if the air is unsaturated, that is, in the presence of water vapor whose partial pressure is lower than the saturation pressure at the temperature of the air. This unsaturated air will be cooled during its ascent at a rate Γ_d equal to the rate for dry air, i.e., 10 K km^{-1}.

If we designate the water vapor mixing rate (that is, the ratio of the mass of water vapor m_V to the mass of air m_a for a given volume, where ρ_V and ρ_a are the respective densities of the water vapor and air) by $r = m_V/m_a = \rho_V/\rho_a$, this ratio will remain constant during ascent of the air. The saturation rate r_e of the air in water can be written as $r_e = 1/\varepsilon(p_0/p)$, where p_0 is the saturated vapor pressure at the temperature of the air, p is the total pressure (equivalent to the pressure of the air), and $\varepsilon = M_V/M_a = 0.622$. A relative humidity rate $r_H = r/r_e$ representing the degree of proximity of humid air to the saturation situation is also often defined.

The rate r_e is a function of the temperature and pressure which can be calculated with the Clapeyron equation; r_e decreases exponentially with the temperature. The capacity of wet air to conserve water vapor thus decreases strongly when the temperature decreases. When $r = r_e$, saturation is attained: the water vapor begins to condense and a cloud forms (Fig. 11.5).

When the wet air is below saturation, it continues its ascent into the atmosphere, preserving a constant rate r: $r = r_0$. When it attains an altitude corresponding to the saturation limit, r_e, the excess water vapor $\Delta r = r_0 - r_e$ is condensed and there is latent heat release $\delta q = L\Delta r$: convective clouds form. The vapor–liquid phase transition is obviously at the origin of cloud formation. The preceding equations must be modified as a consequence.

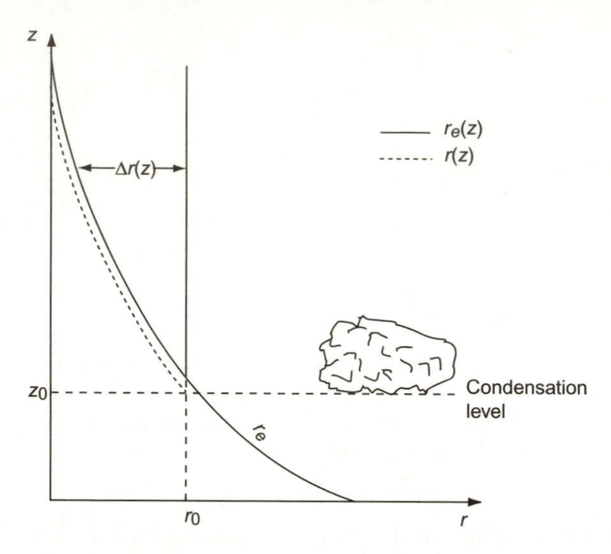

Fig. 11.5. Mechanism of cloud formation. The saturation rate r_e and the mixing rate r are functions of the altitude. When an air mass rises with an initial rate $r = r_0$ to altitude z_0 where, $r = r_e$, saturation is attained: the water vapor begins to condense, and a cloud is formed. For $z > z_0$, $r_e < r$, the excess vapor $\Delta r(z)$ is condensed (M. L. Salby, *Fundamentals of Atmospheric Physics*, Academic Press (1996)).

The temperature at which the air must be isobarically cooled to attain saturation is called the **dew point**.

Considering 1 g of dry air with x g of water, where r g is in the form of vapor and $(x-r)$ g is in the form of liquid, L is the latent heat of vaporization, and p_0 is the saturated vapor pressure, the total entropy is then:

$$S = (C_p + xc)\ln T - \frac{R}{M_a}\ln(p - p_0) + \frac{Lr}{T} + C^{te} \tag{11.10}$$

where c is the specific heat of the liquid water.

Knowing that the hydrostatic equation is written for humid air as:

$$\mathrm{d}p = -g\rho_a(1 + x)\mathrm{d}z \tag{11.11}$$

with

$$r = \frac{\rho_V}{\rho_a} = \left(\frac{p_0}{p - p_0}\right)\frac{M_V}{M_a} \approx \frac{p_0\varepsilon}{p} \tag{11.12}$$

One has treated the air and water vapor masses as ideal gases where M_V is the molar mass of water vapor of density ρ_V, and one knows that $p_0 < p$.

By differentiation of (11.10), replacing dp by (11.4), in adiabatic conditions, we can write:

$$(C_p + xc)\frac{\mathrm{d}T}{T} + \mathrm{d}\frac{Lr}{T} + \frac{R}{M_a(p - p_0)}\frac{\mathrm{d}p_0}{\mathrm{d}T}\mathrm{d}T + \frac{g}{T}(1 + x)\mathrm{d}z = 0 \tag{11.13}$$

In addition, we have from (11.12):

$$\frac{\mathrm{d}r}{\mathrm{d}z} = \frac{\varepsilon}{p}\frac{\mathrm{d}p_0}{\mathrm{d}T}\frac{\mathrm{d}T}{\mathrm{d}z} - \varepsilon\frac{p_0}{p^2}\frac{\mathrm{d}p}{\mathrm{d}z} \tag{11.14}$$

At saturation, we have $x = r$ and using the Clapeyron equation, we can write:

$$\frac{\mathrm{d}p_0}{\mathrm{d}T} = Lp_0\frac{M_V}{RT^2} \tag{11.15}$$

finally with $-(\mathrm{d}T/\mathrm{d}z) = \Gamma_S$:

$$\Gamma_S = \frac{g}{C_p}\,\frac{(1 + Lp_0M_V/pRT)(1 + \varepsilon p_0/p)}{1 + \varepsilon p_0/pC_p(c + \mathrm{d}L/\mathrm{d}T) + \varepsilon p_0L^2M_V/C_ppRT^2} \tag{11.16}$$

Introducing the numerical values for the quantities in this equation, it simplifies significantly and it is written as:

$$\Gamma_S = \Gamma_d\big(1 + \frac{Lp_0M_V}{pRT}\big)\big(1 + \frac{LpM_V}{pRT}\frac{\varepsilon L}{C_pT}\big)^{-1} \tag{11.17}$$

The condensation phenomenon associated with latent heat release thus lowers the adiabatic temperature gradient because we have $\Gamma_S < \Gamma_d$. In the atmosphere, we have $0.3\Gamma_d < \Gamma_S < \Gamma_d$, with $\Gamma_S \approx 6.5$ K km $^{-1}$. Conditions of stability of the humid air column will be found if $-\mathrm{d}T/\mathrm{d}z < \Gamma_S$. On the contrary, if $-\mathrm{d}T/\mathrm{d}z > \Gamma_S$, the column is unstable.

In the preceding calculation, it was assumed that the condensed water in liquid form remained in the air packet when it rises, a condition which is not necessarily verified. However, one can show that the result of the calculation is slightly modified if we assume that all of the condensed liquid water falls into the air column.

11.2.2 Thermodynamic Behavior of Humid Air

There are graphs that allow determining the thermodynamic behavior of a volume of humid air when it is moving up. Below the condensation point, the air describes an adiabatic "trajectory" represented by the Poisson equation:

$$Tp^{-K} = \mathrm{C^{te}} \tag{11.18}$$

where $K = R/C_p = 0.286$.

A reference temperature θ corresponding to the temperature that the air would have if it was adiabatically compressed to pressure $p_0 = 10^2$ kPa (1 atm = 1000 mb) is defined.

$$\theta = T\Big(\frac{p_0}{p}\Big)^K \tag{11.19}$$

Up to the condensation point, the air moves adiabatically on the curve $\theta = \mathrm{C^{te}}$.

When there is heat release δQ, the process is no longer adiabatic and globally we have:

$$\delta Q = C_p T \mathrm{d} \ln \theta \tag{11.20}$$

As $\delta Q = L\delta r_e$, by integration of (11.20), we obtain

$$\frac{\theta_e}{\theta} = e^{Lr_e/C_pT} \tag{11.21}$$

where θ_e is the **equivalent potential temperature** representing the value that θ would have if all of the vapor were condensed ($r_S = 0$). Above the condensation point, the humid air will move on a pseudo-adiabatic $\theta_e = \mathrm{C^{te}}$.

In Fig. 11.6 showing a graph with adiabatics ($\theta = \mathrm{C^{te}}$) and pseudo-adiabatics ($\theta_e = \mathrm{C^{te}}$) as a function of the pressure and temperature, a process starting from the initial situation (i) at $p = 900$ mb, $T = 15°\mathrm{C}$, and $r = 6$ g kg^{-1}, is described with the air. The condensation point corresponds to $r = r_e$ (point 1 or dew point). When the humid air rises below saturation, it describes an adiabatic corresponding to $\theta = 297$ K up to point (2), where $p = 770$ mb. Above the condensation point, the humid air describes the pseudo-adiabatic $\theta_e = 315$ K up to point (3), where $\theta = 303$ K, which corresponds to $r = 4$ gK^{-1}.

Clouds play a critical role in all atmospheric and climatic phenomena. The water droplets in clouds have diameters of less than 200 μ. They are formed by nucleation. Nucleation is homogeneous if it is induced by spontaneous condensation of water molecules by collision; they form a stable liquid nucleus which grows. This nucleation can be heterogeneous if condensation occurs in extraneous molecules or when electric charges are present in the atmosphere.

It is also necessary to note that the latent heat of condensation is the source of energy for a hurricane. Hurricanes are meteorological phenomena, occuring in intertropical regions, which are initiated by a liquid–vapor phase transition (oceanic surface water temperature should exceed 26°C). Vaporisation of warm surface water produces a humid air which rises into the high atmosphere. As soon as it has reached a low temperature zone, it condenses and gives up its latent heat, this thermal energy increases rapidly the wind kinetic energy. Air in the neighborhhod is drawn into the centre of the hurricane, thus creating a vortex which is going up; wind speed can reach 300 km/h.

The droplets grow by condensation of additional water molecules or by coalescence on collisions between drops in the atmosphere. The water droplets that constitute the clouds are not only of meteorological importance: they are certainly at the origin of rain or snow when they crystallize, but they also have a long-term influence on the planet's climate. The snow cover and its behavior are an important parameter in the models that attempt to predict the evolution of the climate, but their influence is far from understood. The water drops in effect reflect incident solar radiation (in particular, the one with the short wavelength in the visible spectrum), because their albedo is

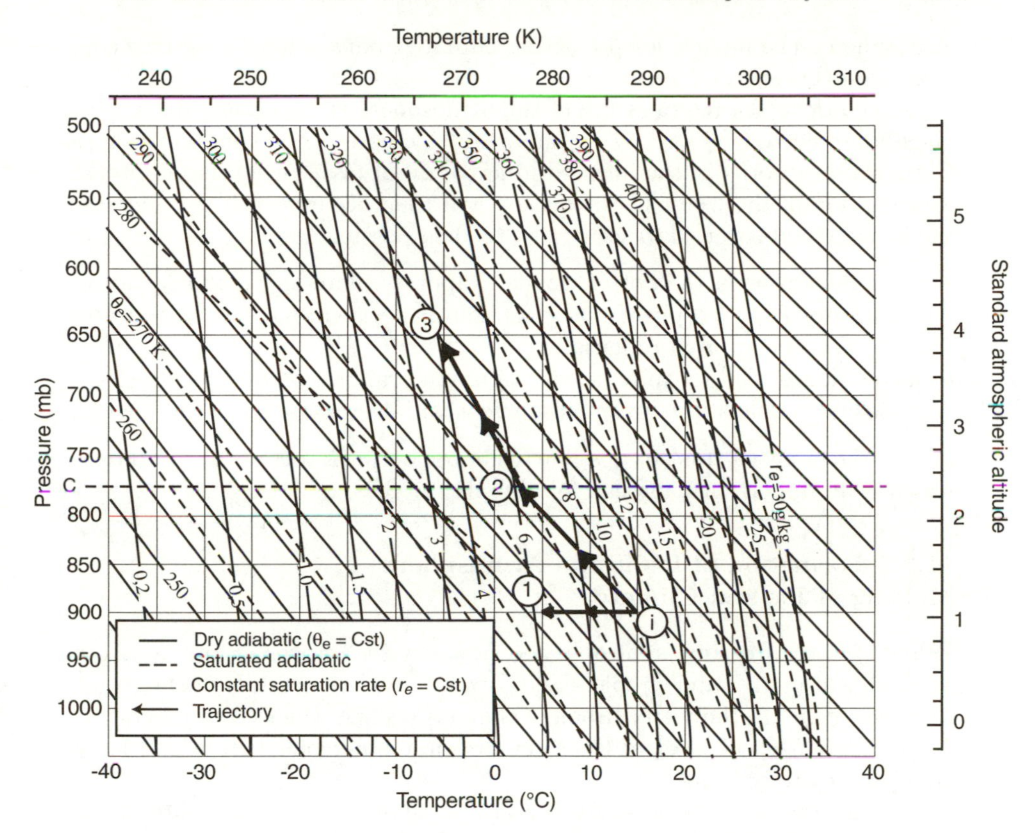

Fig. 11.6. Pseudo-adiabatic chart. Dry ($\theta = \mathrm{c^{te}}$) and saturated ($\theta_e = \mathrm{c^{te}}$) adiabatics and constant saturation rate curves ($r_e = \mathrm{c^{te}}$) in the (p, T)-plane are shown. An isobar up to point (1) (saturation point $r_e = r$, dew point) is described with an air packet from the initial situation (i). If the air rises, it follows a dry adiabatic up to condensation level C at point (2) corresponding to $r_e = r$. Above level C, the air describes a pseudo-adiabatic, for example, up to (3), where the mixing rate r is lowered (M. L. Salby, *Fundamentals of Atmospheric Physics*, Academic Press (1996)).

generally high: it can attain 0.9 for **strati**, that is, clouds in stratified layers. The reflection of solar radiation outside of the atmosphere tends to decrease its temperature. On the contrary, clouds with a high water content absorb radiation of longer wavelength (in the infrared), which is re-emitted by the surface of the earth and contributes to increasing the temperature of the atmosphere (this is the **greenhouse effect**).

Satellite observations seem to indicate that the energy balance of these radiative mechanisms is almost zero in tropical zones. Outside of these zones, this balance is negative: clouds globally contribute to the cooling of the earth's atmosphere. The models for predicting climatic changes must take into consideration this energy factor.

Moreover, if the temperature of the atmosphere is increasing, due to the greenhouse effect (induced by an important increase in the concentration of CO_2, which absorbs in the infrared, due to use of fossil fuels), more intense vaporization of the water in the oceans will then occur, which will increase the heavy cloud cover. Two scenarios are then possible:

- the water drops in the hotter clouds evaporate more and discharge water vapor into the troposphere: the greenhouse effect will increase since water vapor absorbs infrared radiation;
- hotter clouds tend to cause heavier rain and evaporation of droplets to the troposphere is less: the greenhouse effect will decrease.

For the moment, it is not possible to choose between these two hypotheses, which is an important handicap for climatic models.

11.2.3 Formation of Ice in the Atmosphere – Melting of Ice and Climate

Below 0°C, the freezing point of water, new physical processes intervene due to solidification of water droplets and formation of ice crystals in the cloud. However, rare are the water droplets that crystallize when the temperature drops below 0°C, as most of them remain in a supercooled state and ice is only progressively formed.

When a temperature of −39°C is reached, all of the remaining water drops in the cloud crystallize since the supercooling limit is attained. Above this temperature, ions and molecules other than air and water, serve as nucleation sites on which ice crystals form. Dust from the ground, aerosols, inorganic compounds (sulfates and sulfuric acid) or organic compounds originating in decomposition of marine phytoplankton (such as DMS or dimethyl sulfide) are such nucleation sites.

In the phase diagram, note that the saturated vapor pressure above an ice surface is lower than this pressure above liquid water below 0°C, and the water vapor in equilibrium with a liquid water surface will thus always be in a state of supersaturation with respect to the ice below 0°C. The difference between the two vapor pressures is maximum at −12°C. In a cloud where water droplets and ice crystals are present, the ice crystals will grow at their cost. This process is maximum in the vicinity of −12°C and is called the **Bergeron process**. It is characteristic of small cumulus or stratocumulus clouds in which the liquid water content is relatively low.

Ice crystals can enlarge in clouds by **aggregation** of the crystals: the film of supercooled water attached to a crystal can serve as "binder" with another crystal and by solidifying, can lead to the formation of a single crystal. This process is maximally effective between 0°C and −70°C, a temperature range where the supercooled water content is highest; it contributes to the formation of snow due to an **accretion** mechanism: the water droplets crystallize by impact on the ice crystals. This mechanism especially prevails in cumulonimbus clouds, cloud forms with a high liquid water content. **Frost formation** is a phenomenon induced by impact of water droplets and accretion on the surface. The ice microcrystals in the polar stratosphere at low temperature in the presence of fluorocarbons are active sites for chemical reactions involving molecules of ozone, causing the formation of an ozone hole above the polar regions in this way.

Hail formation is also due to accretion of liquid water on ice crystals in clouds with a high water content, particularly in certain cumulonimbus regions.

Finally, it is necessary to note that melting of ice (the ice of glaciers and ice caps in the Arctic and Antarctic) plays a very important role in climatology. In effect, oceanic circulation, one of the engines of climate (the role of the Gulf stream) because it transports the heat of intertropical zones in the North Atlantic to high latitudes is a critical function of small differences in density in salt water (it is a function of the salinity and the temperature): this is called **thermohaline circulation**. The hot surface water flows to the north in the Atlantic, releasing its heat into the atmosphere, while the cold water from the north (denser) circulates in the depths (2 km) to the south. This is in a way the equivalent of a conveyor belt.

An influx of fresh water, which is less dense, can seriously perturb oceanic circulation and even stop it totally according to some models. This phenomenon can occur if there is important melting of ice from the Arctic, northern Canada, and Greenland following heating of the atmosphere due to the greenhouse effect. This was perhaps the case in the interglacial periods when heating of the atmosphere induced partial melting of glaciers. In other words, the greenhouse effect could lead to local cooling due to strong perturbation of circulation, which is an unstable mechanism. This cooling would not necessarily be compensated by the greenhouse effect itself.

Finally, it is necessary to note that the characteristics of the surface temperature of the ocean (which can fluctuate) are communicated to the atmosphere with which it is linked by evaporation/precipitation mechanisms; this in return leads to changes in the temperature, level of precipitation, and frequency of storms in certain regions, in Europe, for example. The advances in weather prediction are thus dependent on advances in prediction of the surface temperatures of the ocean.

11.3 Phase Transitions in Technical Systems

Heat engines, turbines, oil installations (oil wells, refineries) are technical systems which are often the site of phase transitions. The liquid–vapor phase transition is a key phenomenon in these systems, as mechanical work is obtained by the expansion of the steam which is produced in a steam generator. In an oil well in an oil field, if the thermodynamic conditions (pressure, temperature) are favorable, a liquid–gas transition can be observed in the hydrocarbon mixture when it rises to the surface. The phase transition is in this case a serious drawback which can very seriously perturb operation if a two-phase "plug" forms in the well. Finally, the third type of situation, in a tanker carrying liquefied natural gas (at low temperature) by sea: vaporization of the liquid methane in contact with a body at a higher temperature (salt water, for example) must be avoided at all costs, because if this vaporization is too sudden, it could cause a thermal explosion with catastrophic consequences.

11.3.1 Vaporization in Heat Engines

In a classic steam engine, as well as in a nuclear reactor, steam is the fluid that activates the piston of a cylinder or a steam turbine.

In French and American nuclear power plants of the PWR type (pressurized water reactor), the water circulating in liquid form at high pressure ($p = 155$ bar and $T = 320°$C) removes the heat produced in the reactor core by fission of nuclear fuel. The calories from this hot water are then transferred to another fluid (always water in "classic" reactors) which circulates in a secondary circuit; it is then vaporized in a steam generator (Fig. 11.7).

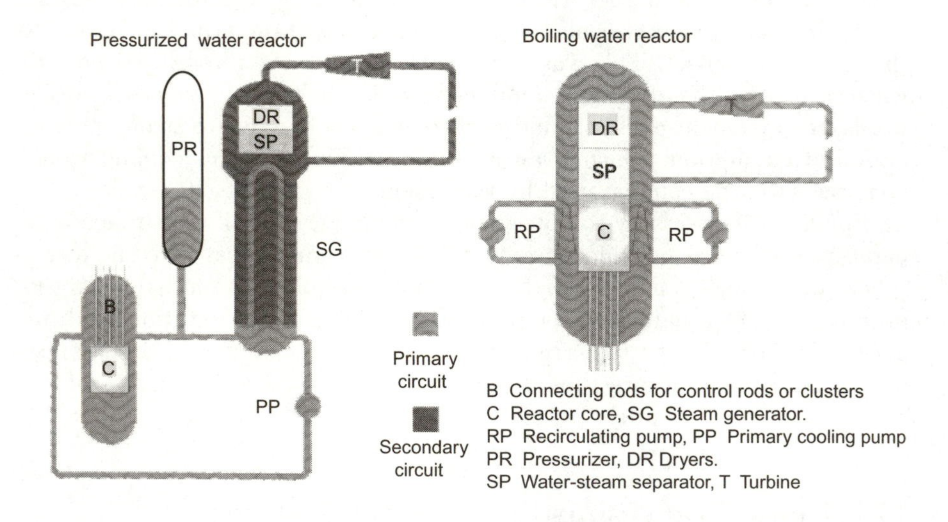

Fig. 11.7. Diagrams of nuclear reactors. Two types of reactors are shown: pressurized water; boiling water (source: C.E.A. France).

In an internal combustion heat engine (classic internal combustion engine or gas turbine, or cryogenic rocket engine of the Ariane type), the liquid–gas phase change results from the combustion of a liquid fuel with an oxidizer (atmospheric oxygen or liquid oxygen in a cryogenic engine). The fuel is first totally or partially vaporized before going into the combustion chamber.

The diagram of a steam turbine is shown in Fig. 11.8a. The water is vaporized in steam generator (1) and the steam, generally at high pressure, feeds turbine (2) (which is coupled with electric generator (3), for example). It is condensed in condenser (4). In the installation shown in this figure, water is delivered to the steam generator by pump (5).

A diagram representing its thermodynamic cycle is associated with each type of installation. The **classic Carnot cycle** is not applied in practice. We note that this cycle describes a system in which wet steam (that is, steam saturated with suspended water droplets) is used. On the contrary, installations described by a **Rankine cycle** are frequently used. A temperature–entropy diagram describing a Rankine cycle is shown in Fig. 11.8a. In a plant operating with this type of cycle, the fluid works between pressures p_1 and p_2. Water

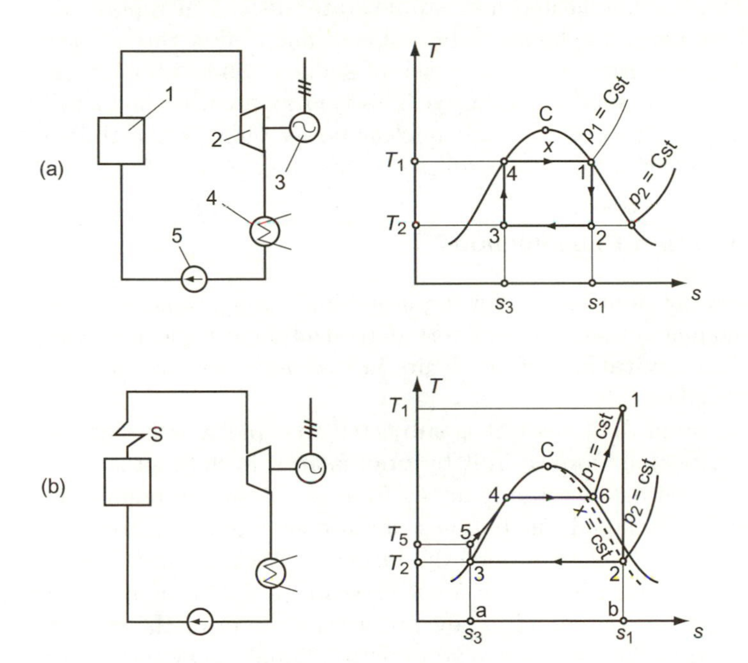

Fig. 11.8. Block diagrams of a heat engine. (**a**) Steam is removed from the boiler (1) and delivered to turbine (2) which drives electric generator (3). It is condensed in condenser (4). The wet steam is introduced in compressor (5). (**b**) In this engine, a superheater (S) has been added in which the steam is brought to a temperature above the saturation temperature for a given pressure. The corresponding Rankine cycle is shown.

is isentropically compressed from p_2 to p_1 (segment 3–4 in the diagram) and isobarically vaporized in the steam generator at p_1 (segment 4–1). Here x is the relative dryness.

To increase the thermodynamic efficiency of the Rankine cycle, **steam superheating** is used: the steam is brought to a temperature above the saturation temperature for the working pressure p_1. Such a cycle is shown in Fig. 11.8b. As the fluid works at a higher temperature, the theoretical thermodynamic efficiency (Carnot efficiency) is greater than for a classic cycle without superheating.

In the PWR type of nuclear reactors, the fission heat is extracted from the reactor core by water at high pressure which circulates in the primary circuit of the reactor. A possible accident scenario considered by the designers of this type of installation is depressurization of the fluid. If, for some reason, a leak occurs in the primary circuit piping, the water pressure will suddenly drop, which should induce vaporization of the liquid and formation of a two-phase liquid–vapor mixture forming a "plug" that limits flow of the fluid. An alternative scenario could consist of passing the liquid undergoing depressurization into a superheated metastable state: instead of vaporizing, the water would remain in the liquid state and continue to flow through the breach. The second scenario involves the risk of sudden vaporization of the superheated liquid which could be explosive. This is probably what happened in the accident at the Three Mile Island nuclear power plant in the United States in 1979.

11.3.2 The Cavitation Phenomenon

The formation of vapor bubbles in a flowing liquid undergoing rapid pressure changes is a phenomenon that can have very detrimental consequences. This phenomenon, called **cavitation**, often occurs in hydraulic systems: pumps, turbines, ship propellers.

We say that a liquid cavitates if it is subjected to rapid pressure drops: vapor bubbles (cavities) are formed in it by **nucleation**. In most situations, nucleation is heterogeneous because it arises from dissolved gas bubbles or impurities present in the liquid; the bubbles can also enlarge by coalescence.

Some bubbles in the liquid can implode during cavitation in the liquid–vapor phase transition and thus cause local pressure peaks, forming shock waves that propagate in the liquid. They are at the origin of the erosion phenomena observed in the walls of a system with a liquid in cavitation (for example, the blades of a dam turbine or a ship's propellers). The cavitation phenomenon also reduces the efficiency of a system in which it occurs (a pump, for example).

Cavitation can also be induced by vibrations maintained in a system. For example, the liquid coolant in some diesel engines can be partially vaporized under the effect of the vibrations the engines produce, and cavitation can cause erosion of engine housing materials in this case. A cavitation phenomenon can be associated with light-emission by imploding bubbles. This phenomenon, called sonoluminescence when it caused by ultrasonic waves, could be explained by the partial ionisation of vapor molecules raised at high temperature (several thousand K) during cavitation; these excited molecules would emit light photons when they get de-excited.

Cavitation also has practical applications in different industrial processes. The principal ones include:

- it assists in ultrasonic machining: a tool is made to vibrate a short distance from the part treated, where the part is immersed in a grit-containing liquid, and the cavitation produced contributes to machining, for example, drilling holes;
- cleaning objects with ultrasound: ultrasound induces cavitation in the liquid in which the object treated is immersed and the shock waves produced clean it (process used for hard objects such as jewelry and clock and watch making);
- fabrication of emulsions and droplets: cavitation causes fragmentation of drops or bubbles.

Cavitation induces surface erosion of a material by shock-wave impact, but we are still far from understanding the mechanism responsible for local degradation of the surface state of the material. When water surrounding an immersed object is completly vaporized through cavitation, one has a **supercavitation** phenomenon which can be used to propagate high speed vehicles (torpedoes for example) undersea.

11.3.3 Boiling Regimes

The nature of boiling of a pure liquid observed in a heating installation will be a function of the thermal conditions to which the liquid is exposed. Nukiyama demonstrated the boiling regimes of a liquid in a series of experiments conducted in 1934. Other experiments have since been conducted, and boiling is characterized by a Nukiyama curve, or boiling curve, which represents the heat flux in the liquid as a function of superheating $\Delta T_{sat} = T_w - T_{sat}$ imposed, where T_w is the heating wall temperature (the wall of a metal pipe or plate, for example), and T_{sat} is the temperature of the liquid maintained at saturation (i.e., 100°C at 10^5 Pa for water).

The curve shown in Fig. 11.9 describes the different boiling regimes while working at a given heating temperature.

Boiling begins at point A with a nucleating, isolated-bubble boiling regime and continues with part BC of the curve by formation of vapor pockets resulting from bubble aggregation: this is **pocket nucleation boiling** or **steam column nucleation boiling**. At point C, the heat flux is maximum and the volume of steam produced is such that the liquid can no longer ensure cooling of the heating wall. In effect, the liquid that falls back down cools the

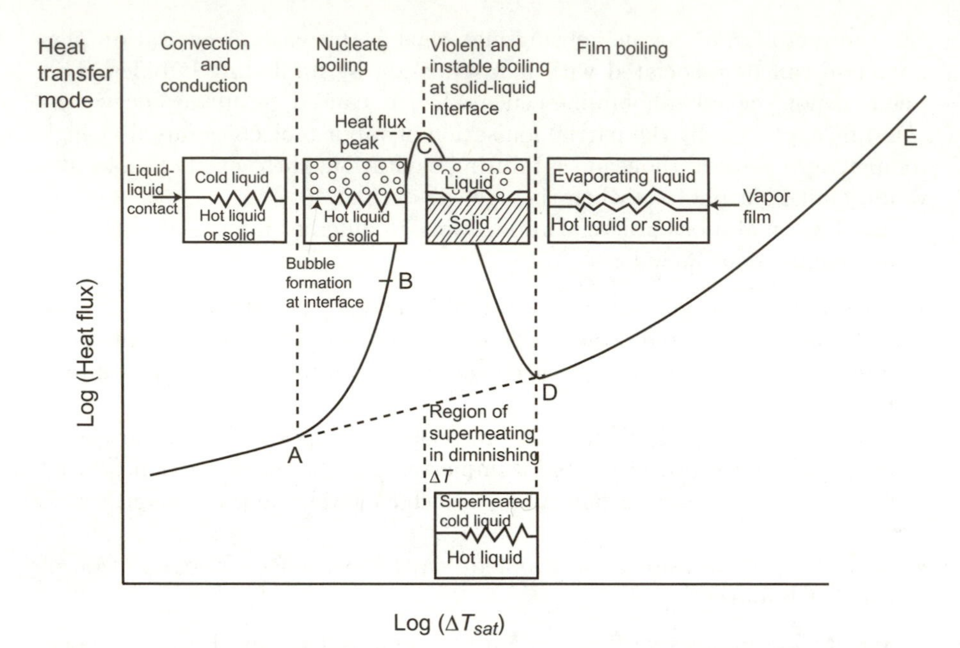

Fig. 11.9. Heat transfer mechanisms at liquid–liquid and liquid–solid interfaces and boiling at the interface. Dashed line AD represents the heat transfer between a stable liquid and a liquid superheated without boiling (D. L. Kate and C. M. Sliepcevich, LNG/Water explosions: cause and effect, *Hydrocarbon Process*, **50**, 240 (1971)).

wall, and if steam production becomes too important, its rate of ascent in the column attains a value that prevents the liquid from falling back down: this is the critical heat flow (Fig.11.10).

This is called **burn-out, critical heat flux**, or **first boiling transition**, and this corresponds to the disappearance of any contact between the liquid and the heating wall. Segment CD of the curve corresponds (Fig. 11.9) to unstable violent boiling. Finally, the film boiling regime is reached from point D (Leidenfrost point) on segment DE, where the heating wall (or metal wire immersed in the liquid) is surrounded by a vapor sheath or vapor film from which vapor bubbles escape.

Nucleation boiling is triggered by a superheating ΔT which is given by the relation (similar to (2.24)

$$\Delta T = \frac{2\gamma T_{sat}}{r_{cav} L \rho_V} \tag{11.22}$$

where γ is the surface tension, L is the specific latent heat of vaporization, ρ_V is the volume density of the vapor, and r_{cav} is the radius of the cavity that will give rise to bubbles. For water, the order of magnitude of a cavity on a smooth metal surface is $r_{cav} \approx 5\mu$ and it is of the order of 0.1–0.3 μ for

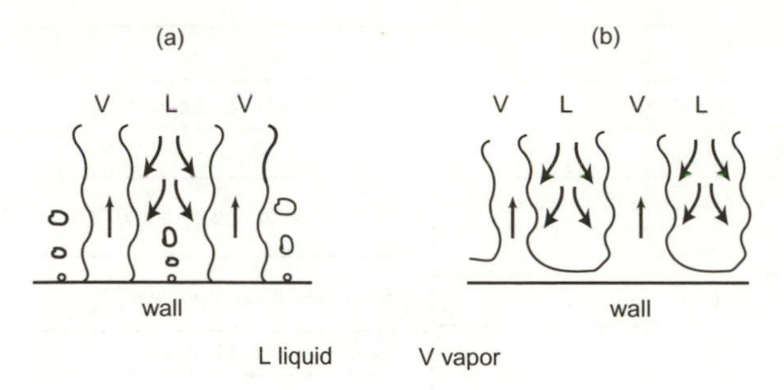

Fig. 11.10. Mechanism of the critical heat flux. Transition from nucleation boiling with vapor columns (**a**) to critical heat flux where the liquid no longer falls back down on the wall (**b**).

cryogenic liquids on a copper or aluminum surface. The more important the superheating is, the smaller the radii of the nucleation sites are.

The theoretical models allow calculating the heat flux density corresponding to the critical heat flux. This critical heat flux q_C corresponding to point C in Fig. 11.9 can be written as follows, for a horizontal boiling plate:

$$q_c = 0.419 L \rho_V^{1/2} [\gamma(\rho_1 - \rho_V) g]^{1/4} \tag{11.23}$$

where ρ_l is the density of the liquid and g is the acceleration due to gravity.

For a horizontal cylinder of radius r, q_C is given by the Sun–Lienhard equation:

$$q_c = \left[0.116 + 0.3 \exp(-3.44 r^{*\,1/2}\right] L \rho_V^{1/2} [\gamma(\rho_l - \rho_v) g]^{1/4} \tag{11.24}$$

where

$$r^* = r \left[\frac{\gamma}{g(\rho_l - \rho_V)}\right]^{-1/2}$$

The evolution of the boiling regime in a vertical heating tube is different from what is found in a horizontal coil or on a plate. The situation is shown in Fig. 11.11.

There is thus a succession of regimes from liquid state to dry steam, passing through the intermediate stages:

- undersaturated boiling (bubble formation);
- saturated nucleate boiling (formation of plugs);
- convection forced by the liquid film (annular flow).

When evaporation of the liquid film on the heating wall is complete, drying takes place, which can be induced by sudden rise in the temperature.

All of these boiling problems are very important in nuclear engineering. In nuclear fission reactions, as we indicated, the nuclear energy released by the fuel rods (enriched uranium) is evacuated by the primary circuit water. Two main systems are used in the world (Fig. 11.7):

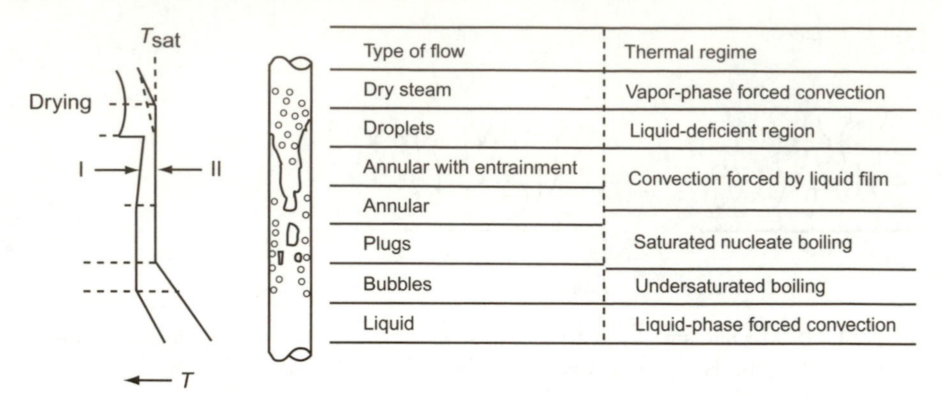

Fig. 11.11. Flow and thermal regimes in a heating tube. The temperatures of the wall (I) and water–steam mixture (II) are shown. The liquid is introduced in the base in an undersaturated state ($T_l < T_{sat}$), and the temperature of the liquid and the wall progressively rise when going up.

- pressurized water reactors, PWR (French series);
- boiling water reactors (BWR), primarily used in the United States.

In the BWR line, the primary circuit water boils in heat transfer and the system is thus a two-phase one; it works at lower pressure and higher temperature than a PWR reactor. With this line, it is necessary to avoid a "critical heat flux" which would partially or totally dry the fuel rods.

Finally, we note that most liquids can be superheated at atmospheric pressure to a temperature of $T \approx 0.9T_C$ (critical temperature). As the volume of the vapor phase is much larger than the volume of the liquid, sudden vaporization is equivalent to an explosion, which can occur when a cold liquid suddenly comes into contact with a hot liquid (the case of methane, for example).

11.3.4 Phase Transitions and Energy Storage

The possibility of storing energy to recover it with a good yield (that is, without too great losses) at a time when its use is needed is a major problem in energetics.

Storage of energy in the thermal form is one possible route: thermal energy in the form of latent heat in a first order liquid–solid phase transition can be used to store energy. Indeed, the thermal energy that must be provided to melt a solid can be recovered during the inverse solidification transformation; that is an amount of energy equivalent to the latent heat of the transition is stored. Glauber's salt ($Na_2SO_4 - 10H_2O$) has long been used as a recoverable material in solidification (the corresponding heat is 0.334 kJ g^{-1}, equivalent to the heat of water but in an eight times smaller volume).

At high temperature, energy can be stored in "heat accumulators" using a salt such as LiF ($T_f = 850°C$ and $L_f = 1.04$ kJ g^{-1}) or the LiF–NaF eutectic mixture ($T_f = 652°C$, $L_f = 20.836$ kJ g^{-1}). These devices are used in wet suits for divers. Hydrates such as $CaCl_2$ $6H_2O$ ($T_f = 29°C$) are also used for storage at ambient temperature. Paraffins which are saturated hydrocarbon mixtures (alkanes), are also used in thermal energy storage systems. They have a melting temperature between 30°C and 90°C when C atoms are in the range 18–50.

At temperatures close to the ambient temperature, the liquid–solid transition of water is also used in air conditioning installations. In this type of device, water cooled by "pumping" heat during the night and using electric power at the night rate is solidified. The ice formed in this way and stored in tanks is used during the day to cool the air circulating in an air conditioning unit, where the latent heat necessary for melting is supplied by the hot air.

In all storage devices utilizing the solid–liquid transition, it is obviously necessary to prevent supercooling of the liquid during recovery of the energy because this would slow release of the stored energy in the form of latent heat.

In nature there is one, probably very important, way of storing energy in the form of crystal phases: gas hydrates. These hydrates are solid phases composed of water molecules forming a three-dimensional lattice with gas molecules such as methane trapped in it at low temperature and low pressure. These gas hydrates are found in the Arctic, under the permafrost in Siberia, and in the depths of the oceans in sediments on the margins of the continental plates (for example, offshore of the United States). Gas hydrates can be considered a form of stored energy, where frozen hydrocarbons can be recovered in gaseous form when the solid melts. It is believed that the resources stored in this way are very important: they could be greater than all of the currently estimated oil and gas resources.

Problems

11.1. Simple Model of the Climate

Let us model the climate change by studying the change in the average air temperature on the surface of the earth T_s. F_{IR} designates the average infrared radiation flux radiated into space, S is the solar constant (flow of solar energy through a plane perpendicular to the solar rays at the top of the atmosphere), α is the albedo, and c is the specific heat of the atmosphere per unit of surface area.

1. Write a simple equation for the evolution of T_s.
2. Rewrite this equation, assuming that $F_{IR} = A + BT_s$, $\alpha = \alpha_0 - \alpha_1 T_s$, and determine T_s at equilibrium.
3. Find the relaxation time τ at equilibrium of T_s.

4. Numerical application: $S = 1370$ W m^{-2}, c $= 4 \times 10^8$ J m^{-2} K^{-1}, $B = 2$ Wm^{-2} K^{-1}, $\alpha = 0.3$, and $\alpha_l = 0.0025$ K^{-1}. What is the significance of τ?

11.2. Snow Balls and Surface Tension
Consider the formation of snow balls from snow crystals under the influence of pressure. $L_f = 6$ kJ/mole for ice and $\Delta V/V = -9.06 \times 10^{-2}$ at the liquid/solid transition.

1. Calculate the change in the melting point associated with a pressure of 15 bar. Knowing that we want to make a snow ball at $-10°$C, what can you conclude from this?
2. What physical phenomenon can intervene in explaining the formation of a snow ball in these conditions?

11.3. Bubble in a Superheated Liquid
A bubble of radius r is held in a liquid superheated to temperature T_0. The liquid/vapor surface tension is designated by γ, the pressure of the liquid imposed from the outside is designated by p_0, and its volume is v_β. $P_s(T_0)$ is the saturated vapor pressure at T_0. Assume that the vapor behaves like an ideal gas.

1. Write the equilibrium condition with the pressure inside the bubble.
2. Express the energy ΔE of formation of the bubble as a function of r and r^*, the critical radius at equilibrium.
3. Discuss the conditions of stability of the bubble. Numerical application $\gamma = 7.5 \times 10^{-2}$ N m^{-1}.

Answers to Problems

1.1. Fluctuations

$$U = \langle E \rangle = \sum_i E_i \exp(-\beta E_i)/Z = -\left(\frac{\partial Z}{\partial \beta}\right)_V, \langle E^2 \rangle = \frac{1}{Z}\left(\frac{\partial^2 Z}{\partial \beta^2}\right)_V$$

hence

$$< \delta E^2 > = < E^2 > - < E >^2 = \left(\frac{\partial^2 \log Z}{\partial \beta^2}\right)_V,$$

$$C_v = (\frac{\partial U}{\partial T})_V = k\beta^2 (\frac{\partial^2 \log Z}{\partial \beta^2})_V, \langle \delta E^2 \rangle = \frac{C_v}{k\beta^2}$$

1.2. Fluctuations and Compressibility

$$\langle N \rangle = kT\left(\frac{\partial \log \Xi}{\partial \mu}\right)_{T,V}, \ \langle N^2 \rangle = \frac{1}{\beta^2 \Xi}\left(\frac{\partial^2 \Xi}{\partial \mu^2}\right)_{T,V}$$

$$\langle \Delta N^2 \rangle = kT^2\left(\frac{\partial^2 \log \Xi}{\partial \mu^2}\right)_{T,V} = kT\left(\frac{\partial \langle N \rangle}{\partial \mu}\right)_{T,V}$$

μ and p are intensive variables and thus homogeneous functions of order 0. We then have the relations:

$$V\left(\frac{\partial p}{\partial V}\right)_{N,T} + \langle N \rangle \left(\frac{\partial p}{\partial \langle N \rangle}\right)_{T,V} = 0$$

$$\left(\frac{\partial \mu}{\partial V}\right)_{N,T} = -\left(\frac{\partial p}{\partial \langle N \rangle}\right) = \frac{V}{\langle N \rangle}\left(\frac{\partial p}{\partial V}\right)_{N,T}$$

$$V\left(\frac{\partial \mu}{\partial V}\right) + \langle N \rangle \left(\frac{\partial \mu}{\partial \langle N \rangle}\right)_{T,V} = 0$$

Hence: $\langle \Delta N^2 \rangle = \langle N^2 \rangle (kT/V)\kappa_T = \langle N \rangle (\kappa_T/\kappa_T^0)$.

1.3. Free Enthalpy of a Binary Mixture

The $G_m(x)$ diagram is plotted. The free enthalpy G_m of the mixture is given by: $G_m = x_A\overline{G_A} + x_B\overline{G_B}$. The Gibbs–Duhem relation is written as: $x_A \mathrm{d}\overline{G_A} + x_B \mathrm{d}\overline{G_B} = 0$ with $x_A + x_B = 1$. If x_B is selected as variable, $\mathrm{d}G_m = \overline{G_A}\mathrm{d}x_A + \overline{G_B}\mathrm{d}x_B$ and $(\mathrm{d}G_m/\mathrm{d}x_B) = \overline{G_B} - \overline{G_A}$.

1.4. Fluctuations and Scattering of Light

1. $\delta\varepsilon(\boldsymbol{r},t) = (\frac{\partial\varepsilon}{\partial T})\delta T(\boldsymbol{r},t)$, $\quad I \propto (\frac{\partial E}{\partial T})_p^2\langle|\delta T(\boldsymbol{k},t)|^2\rangle$. That is

$$I(\boldsymbol{k},\omega) \propto \left(\frac{\partial E}{\partial T}\right)_p^2 \int_{-\infty}^{+\infty} \langle \delta T(\boldsymbol{k},t+\tau)\delta T^*(\boldsymbol{k},t)\rangle e^{\mathrm{i}(\omega-\omega_0)\tau}\mathrm{d}\tau$$

2. $\frac{\mathrm{d}\delta T(\boldsymbol{r},t)}{\mathrm{d}t} = \frac{\lambda}{\rho C_p}\nabla^2\delta T(\boldsymbol{r},t)$
   i.e., by Fourier transformation $\frac{\mathrm{d}\delta T(\boldsymbol{k},t)}{\mathrm{d}t} = -\frac{\lambda}{\rho C_p}\boldsymbol{k}^2\delta T(\boldsymbol{k},t)$. This equation has one solution: $\delta T(\boldsymbol{k},t) = \delta T(\boldsymbol{k},0)\exp -\big(\frac{\lambda}{\rho C_p}k^2 t\big)$, that is

$$\langle \delta T(\boldsymbol{k},t+\tau)\delta T^*(\boldsymbol{k},t)\rangle = \exp -\big(\frac{\lambda}{\rho C_p}k^2\tau\big)\langle|\delta T(\boldsymbol{k},t)|^2\rangle \quad \text{and}$$

$$I = \frac{\langle|\delta T(\boldsymbol{k},t)|^2\rangle}{(\omega-\omega_0)^2 + \Gamma_e^2} I_0\big(\frac{\partial\varepsilon}{\partial T}\big)^2\Gamma_e,\ \ I_0 = E_0^2, \quad \Gamma_e = \frac{\lambda}{\rho C_p}k^2$$

3. The spectrum has the form of a Lorentzian centered around ω_0, of width Γ_e. This is the Rayleigh line.

1.5. Density Fluctuations in a Heterogeneous Medium

1. $\bar{F}$ is the equilibrium value of F, $\Delta F_t = \int\big[(a/2)\delta\rho^2 + (b/2)|\nabla\delta\rho|^2\big]\mathrm{d}\boldsymbol{r}$. With $\delta\rho_{\boldsymbol{k}} = \frac{1}{V}\int\delta\rho(\boldsymbol{r})e^{-\mathrm{i}\boldsymbol{kr}}\mathrm{d}\boldsymbol{r}$; $\quad \Delta F_t = \frac{V}{2}\sum_k(a+b\boldsymbol{k}^2)|\delta\rho_{\boldsymbol{k}}|^2$, and then $\Delta F_{\boldsymbol{k}} = V(a+b\boldsymbol{k}^2)|\delta\rho_{\boldsymbol{k}}|^2$.
2. $w_k \propto \exp -\big(\frac{\Delta F_{\boldsymbol{k}}}{kT}\big)$, let $w_k = \exp -\big[\frac{V}{kT}(a+b\boldsymbol{k}^2)|\delta\rho_k|^2\big]$. Each term $|\delta\rho_k|^2$ enters twice in the summation $(\pm\boldsymbol{k})$.
3.

$$\langle|\delta\rho_{\boldsymbol{k}}|^2\rangle = \frac{kT}{V(a+b\boldsymbol{k}^2)}$$

4. a is $\propto 1/\kappa_T$, where κ_T is the compressibility of the fluid. At T_c, $\kappa_T \to \infty$, $\langle|\delta\rho_{\boldsymbol{k}}|^2\rangle \to \infty$ if $k \to 0$. The fluctuations in a scattering experiment (previous problem) are very important: this is observed as critical opalescence.

2.1. Spinodal Decomposition

1. According to (2.63), we have

$$\frac{\partial\delta c_{\boldsymbol{k}}}{\partial t} = -\tau(\partial\Delta F_k/\partial\delta c_k),\ \ \Delta F_k = \frac{1}{2}\left[\left(\frac{\partial^2 f}{\partial c^2}\right)_{c_0} + K\boldsymbol{k}^2\right]\delta c_k^2(t)$$

and

$$\frac{\partial\delta c_{\boldsymbol{k}}(t)}{\partial t} = -\tau\left[\left(\frac{\partial^2 f}{\partial c^2}\right)_{c_0} + K\boldsymbol{k}^2\right]\delta c_k^2(t)$$

2. $\delta c_k(t) \propto \delta c_k(0)\exp -\tau\big[\big(\frac{\partial^2 f}{\partial c^2}\big)_{c_0} + K\boldsymbol{k}^2\big]t$.

3. In the vicinity of the spinodal, we can have $\left(\frac{\partial^2 f}{\partial c^2}\right)_{c_0} < 0$.
Setting $k_c^2 = -\left(\frac{\partial^2 f}{\partial c^2}\right)_{c_0}/K$, we see that in this region, the fluctuations corresponding to $k < k_c$ are no longer damped: we have spinodal decomposition.

2.2. Soft Mode

1. The oscillator obeys the equation $m\ddot{q} + \gamma\dot{q} = -aq - bq^3$.
2. At equilibrium, we have $aq_e + bq_e^3 = 0$, hence, $m\delta\ddot{q} = -\gamma\delta\dot{q}(t) - a\delta q - 3bq_e^2\delta q(t)$. We have the solutions at equilibrium $q_e = 0$ and $q_e'^2 = -(a/b)$.
3. If $m = 0$, in the vicinity of $q_e = 0$, $\delta q_1(t) = \delta q_1(0)\exp - \left[\frac{\alpha(T-T_c)}{\gamma}t\right]$ and in the vicinity of q_e', $\delta q_2(t) = \delta q_2(0)\exp - \left[\frac{2\alpha(T-T_c)}{\gamma}t\right]$. If $T \to T_c$, the damping factor of $\delta q(t)$ approaches zero: the movements become permanent. We have a soft mode.

2.3. Heterogeneous Nucleation

1. We have $\Delta G_{surf} = 2\pi r^2(1-\cos\theta)\gamma_{sl} + \pi r^2(1-\cos^2\theta)(\gamma_{sc} - \gamma_{lc})$
2. At equilibrium, we have the relation $\gamma_{lc} = \gamma_{sc} + \gamma_{sl}\cos\theta$, which allows eliminating $\gamma_{sc} - \gamma_{lc}$, hence $\Delta G = 4\pi r^2 f(\theta)\gamma_{sl} + \frac{4\pi}{3}f(\theta)r^3\Delta g_V$
3. r^* is given by the solution of $(\partial\Delta G(r)/\partial r) = 0$, that is, $r^* = \frac{16\pi}{3}(\gamma_{sl}^3/\Delta g_V^2)f(\theta)$, which corresponds to (2.29).

2.4. Liquid/Vapor Transition and Interface

1. The new equilibrium conditions are $p + \Delta p$ and $T + \Delta T$ in the convex part and $p + \Delta p + 2\gamma/r$ and $T + \Delta T$ in the concave part with $\mu^\alpha = \mu^\beta$. By limited expansion around (p, T), we then obtain:

$$\left\{\left(\frac{\partial\mu^\alpha}{\partial p}\right)_T - \left(\frac{\partial\mu^\beta}{\partial p}\right)_T\right\}\Delta p$$

$$+\left(\frac{\partial\mu^\alpha}{\partial p}\right)_T\frac{2\gamma}{r} + \left\{\left(\frac{\partial\mu^\alpha}{\partial T}\right)_p - \left(\frac{\partial\mu^\beta}{\partial T}\right)_p\right\}\Delta T = 0$$

$$\text{hence}\quad (v^\alpha - v^\beta)\Delta p + v^\alpha\frac{2\gamma}{r} - \frac{L}{T}\Delta T = 0$$

2. $\Delta T = 0$, $\Delta p = (2\gamma/r)\rho^\beta/(\rho^\alpha - \rho^\beta)$, $\Delta p = 0$, $\Delta T = (2\gamma r/\rho^\alpha L)T$ where ρ^α and ρ^β are the densities and L is the latent heat.

2.5. Homogeneous Nucleation

1. We use (2.23) and (2.28). We find for: $r = 0.5$ nm, $n_r = 4.4 \times 10^{-35}$.
2. r^* is given by (2.24). For $n_r = 1, r^* = 0.6$ nm.
3. $r^* = 35$ nm for $\Delta T = 10$ K.

4. $n_{r*} = 10^{-24}$ nuclei cm^{-3}. Homogeneous nucleation is impossible and would require a greater degree of supercooling or the presence of impurities.

3.1. Richard and Trouton Rule for Melting

1. The total partition function is $Z = \int \exp -(\beta \mathcal{H}) \mathrm{d}\boldsymbol{p} \mathrm{d}\boldsymbol{r}$ where $\mathcal{H}$ is the total Hamiltonian $\mathcal{H} = \sum_i \frac{p_i^2}{2m} + \sum_{i,j} u_2(\boldsymbol{r}_{ij})$. Z is in the form of a product $Z = Q_N(2\pi mkT)^{3N/2}$ with $Q_N = \frac{1}{N!}\int \exp(-\beta \sum_{i,j} u_2(\mathbf{r}_{ij}))\mathrm{d}\boldsymbol{r}$ for the liquid. In the solid permutations are not allowed and hence we have $Q'_N = \int \exp(-\beta \sum_{i,j} u_2(\boldsymbol{r}_{ij}))\mathrm{d}\boldsymbol{r}$. The only contributions to the integrals correspond to terms $r > r_0$ with $u_2 = 0$. We thus have $Q_N = \frac{V^N}{N!}$ and $Q'_N = (\frac{V}{N})^N$.
2. At the solid/liquid transition, we have $\Delta S = -\frac{\Delta F}{T} = k \ln \frac{Q_N}{Q'_N} = kN$.
3. We have $\Delta S = \frac{L_f}{T_m} = kN$, hence $\frac{L_f}{RT_m} = 1$, the Richard-Trouton rule.

3.2. Sublimation

1. The energy of the solid for a vibration state is $E_i = -N_s\varphi + E_{i\nu}$, and we find $Z_s = \sum_i e^{-E_i/kT} = e^{N_s\varphi/kT}(Z_\nu)^{N_s}$; Z_ν is the partition function for an atom. We find $Z_s = E^{N_s\varphi/kT}\left(2sh\frac{h\nu}{2kT}\right)^{-3N_s}$.
2. For an ideal gas, $Z_G = (1/N_G!)(V^{N_G}/\Lambda^{3N_g})$, $\Lambda = \left(\frac{2\pi mkT}{h^2}\right)^{1/2}$.
3. $F = -kT\ln Z_G - kT\ln Z_s$. At equilibrium, knowing that $N = N_s + N_G$
4. $\frac{\partial F}{\partial N_G} = 0$, that is:

$$N_G = \frac{V}{\Lambda^3}\big(2sh\frac{h\nu}{2kT}\big)e^{-\varphi/kT}, \quad p = N_G\frac{kT}{V}$$

3.3. Melting of a Solid with Defects

1. $\Delta S^* = -\left[\frac{\partial \Delta G}{\partial T} - \frac{\partial \Delta G^\nu}{\partial T}\right]$ where the corresponding expressions are given in Sect. 3.3.2. The probability is $w \propto \exp -\left[\Delta S^*/k\right]$.
2. ΔS^* decreases when c increases. For $c = c^*$, by definition $\Delta S^* = 0$, $w = 1$. The probability of any fluctuation becomes equal to 1; the crystal is not stable and melts.

3.4. Melting of a Polymer

1. We have a relation similar to (3.38) with

$$\Delta G = -2n\,b\gamma_e + N\,ablL_f\frac{\Delta T}{T_m^0}, \quad \Delta T = T_m^0 - T.$$

2. At equilibrium, $\Delta G = 0$, that is, $T_m = T_m^0 - 2\gamma(T_m^0/lL_f)$. As $L_f > 0$, the melting point of the crystal will always be less than the melting point T_m^0 of the ideal crystal in the absence of a surface.

4.1. Virial Theorem

1. $\langle \frac{1}{2}\sum_i m\boldsymbol{v}_i^2 \rangle = \langle \frac{1}{2}\sum_i \boldsymbol{p}_i.\boldsymbol{v}_i \rangle = \frac{3}{2}NkT$.
2. By conservation of the virial, we have $\frac{\mathrm{d}}{\mathrm{d}t}\sum_i \langle \boldsymbol{p}_i.\boldsymbol{r}_i \rangle = 0 = \sum_i \langle \dot{\boldsymbol{p}}_i.\boldsymbol{r}_i \rangle + \sum_i \langle \boldsymbol{p}_i.\boldsymbol{v}_i \rangle$ that is $-3NkT = \langle \sum_i \dot{\boldsymbol{p}}_i.\boldsymbol{r}_i \rangle$, where $\dot{\boldsymbol{p}}_i$ is the force on each particle. Let $\langle \sum_i \dot{\boldsymbol{p}}_i.\boldsymbol{r}_i \rangle = \langle \sum_i \boldsymbol{r}_i.\boldsymbol{f}_i \rangle - p\int \boldsymbol{r}.\mathrm{d}\boldsymbol{A}$, where $\mathrm{d}\boldsymbol{A}$ is a surface element perpendicular to the surface of the walls and directed towards the outside. Application of the Gauss theorem for surface integrals allows writing: $-p\int \boldsymbol{r}.\mathrm{d}\boldsymbol{A} = -p\int(\nabla.\boldsymbol{r})\mathrm{d}\boldsymbol{r} = -3pV$, where $-3NkT = -3pV + \langle \sum_i \boldsymbol{r}_i.\boldsymbol{f}_i \rangle$.
3. $\boldsymbol{f}_i = \sum_j \boldsymbol{f}_{ij}(\boldsymbol{r}_{ij})$ where $\langle \sum_i \boldsymbol{r}_i.\boldsymbol{f}_i \rangle = \langle \sum_{i\neq j} \boldsymbol{r}_i.\boldsymbol{f}_{ij} \rangle = \frac{1}{2}\langle \sum_{i\neq j} \boldsymbol{r}_{ij}.\boldsymbol{f}_{ij} \rangle$ The coefficient $1/2$ prevents counting twice. Then one finds $pV = NkT - \frac{1}{6}\langle \sum_{i\neq j} \boldsymbol{r}_{ij}.\boldsymbol{f}_{ij} \rangle$.
4. Replacing the summation by an integral, we have

$$pV = NkT\left[1 - \frac{1}{6kT}\left(\frac{N}{V}\right)\int(\boldsymbol{r}.\nabla \boldsymbol{u})g(\boldsymbol{r})\mathrm{d}\boldsymbol{r}\right]$$

4.2. Law of Corresponding States

1. The total energy is in the form $E = \sum_i \frac{p_i^2}{2m} + \sum_{i,j} \varepsilon_0 f\left(\frac{|\boldsymbol{r}_i - \boldsymbol{r}_j|}{r_0}\right)$. Then introducing $\boldsymbol{r}_i^* = \boldsymbol{r}_i/r_0, \quad T^* = kT/\varepsilon_0$,

$$Z = \frac{1}{N!}\left(\frac{2\pi mkT}{h^2}r_0^2\right)^{3N/2}\int \exp -[\sum_{i,j} f(\boldsymbol{r^*}_i - \boldsymbol{r^*}_j)/T^*]\mathrm{d}\boldsymbol{r}_1^* \ldots \mathrm{d}\boldsymbol{r}_N^*$$

The integral is only a function of the reduced variables and the form of f.
2. $p = -\left(\frac{\partial F}{\partial V}\right)_{T,N} = (kT/\varepsilon_0)(\varepsilon_0/Nr_0^3)\frac{\partial}{\partial V^*}\log\int$ where $V^* = V/(Nr_0^3)$. We thus have the reduced pressure $p^* = (pr_0^3/\varepsilon_0)$ which is a universal function of the reduced variables.
3. The equation of state $p^*V^*/T^* = \frac{V^*}{N}\frac{\partial}{\partial V^*}\log\int$ is a universal function of the reduced variables and has the same form for all fluids described with the same form of potential f.

4.3. Van der Waals Equation of State

1.

$$Z = Z_c\int\ldots\int \exp -\beta(U_N^0 + U_N^1)\mathrm{d}\boldsymbol{r}_1\ldots\mathrm{d}\boldsymbol{r}_N,$$

Z_c is the kinetic part, let $Z = Z_c Z_N^0\langle \exp -\beta U_N^1\rangle_0$, where

$$Z_N^0 = \int\ldots\int \exp -\beta U_N^0 \mathrm{d}\boldsymbol{r}_1\ldots\mathrm{d}\boldsymbol{r}_N$$

corresponding to the configuration partition function of the unperturbed system and $\langle \exp -\beta U_N^1\rangle_0 = (1/Z_n^0)\int\ldots\int \exp -\beta[U_N^0 + U_N^0]\mathrm{d}\boldsymbol{r}_1\ldots\mathrm{d}\boldsymbol{r}_N$ is the ensemble average calculated with the unperturbed system.

2. $\langle \exp -\beta U_N^1 \rangle_0 \approx 1 - \beta \langle U_N^1 \rangle_0 \approx \exp -\beta \langle U_N^1 \rangle_0$ that is,

$$\langle U_N^1 \rangle_0 = \sum_{i,j} \langle u^1(\boldsymbol{r}_i, \boldsymbol{r}_j) \rangle = \frac{N(N-1)}{2} \int u^1(\boldsymbol{r}_1, \boldsymbol{r}_2) p(\boldsymbol{r}_1, \boldsymbol{r}_2) \mathrm{d}\boldsymbol{r}_1 \mathrm{d}\boldsymbol{r}_2$$

$$p(\boldsymbol{r}_1, \boldsymbol{r}_2) = \int \mathrm{d}\boldsymbol{r}_3 \ldots \mathrm{d}\boldsymbol{r}_N \exp -\beta E / \int \mathrm{d}\boldsymbol{r}_1 \mathrm{d}\boldsymbol{r}_2 \ldots \mathrm{d}\boldsymbol{r}_N \exp -\beta E,$$

$$n_2(|\boldsymbol{r}_1 - \boldsymbol{r}_2|) = N(N-1) p(|\boldsymbol{r}_1 - \boldsymbol{r}_2|) = \frac{N^2}{V^2} g(|\boldsymbol{r}_1 - \boldsymbol{r}_2|), \text{hence}$$

$$\langle U_N^1 \rangle_0 = \frac{N^2}{2V^2} \int u^1(|\mathbf{r}_1 - \mathbf{r}_2|) g(|\boldsymbol{r}_1 - \boldsymbol{r}_2|) \mathrm{d}\boldsymbol{r}_1 \mathrm{d}\boldsymbol{r}_2$$

$$= \frac{N^2}{2V} \int u^1(\boldsymbol{r}) g(\boldsymbol{r}) \mathrm{d}\boldsymbol{r}.$$

3. $u_N^0 = \infty$ for $r < r_0$ and $u_N^0 = 0$ for $r > r_0$, thus $g(r) = 0$ for $r < r_0$ and $g(r) = 1$ for $r > r_0$, $\left(n_2 \to N^2/V^2\right)$.
Hence $\langle U_N^1 \rangle_0 = -aN\rho$ with $\rho = N/V$ and $a = -2\pi \int u^1(r) r^2 \mathrm{d}r$ $(a > 0)$, $\langle \exp -\beta U_N^1 \rangle_0 = \exp(\beta a N \rho)$
4. $F = -kT \ln Z$, $p = -\left(\frac{\partial F}{\partial V}\right)_T$. If b designates the volume associated with r_0 (covolume) corresponding to the repulsive part of the potential, $Z_N^0 = (V - Nb)^N$, $F = -kT[\ln Z_N^0 + \ln Z_c + aN\rho\beta]$, $p = kT\left[-\frac{aN^2}{V^2}\frac{1}{kT} + \frac{N}{V-Nb}\right]$ which is the van der Waals equation.

4.4. Spinodal for a Fluid

1. At the metastability limiting point, the isotherm has a horizontal tangent in the (p, V)-plane, and we thus have $(1/\kappa_T) = 0$. Calculating $(\partial p/\partial V)_T$ with the van der Waals equation, we have $-(RT/(V-b)^2) + 2a/V^3 = 0$. Eliminating T and using the van der Waals equation, the spinodal equation is written: $p = a/V^2 - 2ab/V^3$.

4.5. Maxwell's Rule and the Common Tangent

1. The Gibbs–Duhem relation at constant T is written $(\mathrm{d}\mu)_T = V(\mathrm{d}p)_T$. At transition pressure p, we thus have $\mu_1 - \mu_2 = \int_1^2 V \mathrm{d}p = -p(V_2 - V_1) + \int_1^2 p \mathrm{d}V = 0$ hence $p(V_2 - V_1) = \int_1^2 p \mathrm{d}V$.
This equality implies equality of the areas of the surfaces between the isotherm and isobar p and allows situating the liquefaction plateau.
2. $F_T(v)$ is obtained by integration of $p = -\left(\frac{\partial F}{\partial V}\right)_T$. With points 1 and 2 in Fig. 4.4b corresponding to an isobar, the preceding relation implies that points 1 and 2 have the same tangent.

5.1. Unmixing of a Glass

1. As the connectivity of the heterogeneities is important, they correspond to concentration fluctuations in the binary mixture in the glassy state resulting from spinodal decomposition.

2. Conditions for spinodal decomposition exist. The phenomenon can be described with the Cahn-Hilliard model ((2.61), for example). Taking a one-dimensional system with concentration fluctuations of the form $\delta c_k = A\cos kx$, $\delta c(x)$ can be put in the form of (2.82): there is demixing in the binary mixture with phases of different concentration.

5.2. Kauzmann Paradox

1. The entropies of the liquid, glass, and crystal are designated by s_l, s_{gl} and s_{cr}, s_{gl} is the entropy of the glass at $T = 0$ K.

$$s_l(T_g) = s_{gl}(0) + \int_0^{T_g} (c_{p,gl}/T)\mathrm{d}T, \quad s_{cr}(T_g) = \int_0^{T_g} (c_{p,cr}/T)\mathrm{d}T,$$

from which

$$\Delta s(T_g) = s_{gl}(0) + \int_0^{T_g} \frac{(c_{pl} - c_{pr})}{T}\mathrm{d}T, \quad \text{if} \quad c_{pl} \approx c_{pr}, s_{gl}(0)$$

is equal to the excess entropy "frozen" at T_g. We can also write:

$$\left(\frac{\partial \Delta s}{\partial T}\right)_p = \frac{(c_{pl} - c_{p,cr})}{T} = (\Delta C_p/T), \quad \text{and} \quad \Delta s(T_g) = \int_{T_g}^{T} \frac{\Delta c_p}{T}\mathrm{d}T$$

2. $s_l(T) = \int_0^{T_m} (c_{pcr}/T)\mathrm{d}T + \frac{\Delta h_f}{T_m} + \int_{T_m}^{T} (c_{pl}/T)\mathrm{d}T$,
$s_l(T) = s_{gl}(0) + \int_0^{T_g} (c_{pgl}/T)\mathrm{d}T + \int_{T_g}^{T_m} (c_{pl}/T)\mathrm{d}T$, where T_m and Δh_f are the melting point temperature and enthalpy. We thus obtain:

$$s_{gl}(0) = \frac{\Delta h_f}{T_f} + \int_0^{T_g} \frac{(c_{pcr} - c_{pgl})}{T}\mathrm{d}T - \int_{T_g}^{T_m} \frac{\Delta c_p}{T}\mathrm{d}T$$

$$\approx \frac{\Delta h_f}{T_m} - \int_{T_g}^{T_m} \frac{\Delta c_p}{T}\mathrm{d}T.$$

5.3. Prigogine–Defay Relation

1. From (4.25) and (4.26), we directly have $R = \frac{\Delta\kappa_T \Delta c_p}{V_m T_g (\Delta\alpha_p)^2} = 1$, where V_m is the molar volume. This is the Prigogine–Defay relation.
2. If $\frac{\mathrm{d}T_g}{\mathrm{d}p} \ll (\Delta\kappa_T/\Delta\alpha_p)$, (4.26) can be rewritten as

$$(\Delta\kappa_T/\Delta\alpha_p) > V_m T_g(\Delta\alpha_p)/\Delta c_p)$$

$$\text{that is } R = \frac{\Delta\kappa_T \Delta c_p}{V_m T_g (\Delta\alpha_p)^2} \gg 1$$

The Prigogine–Defay coefficient is greater than 1 at the glass transition, which has been verified experimentally. This is not a second order transition in the Ehrenfest sense.

5.4. Supercooling and Glass Transition

1. $\log \eta = C^{st} + b/f$. Hence $\ln(\eta(T)/\eta(T_g)) = b(f^{-1} - f_g^{-1})$ where $f_g^{-1} = \left(v_0 / v_f \right)_g$.
2.

$$\log_{10} \frac{\eta(T)}{\eta(T_g)} = \frac{b}{2.303}(f^{-1} - f_g^{-1}) = -\frac{b'}{f_g} \frac{(T - T_g)}{(T - T_g) + f_g/\Delta\alpha_p}$$

$$= -\frac{c(T - T_g)}{c' + T - T_g}.$$

This is the Williams–Landel–Ferry equation for the viscosity.

6.1. Coil Conformation

1. We have $\langle r^2 \rangle^{1/2} = \left[\int_0^\infty r^2 w(r) \mathrm{d}r\right]^{1/2} = l\sqrt{n}$
The result is similar to the result of random walk of a particle subject to Brownian motion. We have $\langle r^2 \rangle^{1/2} \ll ln$, the chain length, assumed to be linear. The conformation is random coil.
2.

$$n = 10^3, l = 0.3 \text{ nm}, L = nl = 300 \text{ nm}, \langle r^2 \rangle^{1/2} = 9.5 \text{ nm}$$

$$n = 10^4, \quad l = 0.3 \text{ nm}, \quad L = 3000 \text{ nm}, \quad \langle r^2 \rangle^{1/2} = 30 \text{ nm}$$

6.2. Conformation of a Protein

1. The total energy is $E = N_\alpha E_\alpha + N_\beta E_\beta$ with $N = N_\alpha + N_\beta$ and $L(N_\alpha, N_\beta) = N_\alpha a + N_\beta b$.
The canonical partition function Z is written:

$$Z = \sum_{N_\alpha} \frac{N!}{N_\alpha! N_\beta!} \exp\left[-\frac{E - XL}{kT}\right], \text{ that is}$$

$$Z = \left[\exp\left(\frac{Xa - E_\alpha}{kT}\right) + \exp\left(\frac{Xb - E_\beta}{kT}\right)\right]^N$$

2.

$$\overline{L} = N\left[a\, e^{\frac{(Xa - E_\alpha)}{kT}} + b e^{\frac{(Xb - E_\beta)}{kT}}\right]/Z$$

If X is small, a finite expansion can be performed, for which

$$\overline{L} = \frac{(a + be^u)}{(1 + e^u)} + \frac{X}{kT}\left[\frac{(a^2 + b^2 e^u)}{(1 + e^u)} - \frac{(a + be^u)}{(1 + e^u)}\right]$$

with $u = \frac{E_\alpha - E_\beta}{kT}$. This is the behavior observed with keratin in a wool strand.

6.3. Sol/Gel Transition and Percolation

1. By definition (Sect. 6.3.1), $p = \sum_s s n_s, \quad S_m = \sum_s n_s s^2 / \sum_s n_s s$.
2.

$$S_m \propto \sum_s s^{2-\tau} \exp -cs = \int s^{2-\tau} \exp -cs \, \mathrm{d}s,$$

$$S_m = c^{\tau-3} \int Z^{2-\tau} \exp -Z \, \mathrm{d}Z.$$

The integral gives a constant. Hence $S_m = B e^{\tau-3} \propto (p_c - p)^{2\tau-6}$.
As $\tau = 2.5$, $S_m \propto (p_c - p)^{-1}$.
The average size diverges at p_c. We have the equivalent of a critical phenomenon.

6.4. Gelation and Properties of Water
The density and stability of the hydrogen bonds increase when the temperature of the liquid water is decreased. The structure of this system contributes to the increase the specific volume of the liquid phase (decrease in density ρ). This is in contrast to any other liquid, where the decrease in the temperature tends to diminish the specific volume (increase in ρ). The two effects thus vary oppositely with the temperature, where the first moves it to the vicinity of 0°C and the second moves it to high temperatures. A maximum of ρ is observed (at 4°C).

7.1. Predictions of Molecular Fields Models

1. $S_e = -\left(\frac{\partial G_0}{\partial T}\right)_p - (a/2)P_e^2 - \left(\frac{\partial G}{\partial P}\right)_e \left(\frac{\partial P}{\partial T}\right)_e$.
The last term is zero at equilibrium. At T_c, $P_e = 0$, we have $\Delta S = 0$. There is no latent heat: this is a property of a second-order transition.
2. $C_E(E = 0) = -T\left(\partial^2 G/\partial T^2\right) = C_0$ for $T \geq T_c$.

$$C_E(E = 0) = C_0 - T\left[\frac{a}{2}\frac{\mathrm{d}P^2}{\mathrm{d}T} + \frac{b}{4}\frac{\mathrm{d}^2 P^4}{\mathrm{d}T^2} + \ldots\right], \quad T \leq T_c.$$

Using (7.53) at T_c, $\Delta C_E = -T_c(a^2/2b)$. There is discontinuity of specific heat.

7.2. Order/Disorder Transitions under Pressure

1. The free enthalpy G per site is given by

$$G = U_0(a) + V(a)P^2 + pa^3$$
$$+ \frac{kT}{2}\left[(1+P)\log\frac{1+P}{2} + (1-P)\log\frac{1-P}{2}\right] - TS_0.$$

2. At equilibrium, $(\partial G/\partial P) = 0, \quad (\partial G/\partial a) = 0$. As the transition is second order, in the vicinity of T_c, it can be expanded to first order.
We then determine $T_c = -2V(a)/k, \quad p_c = -U_0'(a)/3a^2$.

3. The equilibrium condition $\partial G/\partial a = 0$ is rewritten with linear expansions in δa, which gives $\delta a = -\big(V'(a_0)/(U''(a_0) + 6pa_0)\big)P^2$. Inserting this value in $\partial G/\partial P = 0$, we find

$$P^2 = k(T_c - T)\big[kT_c/3 - 2V'^2(a_0)/(U_0''(a_0) + 6pa_0)\big]^{-1}$$

For the transition to be of second order, the term in brackets must be positive. The transition stops being second order either when this term is zero, or at the pressure p_c^*: $kT_c^*/3 = 2V'^2(a_0)/(U_0''(a_0) + 6p_c^*a_0)$ and $p_c^* = -U_0'(a_0)/3a_0^2$.

7.3. Modeling a Structural Transition

1. We have the equation $m\ddot{q} + \gamma\dot{q} = -\alpha q - \beta q^3$.
2. In the vicinity of equilibrium, $\alpha q_e + \beta q_e^3 = 0$, hence

$$m\delta\ddot{q} + \gamma\delta\dot{q} = -\alpha\delta q(t) - 3\beta q_e^2\delta q(t)$$

with $q_e = 0$ and $q_e^2 = -\frac{\alpha}{\beta}$.
3. $m = 0$, there are two solutions:
$\delta q_1(t) = C^{st}\exp -\frac{a(T-T_c)t}{\gamma}$ and $\delta q_2(t) = C^{st}\exp -2a\frac{(T_c-T)t}{\gamma}$. If $T \to T_c$, the two damping constants approach zero. This the equivalent of a soft mode.

7.4. Universality and Critical Exponents

1. $\lambda^m \frac{\partial G}{\partial \lambda^m H}(\lambda^n\varepsilon, \lambda^m H) = \lambda\frac{\partial G}{\partial H}(\varepsilon, H) \quad (1)$

$$\text{as} \quad \frac{\partial G}{\partial H} = -M(\varepsilon, H), \quad \lambda^m M(\lambda^n\varepsilon, \lambda^m H) = \lambda M(\varepsilon, H)$$

For $H = 0, M(\varepsilon, 0) \propto (-\varepsilon)^\beta$. Then $M(\varepsilon, 0) = \lambda^{m-1}M(\lambda^m\varepsilon, 0)$.
2. Setting $H = 0$ in (1), $M(0, H) = \lambda^{m-1}M(0, \lambda^m, H)$ taking $\lambda^m H = \mathrm{C^{te}}$, we have $\delta = m/(1-m)$.
3. Differentiating (1) with respect to H, we obtain

$$\lambda^{2m}\chi_T(\lambda^n\varepsilon, \lambda^m H) = \lambda\chi_T(\varepsilon, H), \text{using } \lambda \propto (-\varepsilon)^{1/n},$$

$$\chi_T(\varepsilon, 0) = (-\varepsilon)^{(2m-1)/n}\chi_T(-1, 0), \quad \gamma = \gamma' = (2m-1)/n$$

7.5. Piezoelectricity

1. With the notation of (7.54), for a Landau expansion, we can write:

$$G = G_0 + \frac{1}{2}a(T - T_c)P^2 + \frac{b}{4}P^4 + \frac{1}{2}c^p x^2 - \frac{1}{2}\lambda x P^2$$

2. At equilibrium $X = (\partial G/\partial x) = c^p x - \frac{1}{2}\gamma P^2$ if for $X = 0, x = \frac{1}{2}\gamma/c^p P^2$.
3. Substituting x in G, $G = G_0 + \frac{1}{2}a(T - T_c)P^2 + \frac{1}{4}\big(b - \frac{1}{2}\frac{\gamma^2}{c^p}\big)P^4$.
Hence $E = a(T - T_c)P + \big(b - \frac{1}{2}\gamma^2/c^p\big)P^3$, which gives at equilibrium in a zero field ($E = 0$), $\varepsilon = \big[a(T - T_c) + \big(b - \frac{1}{2}\gamma^2/c^p\big)P_e^2\big]^{-1}$ where P_e is the polarization at equilibrium. We find the classic behavior in the neighborhood of T_c.

8.1. Latent Heat of the Nematic Transition

1. At the transition, L, the latent heat of fusion of the nematic phase is equal to $L = T\Delta S = N_0 kT \log\left[z(T_{NI}, \langle s_{NI}\rangle)_{int}\right]$ with the condition $\Delta\mu = 0$, hence $(A/2V^2)\langle s_{NI}\rangle^2 - kT\log[z(T_{NI}, \langle s_{NI}\rangle)_{int}] = 0$, hence $L = N_0(A/2V^2)\langle s_{NI}\rangle^2$.
   With $< s_{NI} >= 0.429$ and $T_{NI} = 0.220(A/kV^2)$, $\quad L = 0.418 N_0 kT_{NI}$ and $(L/T_{NI}) = 0.418$, R $= 3.46$.
2. The existence of latent heat at the transition implies that it is first order.

8.2. Nematic–Smectic A transition

1. We may write: $F(\langle s_0\rangle + \delta s, \rho_1) = F_N(\langle s_0\rangle) - C\rho_1^2\delta s + \frac{1}{2\chi(T)}\delta s^2 + \frac{1}{2}\alpha(T - T_0)\rho_1^2 + \frac{1}{4}u_0\rho_1^4 + \frac{1}{6}u_1\rho_1^6 + \mathcal{O}(\rho_1^8)$.
2. At equilibrium we should have $\partial F/\partial\delta s = 0$, that is: $(1/\chi) < \delta s >_{\rho_1} = C\rho_1^2$ in the presence of smectic order;
3. We can write:
$$F = F_N(\langle s_0\rangle) + \frac{1}{2}\alpha(T - T_0)\rho_1^2 + \frac{1}{4}(u_0 - 2\chi(t)C^2)\rho_1^4 + \frac{1}{6}u_1\rho_l^6 + \mathcal{O}(\rho_l^8)$$
4. Setting $u(T) = u_0 - 2\chi(t)C^2$, one has at equilibrium $\alpha(T - T_0)\rho_1 + u(T)\rho_1^3 + u_1\rho_1^5 + \mathcal{O}(\rho_1^7) = 0$ with $\frac{\partial^2 F}{\partial^2\rho} > 0$. We then find $< \rho_1 >_1 = 0$ for $T > T_0$ and, for $T < T_0 + u(T)^2/(2\alpha u_1)$
$$\langle\rho_1\rangle_2^2 = \frac{-u(T) + \sqrt{u(T)^2 - 4u_1\alpha(T - T_0)}}{2u_1}.$$
   At the "nematic–smectic A" transition, we should have equality of the free energies of the two phases, that is:
$$\alpha(T - T_0) + (1/2)u(T)C^2 < \rho_1 >_2^2 + (1/3)u_1 < \rho_1 >_2^4 = 0$$
   Hence at the transition: $\langle\rho_{1AN}^2\rangle = -(3u/4u_1), T = T_0 + (3u^2/16\alpha u_1)$ and for $u_0 = 0$:
$$\langle\rho_{1AN}\rangle = \frac{3C^2}{2u_1}\chi(T_{AN}), \quad T_{AN} = T_0 + \frac{3C^4}{4\alpha u_1}\chi(T_{AN})^2$$

8.3. Light Scattering in a Liquid Crystal

1. Using (8.61) and (8.65), we have
$$F = F_0 + \frac{1}{2V}\sum_q [(K_3 q_\parallel^2 + K_1 q_\perp^2 + \chi_a\, H^2)|\delta n_\perp(\boldsymbol{q})|^2 + (K_3 q_\parallel^2 + K_2 q_\perp^2 + \chi_a\, H^2)|\delta n_t(\boldsymbol{q})|^2]$$
   where $q_\parallel$ is the component of $\boldsymbol{q}$ in the $\boldsymbol{n}_0$ direction and $q_\perp$ is the perpendicular component.

2. The fluctuation distribution function is $\propto \exp -F/kT$, hence

$$\langle|\delta n_\perp(\boldsymbol{q})|^2\rangle = \frac{kTV}{(K_3 q_\parallel^2 + K_1 q_\perp^2 + \chi_a\ H^2)}, \quad \text{and}$$

$$\langle|\delta n_t(\boldsymbol{q})|^2\rangle = \frac{kTV}{(K_3 q_\parallel^2 + K_2 q_\perp^2 + \chi_a\ H^2)}$$

These fluctuations scatter light and we have for the scattered intensity:

$$I \propto \frac{kTV}{(K_3 q_\parallel^2 + K_1 q_\perp^2 + \chi_a\ H^2)} + \frac{kTV}{(K_3 q_\parallel^2 + K_2 q_\perp^2 + \chi_a\ H^2)}$$

3. One observes a strong opalescence in the nematic phase.

9.1. Droplets in Equilibrium with their Vapor

For the system $G = N_v\mu_v + N_L\mu_L + 4\pi r^2\gamma$, where N_v and N_L are the number of molecules of vapor and liquid $N = N_v + N_L$. We have $N_L = \frac{4\pi}{3}\frac{r^3}{V_0}$, where V_0 is the molecular volume in the liquid phase. At equilibrium, $\mathrm{d}G = 0$. Then $\mu_v - \mu_L - \frac{2\gamma}{r}V_0 = 0$.

Deriving this equation at constant T, $(V - V_0)\mathrm{d}p = 2\gamma V_0 \mathrm{d}(1/r)$. Assuming $V_0 \ll V$ and considering that the vapor behaves like an ideal gas, we can write $kT\mathrm{d}(\ln p) = 2\gamma V_0 \mathrm{d}\left(\frac{1}{r}\right)$, that is $\ln(p/p_\infty) = \ln S = (2\gamma V_0/rkT)$.

9.2. Creation of Holes

1. $\Delta G = nc\varepsilon_f - T\Delta S$, where ΔS is the change in entropy related to the formation of vacancies (entropy of mixing).
   $\Delta S = kN\{c\ln c + (1-c)\ln(1-c)\}$.
2. At equilibrium, $\partial\Delta G/\partial c = 0$ if $\varepsilon_f + kT\ln c/(1-c) = 0$ and thus $c = \exp -\frac{\varepsilon_f}{kT}$.
3. At 300 K, $c = 1.6 \cdot 10^{-17}$ and at 1000 K, $c = 9 \cdot 10^{-6}$.

9.3. Energy of Bloch Walls

1. $W_e = -2JS^2(\cos\varphi - 1) \approx JS^2\varphi^2$.
2. For the total rotation of M achieved by N spins in the Bloch wall, one may take $\varphi = \frac{\pi}{N}$, hence $W_e = JS^2\left(\pi/N\right)^2$. For a line of N spins, we thus have a change of total energy $NW_e = JS^2\left(\pi^2/N\right)$.

9.4. Molecular Films

1. $\gamma = \gamma_0 - \Pi$, where Π is the equivalent of surface tension whose effect is opposite of γ_0.
2. Since the liquid is dilute, $\mu = \mu^0 + RT\ln x$, from which $(\mathrm{d}\gamma/\mathrm{d}\mu) = (\mathrm{d}\gamma/\mathrm{d}x)(\mathrm{d}x/\mathrm{d}\mu) = -(bx/kT)$ with $\gamma = \gamma_0 - bx$, that is $\Gamma = (bx/RT) = (\Pi/RT) = (n_s/A)$, $(n_s/A) = a$, is the area per adsorbed mole. Hence $\Pi a = RT$.
   This is an equation of the ideal gas type for adsorbed molecules.

11.1. Simple Model of the Climate

1. $C(\mathrm{d}T_s/\mathrm{d}t) = (S/4)(1-\alpha) - F_{IR}$, the solar flux should be averaged over the entire surface of the earth: $(\pi R^2 S/4\pi R^2) = \frac{S}{4}(1-\alpha)$ is absorbed.
2. Placing F_{IR} in this equation, it is written as
$$\frac{\mathrm{d}T_s}{\mathrm{d}t} + \frac{B - S\alpha_1/4}{C} T_S = \frac{S}{4}\frac{(1-\alpha_0)}{C} - \frac{A}{C}$$
At equilibrium,
$$\frac{\mathrm{d}T_s}{\mathrm{d}t} = 0, \text{ thus } T_S = \frac{(S/4)(1-\alpha_0) - A}{B - \alpha_1(S/4)}.$$
3. The general solution of the differential equation is:
$$T_S(t) = T_S(0)e^{-t/\tau} \text{with } \tau = \frac{C}{B - \alpha_1(S/4)}.$$
4. One finds $\tau \approx 10$ years. τ measures the response time to a climatic perturbation.

11.2. Snow Balls and Surface Tension

1. $\mathrm{d}p/\mathrm{d}T = 0.0074°\mathrm{C}\mathrm{bar}^{-1}$ if the pressure is changed by 15 bar, the melting point is reduced by 0.1°C. This pressure is insufficient to induce local melting of snow and to cause "welding" of snow crystals to form a snow ball at −10°C.
2. Two ice particles in contact exhibit regions with strong curvature (radius $\approx 10-100$ nm). These concave regions are the site of high overpressures $\approx (2\gamma/r)$ ($\gamma \approx 75$ dynes cm^{-1}) which can reach $10^3 - 10^4$ bar and thus cause local melting of ice.

11.3. Bubble in a Superheated Liquid

1. By integration along an isotherm, we obtain the chemical potential of the pure liquid: $\mu_\beta = \mu_S(T) + v[p - p_S(T)]$, where $\mu_S(T)$ is the saturation potential. Integrating $\left(\frac{\partial\mu}{\partial T}\right)_T = v = RT/p$, the chemical potential of the vapor is $\mu_\alpha = \mu_S + RT\log(p/p_S(T))$.
Hence:
$$v_\beta[p_0 - p_S(T_0)] = RT_0 \log\frac{p_\alpha^*}{p_S(T_0)}$$
2. Radius r is different from critical radius r^* at equilibrium, given by $p_\alpha^* - p_0 = 2\gamma/r^*$, where $\Delta E = -(p_\alpha^* - p_0)v_\alpha + A\gamma$ is the surface area of the bubble and v_α is its volume, so $\Delta E = \frac{4\pi}{3}r^3\left(3\gamma/r + p_0 - p_\alpha^*\right)$ where $\Delta E = \frac{4\pi}{3}r^3\left(3/r - 2/r^*\right)$.
3.
$$\frac{\partial\Delta E}{\partial r} = 8\pi r^2\gamma\left(1/r - 1/r^*\right), \frac{\partial^2\Delta E}{\partial r^2} = 8\pi r\gamma\left(1/r - 2/r^*\right).$$

The equilibrium is stable at $r = 0$ and unstable for $r = r^*$. The limited expansion can be written as:

$$\Delta E = \frac{4\pi}{3} {r^*}^2 \gamma - 4\pi\gamma (r - r^*)^2 + \dots.$$

A. Conditions for Phase Equilibrium

To write the conditions for equilibrium between phases, it is necessary to utilize thermodynamic potentials (the chemical potentials, for example) and constraints which are a function of the variables used to describe the system. They depend on the type of experiments being conducted.

To describe a mixture of several constituents, three, for example, the mole fractions n_1, n_2, n_3 and corresponding concentrations x_2, x_3 are introduced, where n is the total number of moles; we then have:

$$n_1 = n(1 - x_2 - x_3),\ n_2 = nx_2,\ n_3 = nx_3 \tag{A.1}$$

with:

$$n = n_1 + n_2 + n_3$$

i.e. also

$$x_2 = \frac{n_2}{n_1 + n_2 + n_3},\ x_3 = \frac{n_3}{n_1 + n_2 + n_3} \tag{A.2}$$

The differential of a function f, for example, with respect to n_1, is written:

$$\frac{\partial f}{\partial n_1} = \frac{\partial f}{\partial n} - \frac{x_2}{n}\frac{\partial f}{\partial x_2} - \frac{x_3}{n}\frac{\partial f}{\partial x_3} \tag{A.3}$$

We have similar relations for $\partial/\partial n_2$ and $\partial/\partial n_3$. This leads to the sequence of differential equations using (A.2):

$$\begin{aligned}
\frac{\partial}{\partial n_1} &= \frac{\partial}{\partial n} - \frac{x_2}{n}\frac{\partial}{\partial x_2} - \frac{x_3}{n}\frac{\partial}{\partial x_3} \\
\frac{\partial}{\partial n_2} &= \frac{\partial}{\partial n} + \frac{1 - x_2}{n}\frac{\partial}{\partial x_2} - \frac{x_3}{n}\frac{\partial}{\partial x_3} \\
\frac{\partial}{\partial n_3} &= \frac{\partial}{\partial n} - \frac{x_2}{n}\frac{\partial}{\partial x_2} + \frac{1 - x_3}{n}\frac{\partial}{\partial x_3}
\end{aligned} \tag{A.4}$$

Applying these equations to the definition of the chemical potentials μ_i $(i = 1, 2, 3)$ for a ternary mixture, we obtain the following expressions:

$$\begin{aligned}
\mu_1 &= \frac{\partial G}{\partial n_1} = G_m - x_2\frac{\partial G}{\partial x_2} - x_3\frac{\partial G}{\partial x_3} \\
\mu_2 &= \frac{\partial G}{\partial n_2} = G_m + (1 - x_2)\frac{\partial G}{\partial x_2} - x_3\frac{\partial G}{\partial x_3}
\end{aligned} \tag{A.5}$$

$$\mu_3 = \frac{\partial G}{\partial n_3} = G_m - x_2 \frac{\partial G}{\partial x_2} + (1 - x_3) \frac{\partial G}{\partial x_3}$$

$$G_m \equiv \frac{G}{n} = \frac{\partial G}{\partial n} \quad \text{is the Gibbs molar function.}$$

We easily determine from equations (A.5):

$$\mu_2 - \mu_1 = \frac{\partial G}{\partial x_2}, \quad \mu_3 - \mu_1 = \frac{\partial G}{\partial x_3} \tag{A.6}$$

As a function of the choice of independent variables (A.1) or (A.2), we can write the conditions of equilibrium between phases either with (A.5), $\mu_1 = \mu_2 = \mu_3$, or with equalities (A.6).

B. Percus–Yevick Equation

We should be able to completely determine the equation of state of a fluid if we can calculate the partition function of a system with a very large number of particles (the molecules in the fluid). This is a multiple integral in position and time variables which is a function of the pressure and the temperature. This calculation can only succeed by using an approximation method. The methods used are of three types: series expansions of the density, integral equations, and perturbation methods.

The density expansions have the advantage of giving the exact expressions for the virial coefficients which can be compared with the experimental results if the interaction forces are known. On the other hand, they do not allow predicting phase transitions, since truncated series expansions do not have singularity. The so-called Padé approximation nevertheless allows performing the calculation for the rest of the series. Kirkwood on one hand and Born–Green–Yvon on the other proposed integral equations for the radial distribution function $g(\boldsymbol{r})$ which is then used to obtain the equation of state. For this purpose, a hierarchy of functions $G^n (n = 2, 3 \ldots)$ which are finite in number is established. A new class of integral equations was subsequently proposed based on a calculation performed by Ornstein and Zernike (1914) to take into account the phenomenon of critical opalescence.

The distribution function for n particles in a system of N particles can be written as:

$$\rho_N^{(n)}(\boldsymbol{r}_1, \ldots, \boldsymbol{r}_n) = \frac{N!}{(N-n)!} \int \cdots \int \mathrm{e}^{-\beta U_N} \mathrm{d}\boldsymbol{r}_{n+1} \ldots \mathrm{d}\boldsymbol{r}_N / Z_N \tag{B.1}$$

In an open system (with a variable number of particles), the probability of observing n molecules in volume $\mathrm{d}\boldsymbol{r}_1 \ldots \mathrm{d}\boldsymbol{r}_\mathrm{n}$ at the point $(\boldsymbol{r}_1, \ldots \boldsymbol{r}_n)$ is given by:

$$\rho^{(n)} = \sum_{N \geq n} \rho_N^{(n)} P_N \tag{B.2}$$

This function is independent of N. The probability P_N is written in the classic manner in statistical mechanics:

$$P_N = \frac{z^N Z_N}{N! \Xi} \tag{B.3}$$

Z_N is the partition function for N particles and Ξ is the grand partition function with

$$z \equiv e^{\beta\mu}\left(\frac{2\pi mkT}{h^2}\right)^{3/2}$$

Carrying (B.3) into (B.2) for $n = 2$ and after integration in $\boldsymbol{r}_1$, and $\boldsymbol{r}_2$, we obtain:

$$\int\int \rho^{(2)}(\boldsymbol{r}_1, \boldsymbol{r}_2)\mathrm{d}\boldsymbol{r}_1\mathrm{d}\boldsymbol{r}_2 = \langle\frac{N!}{(N-2)!}\rangle = \langle N(N-1)\rangle = \overline{N^2} - \overline{N} \quad \text{(B.4)}$$

and for $n = 1$, we have:

$$\int\int \rho^{(1)}(\boldsymbol{r}_1)\rho^{(1)}(\boldsymbol{r}_2)\mathrm{d}\boldsymbol{r}_1\mathrm{d}\boldsymbol{r}_2 = (\overline{N})^2 \quad \text{(B.5)}$$

Subtraction of these two equations gives:

$$\int\int [\rho^{(2)}(\boldsymbol{r}_1, \boldsymbol{r}_2) - \rho^{(1)}(\boldsymbol{r}_1)\rho^{(1)}(\boldsymbol{r}_2)]\mathrm{d}\boldsymbol{r}_1\mathrm{d}\boldsymbol{r}_2 = \overline{N^2} - \overline{N}^2 - \overline{N} \quad \text{(B.6)}$$

where $\langle N\rangle = \overline{N}$ and $\langle N^2\rangle = \overline{N^2}$. We know that:

$$\overline{\Delta N^2} = \langle(N - \overline{N})^2\rangle = kT\kappa_T\frac{\overline{N}}{V} = \overline{N^2} - \overline{N}^2 \quad \text{(B.7)}$$

with $\rho = (N/V)$ and $\kappa_T = \rho^{-1}\left(\frac{\partial\rho}{\partial p}\right)_T$.

Setting $\rho^{(2)}(\boldsymbol{r}_1, \boldsymbol{r}_2) = \rho^2 g(\boldsymbol{r}_1, \boldsymbol{r}_2)$ and $\rho^{(1)} = \rho^{(2)} = \rho$, i.e. assuming that the fluid is invariant under translation, and utilizing (B.7), (B.6) is rewritten as:

$$\rho^2 V\int [g(\boldsymbol{r}) - 1]\mathrm{d}\boldsymbol{r} = \langle\Delta N^2\rangle - \overline{N} \quad \text{(B.8)}$$

that is

$$\rho kT\kappa_T = 1 + \rho\int [g(\boldsymbol{r}) - 1]\mathrm{d}\boldsymbol{r} \quad \text{(B.9)}$$

Function $h(\boldsymbol{r}) = g(\boldsymbol{r}) - 1$ is called the **pair correlation function** and it can be obtained from scattering experiments (light, neutrons). It is a measure of the influence of molecule 1 on molecule 2.

The **structure function** $S(\boldsymbol{k})$ is defined using the Fourier transform of $g(r)$ (Fig. B.1) by the equation:

$$S(\boldsymbol{k}) = 1 + \rho\int \exp(-i\boldsymbol{k}\boldsymbol{r})g(\boldsymbol{r})\mathrm{d}\boldsymbol{r} \quad \text{(B.10)}$$

Ornstein and Zernike proposed separating the influence exercised by each molecule into two parts. The first term represents the direct interaction of a molecule with its neighbors, and the second is the contribution of the indirect influence, which is the force transmitted via an intermediate molecule. The direct correlation corresponding to the first term is the function $c(\boldsymbol{r}_{12})$,

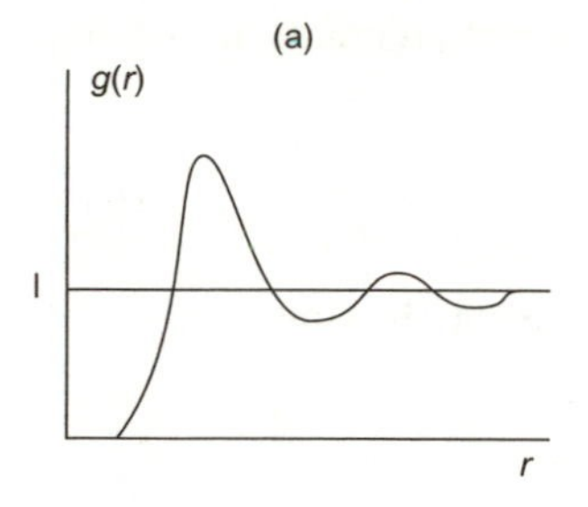

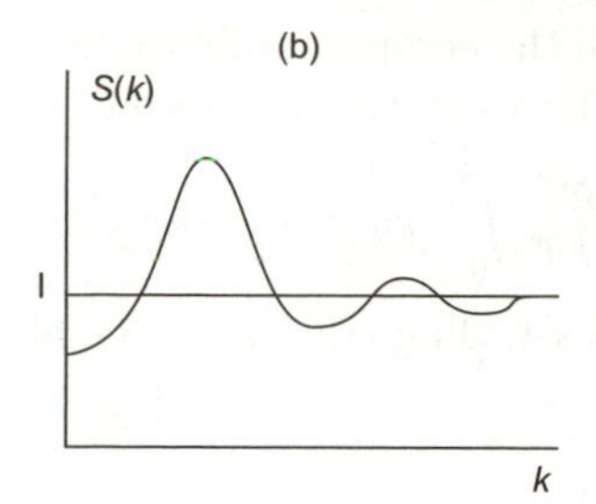

Fig. B.1. Pair correlation function (**a**) and structure function (**b**) for a liquid in normal conditions.

the second contribution to $h(\boldsymbol{r}_{12})$ is an average over all positions of the intermediate molecule $\boldsymbol{r}_3$:

$$h(\boldsymbol{r}_{12}) = c(\boldsymbol{r}_{12}) + \int c(\boldsymbol{r}_{13})\, h(\boldsymbol{r}_{32}) \mathrm{d}\boldsymbol{r}_3 \tag{B.11}$$

This equation in fact defines function $c(\boldsymbol{r})$. It can be solved by taking the Fourier transform and using the properties of the convolutions.

$$H(\boldsymbol{k}) = c(\boldsymbol{k}) + \rho\, H(\boldsymbol{k})\, c(\boldsymbol{k}) \tag{B.12}$$

that is:

$$H(\boldsymbol{k}) = \frac{c(\boldsymbol{k})}{1 - \rho\, c(\boldsymbol{k})} \tag{B.13}$$

Substituting $H(\boldsymbol{k})$ in (B.9), we then have:

$$kT\big(\frac{\partial \rho}{\partial p}\big)_T = 1 + \rho\, H(0) = \frac{1}{1 - \rho\, c(0)} = \frac{1}{1 - \rho \int c(\boldsymbol{r}) \mathrm{d}\boldsymbol{r}} \tag{B.14}$$

We must still determine $c(r)$. This can be done by taking an approximation that describes short-range interactions. Here $c(r)$ is the difference between the function $g_{total}(r)$ corresponding to the effective interaction represented by potential $w(r)(g(r) = \mathrm{e}^{-\beta w(r)})$, and the indirect interaction which is the remainder of the potential when the direct interaction is eliminated:

$$c(r) = g_{total}(r) - g_{indirect}(r) = \mathrm{e}^{-\beta w(r)} - \mathrm{e}^{-\beta[w(r) - u(r)]} \tag{B.15}$$

This is the Percus–Yevick approximation for function $c(r)$. Using the Ornstein-Zernicke method, we can then solve the system of equations.

Assuming $y(r) = \mathrm{e}^{\beta u(r)} g(r)$, we can rewrite (B.11):

$$y(\boldsymbol{r}_{12}) = 1 + \rho \int [\mathrm{e}^{-\beta u(\boldsymbol{r}_{13})} - 1] y(\boldsymbol{r})_{13}) h(\boldsymbol{r}_{32}) \mathrm{d}\boldsymbol{r}_3 \tag{B.16}$$

This is the Percus–Yevick equation. As $h = g - 1 = \mathrm{e}^{\beta u} y - 1$, this equation determines $y(\mathbf{r})$ and thus $g(\mathbf{r})$. It was solved for hard spheres. The equation for p and thus the equation of state is not the same as a function of whether it is determined with the compressibility (B.9) or by direct calculation utilizing the equation for the virial (Problem 4.1):

$$\frac{p}{kT} = \rho - \frac{\rho^2}{6kT} \int_0^\infty r u(r) g(r) 4\pi r^2 \mathrm{d}r \tag{B.17}$$

The Carnahan–Starling equation is established using the first method.

C. Renormalization Group Theory

Take the Ising Hamiltonian $\mathcal{H}_0$ for a magnetic system

$$\mathcal{H}_0 = -J_0 \sum_{i,j} \sigma_i \sigma_j - \mu_0 H_0 \sum_i \sigma_i \tag{C.1}$$

The free energy F is the sum of the singular part near the critical point F_S and the regular part F_R corresponding to nonmagnetic lattice contributions. It is written (7.74):

$$\mathrm{e}^{-F/kT} = \big[\sum_{(\Omega_1)} \mathrm{e}^{-\beta \mathcal{H}_1}\big] \mathrm{e}^{-F_{R1}/kT} \tag{C.2}$$

after taking only one out of two spins in the summation Σ, where (Ω_1) represents the new configurations.

$\mathcal{H}_1$ has the same form as $\mathcal{H}_0$ but with half the spins and with new values of exchange constant J_1 and applied field H_0. Introducing the reduced variables $K = J/kT$ and $b = \mu_0 H_0/kT$, we have:

$$K_1 = f(K_0) \quad \text{and} \quad b_1 = g(b_0) \tag{C.3}$$

This process is iterated several times, doubling the scale with each operation, and a sequence of new variables is obtained: $K_1, K_2 \ldots b_1, b_2$. The critical point will be a fixed point corresponding to variables K_C and b_C so that:

$$K_C = f(K_C) \quad \text{and} \quad b_C = g(b_C) \tag{C.4}$$

because no scale change can modify the effective Hamiltonian any further.

For finding T_C, we operate in the vicinity of this temperature after n renormalization operations $K_n = f(K_{n-1})$ and $K_C = f(K_C)$, that is:

$$\delta K_n = K_C - K_n = f(K_C) - f(K_{n-1}) = f'(K_C - K_{n-1}) = f' \delta K_{n-1} \tag{C.5}$$

At T_C, by definition $K_C = J_0/kT_C$ (renormalization no longer alters the form of the Hamiltonian). Moreover, $b_n = g' b_{n-1}$ and at the critical point, $H = b_C = 0$ (C.5) is rewritten:

$$K_n = K_C - \delta K_n = K_C - f' \delta K_{n-1} \tag{C.6}$$

knowing that

$$\delta K_0 = K_C - K_0 = J_0/kT_C - J_0/kT = \frac{J_0}{kT_C}\varepsilon \tag{C.7}$$

$$K_n = \frac{J_0}{kT_C}[1 - f'^n\varepsilon]; \quad b_n = g'^n \frac{\mu_0 H_0}{kT_C}$$

After n renormalizations, if d is the dimension of the system and n is the dimension of order parameter σ, the total number of remaining spins for N initial sites is $N/(2^d)^n$. The free energy by spins F_{sn} is a function of reduced variables K_n and b_n and thus of $(f'^n)\varepsilon$ and $g'^n H_0$, with $\varepsilon = (T_C - T)/T$. As J_0, T_C, and H_0 are constants, we have:

$$F_{sn}2^{dn} = F(f'^n\varepsilon, g'^n H) \tag{C.8}$$

If we put $\lambda = 2^{dn}, f' = (2^d)^p, g' = (2^d)^q$, we find an equation similar to (7.69):

$$\lambda F_{sn} = F(\lambda^p\varepsilon, \lambda^q H) \tag{C.9}$$

We have a scaling law here which allows calculating the critical exponents. A Hamiltonian representing the energy, which has a Landau-type form, is then introduced:

$$\mathcal{H} = \int (tP(x)^2 + u_4 P^4(x) + a|\nabla P(x)|^2 + \ldots)\mathrm{d}^d x \tag{C.10}$$

where t, u_4, and a are functions of T, but where t does not necessarily have the form given it in the classic Landau expansion (t can be different from $(T - T_C)$). Application of renormalization represented by the operator R results in an effective Hamiltonian and a fixed critical point Π_C^* defined in the space of parameters (t, u_4, a):

$$\Pi_C^* = R(\Pi_C^*) \tag{C.11}$$

Bibliography

Chapter 1

Binney, J. J., Dowrick, A. J., Fisher, A. J., Newman, E. J.: *The Theory of Critical Phenomena* (Clarendon Press, Oxford 1995).
Boccara, N.: *Symétries brisées* (Hermann, Paris 1976).
Callen, H. B.: *Thermodynamics and Introduction to Thermostatics* (Wiley, New York 1985).
Cahn, R. W.: *The Coming of Materials Science* (Pergamon, Oxford, 2001).
Chaikin, P. M., Lubenski, T. C.: *Principles of Condensed Matter Physics* (Cambridge University Press, Cambridge 1995).
Kittel, C., Kroemer, H.: *Thermal Physics*, 2nd edn. (W.H. Freeman, New York 1984).
Kurz, W., Mercier, J. P., Zambelli, G.: *Introduction à la science des matériaux* (Presses Polytechniques et Universitaires Romandes, Lausanne 1991).
Papon, P., Leblond, J.: *Thermodynamique des états de la matière* (Hermann, Paris 1990).
Ragone, D. V.: *Thermodynamics of Materials*, Vols. I and II (Wiley, New York 1995).
Reisman, A.: *Phase Equilibria* (Academic Press, New York 1970).
Yeomans, J. P.: *Statistical Mechanics of Phase Transitions* (Clarendon Press, Oxford 1992).
Zemansky, M.: *Heat and Thermodynamics* (McGraw-Hill, New York 1968).

Chapter 2

Ashby, M. F., Jones, D. R. H.: *Engineering Materials*, Vol. 2, 2nd edn. (Butterworth–Heinemann, Oxford 1998).
Doremus, R. H.: *Rates of Phase Transformations* (Academic Press, Orlando 1985).
Haasen, P. (ed.): *Phase Transformations in Materials. Materials Science and Technology*, Vol. 5 (VCH, Weinheim 1991).
Kelton, K. P.: *Crystal Nucleation in Liquids and Glasses*, Solid State Physics, Vol. 45 (Academic Press, Orlando 1991).

Kurz, W., Mercier, J. P., Zambelli, G.: *Introduction à la science des matériaux* (Presses Polytechniques et Universitaires Romandes, Lausanne 1991).

Liu, S. H.: Fractals and their applications in condensed matter physics, in *Solid State Physics*, Vol. 39 (Academic Press, Orlando 1991).

Ragone, D. V.: *Thermodynamics of Materials*, Vol. II (Wiley, New York 1995).

Romano, A.: *Thermomechanics of Phase Transitions in Classical Field Theory* (World Scientific, Singapore 1993).

Stauffer, D.: *Introduction to Percolation Theory* (Taylor & Francis, London 1985).

Zallen, R.: *The Physics of Amorphous Solids* (Wiley, New York 1983).

Chapter 3

Ashby, M. F., Jones, D. R. H.: *Engineering Materials*, Vol. 2, 2nd edn. (Butterworth–Heinemann, Oxford 1998).

Bernache-Assolant, B. (ed.): *Chimie-physique du frittage* (Hermès, Paris 1995).

Daoud, M., Williams, C. (eds.): *Soft Matter* (Springer-Verlag, Berlin, Heidelberg, New York 1999).

Dash, J., Haying Fu, Wettlaufer, J. S.: The premelting of ice and its environmental consequences, *Rept. Progr. Phys.* **58**, 115–167 (1995).

Israelachvili, J.: *Molecular and Surface Forces* (Academic Press, New York 1992).

Kurz, W., Fisher, D. J.: *Fundamentals of Solidification* (Trans. Tech. Publications, Aedermannsdorf 1992).

Kurz, W., Mercier, J. P., Zambelli, G.: *Introduction à la science des matériaux* (Presses Polytechniques et Universitaires Romandes, Lausanne 1991).

Tiller, W. A.: *The Science of Crystallization* (Cambridge University Press, Cambridge 1992).

Walton, J.: *Three Phases of Matter* (McGraw-Hill, New York 1976).

Young, R. P. J., Lovell, P. A.: *Introduction to Polymers* (Chapman and Hall, London 1995).

Chapter 4

Cyrot, M., Pravuna, D.: *Introduction to Superconductivity and High T_C Materials* (World Scientific, Singapore 1992).

Debenedetti, P. G.: *Metastable Liquids* (Princeton University Press, Princeton 1996).

Domb, C.: *The Critical Point: A Historical Introduction to the Modern Theory. Theory of Critical Phenomena* (Taylor & Francis, Bristol 1996).

Hansen, J. P., McDonald, I. R.: *Theory of Simple Liquids* (Academic Press, London 1976).

Huang, K.: *Statistical Mechanics*, 2nd edn. (Wiley, New York 1987).

Kauzmann, K. W., Eisenberg, D.: *The Structure and Properties of Water* (Oxford University Press, Oxford 1969).

Lawrie, D., Sarbach, S.: Theory of tricritical points, in *Phase Transitions and Critical Phenomena*, Domb, C., Lebowitz, J. L. (eds.), Vol. 9 (Academic Press, London 1984) pp. 1–161.

Levelt Sengers, J. M. H.: Critical behavior of fluids, in *Supercritical Fluids*, Kiran, E., Levelt Sengers, J. M. H. (eds.) (Kluwer, Dordrecht 1994) pp. 3–38.

Levelt Sengers, J. M. H.: *How fluids unmix* (Royal Netherlands Academy of Arts and Sciences, Amsterdam 2002).

Moldover, M. R., Rainwater, J. C.: Thermodynamic models for fluid mixtures near critical conditions, in *Chemical Engineering at Supercritical Fluid Conditions*, Chap. 10, Gray Jr., R. D., Paulaitis, M. E., Penninger, J. M. L., Davidson, P. (eds.) (Ann Arbor Science Publications, Ann Arbor, Michigan 1983).

Papon, P., Leblond, J.: *Thermodynamique des états de la matière* (Hermann, Paris 1990).

Plischke, M., Bergersen, B.: *Equilibrium Statistical Physics* (World Scientific, Singapore 1994).

Reid, R. C., Prausnitz, J. M., Poling, B. E.: *The Properties of Gases and Liquids* (McGraw-Hill, New York 1987).

Scott, R. L., Van Konynenburg, P. H.: Critical lines and phase equilibria in binary van der Waals mixtures, *Philos. Trans. Roy. Soc.* **298**, 495–594 (1980).

Vidal, J.: *Thermodynamique, application au génie chimique et à l'industrie pétrolière* (Technip, Paris 1997).

Chapter 5

Angell, C. A.: Formation of glasses from liquids and biopolymers, *Science*, **267**, 1924–1935 (1995).

Debenedetti, P. G.: *Metastable Liquids* (Princeton University Press, Princeton 1996).

Elliot, S. R.: *Physics of Amorphous Materials* (Longman, Essex 1990).

Götze, W., Sjögren, L.: Relaxation processes in supercooled liquids, Rep. Prog. Phys. **55**, 241–376 (1992).

Zallen, R.: *The Physics of Amorphous Solids* (Wiley, New York 1983).

Zarzycki, J.: *Les verres et l'état vitreux* (Masson, Paris 1982).

Chapter 6

Clark, A., Ross-Murphy, S. B.: Structural and mechanical properties of biopolymer gels, *Adv. Polym. Sci.* **83**, 57–1923 (1987).

Djabourov, M.: Architecture of gelatin gels, in *Contemp. Phys.*, **29**, No. 3, 273–297 (1988).
Djabourov, M.: Gelation, a review, *Polymer Int.*, **25**, 135–143 (1991).
Donald, A.M.: Physics of foodstuff, *Rep. Prog. Phys.* **57**, 1081–1135 (1994).
de Gennes, P. G.: *Scaling Concepts in Polymer Physics* (Cornell University Press, Ithaca 1985).
Guenet, J. M.: *Thermoreversible Gelation of Polymers and Biopolymers* (Academic Press, London 1992).
Martin, J. E., Adolf, D.: The sol–gel transition in chemical gels, *Ann. Rev. Phys. Chem.*, **42**, 311–339 (1991).
Nishinari, K.: Rheology of physical gels and gelling processes, *Rep. Prog. Polymer Phys. Japan* **43**, 163 (2000).

Chapter 7

Binney, J. J., Dowrick, N. J., Fisher, A.J., Newman, M. E. J.: *The Theory of Critical Phenomena* (Oxford Science Publications, Oxford 1995).
Brout, R. H.: *Phase Transitions* (W.A. Benjamin, New York 1965).
Chaikin, P. M., Lubensky, T. C.: *Principles of Condensed Matter Physics* (Cambridge University Press, Cambridge 1995).
Domb, L., Lebowitz, J. L. (eds.): *Phase Transition and Critical Phenomena* Vol. 9 (Academic Press, London 1984).
Gerl, M., Issi, J. P.: *Physique des matériaux* (Presses Polytechniques et Universitaires Romandes, Lausanne 1997).
Kittel, C.: *Introduction to Solid State Physics*, 7th edn. (Wiley, New York 1996).
Huang, K.: *Statistical Mechanics* (Wiley, New York 1963).
Kubo, R.: *Statistical Mechanics* (North Holland, Amsterdam 1965).
Papon, P., Leblond, J.: *Thermodynamique des états de la matière* (Hermann, Paris 1990).
Stanley, H. E.: *Introduction to Phase Transitions and Critical Phenomena* (Clarendon Press, Oxford 1971).
Swalin, R.A.: *Thermodynamics of Solids* (Wiley, New York 1972).
Toledano, J. C., Toledano, P.: *The Landau Theory of Phase Transition* (World Scientific, Singapore 1987).
White, R. M., Geballe, Th. H.: *Long Range Order in Solids* (Academic Press, New York 1979).

Chapter 8

Chaikin, P. M., Lubenski, T. C.: *Principles of Condensed Matter Physics* (Cambridge University Press, Cambridge 1995).
de Gennes, P. G., Prost, J.: *The Physics of Liquid Crystals*, 2nd edn. (Oxford Science Publications, Oxford 1995).

Heppke, G., Moro, D.: Chiral order from achiral molecules, *Science* **279**, 1872–1873 (1998).

Huang, K.: *Statistical Mechanics* (Wiley, New York 1963).

Liebert, L. (ed.): *Liquid Crystals*, Solid State Physics, Suppl. Vol. 14 (Academic Press, New York 1978).

Rice, R. W.: Ceramic tensile strength–grain size relations, *J. Mater. Sci.* **32**, 1673–1692 (1997).

Sachdev, S.: *Quantum Phase Transitions* (Cambridge University Press, Cambridge 2001)

Tilley, D. R., Tilley, J.: *Superfluidity and Superconductivity* (Adam Hilger, Bristol 1990).

Wright, D., Mermin, D.: Crystalline liquids: the blue phases, *Rev. Mod. Phys.* **385** (April 1989).

Chapter 9

Ashby, M. F., Jones, D. R. H.: *Engineering Materials*, Vol. 1, 2nd edn. (Butterworth–Heinemann, Oxford 1996)

Daoud, M., Williams, C. (eds.): *Soft Matter* (Springer-Verlag, Berlin, Heidelberg, New York 1999).

Fujita, F. E.: *Physics of New Materials* (Springer-Verlag, Berlin, Heidelberg, New York 1994).

Gerl, M., Issi, J.P.: *Physique des matériaux*, Vol. 8 (Presses Polytechniques et Universitaires Romandes, Lausanne 1997).

Israelachvili, J.: *Intermolecular and Surface Forces* (Academic Press, London 1992).

Moriarty, Ph.: Nanostructural materials, *Rep. Prog. Physics* **63**(3), 29–381 (2001).

Rice, R. W.: Ceramic tensile strength–grain size relations, *J. Mater. Sci.* **32**, 1673–1692 (1997).

Timp, G. (ed.): *Nanotechnology* (Springer-Verlag, Heidelberg 1998).

Chapter 10

Israelachvili, J.: *Intermolecular and Surface Forces* (Academic Press, London 1992).

Knobler, C. M., Desai, R.C.: Phase transitions in monolayers, *Ann. Rev. Phys. Chem.* **43**, 207–236 (1992).

Mühlwald, H.: Surfactant layers at water surfaces, *Rep. Prog. Phys.* **56**, 653–685 (1993).

Chapter 11

Debenedetti, P. G.: *Metastable Liquids* (Princeton University Press, Princeton 1996).

Delhaye, J. M., Giot, M., Riethmuller, M. L.: *Thermohydraulics of Two-Phase Systems for Industrial Design and Nuclear Engineering* (McGraw-Hill, New York 1980).

Holzapfel, W. B.: Physics of solids under strong compression, *Rep. Prog. Phys.* **59**, 29–90 (1996).

Houghton, J. T.: *The Physics of Atmosphere* (Cambridge University Press, Cambridge 1995).

Leggett, A. J.: Bose–Einstein condensation in the alkali gases, *Rev. Mod. Phys.* **73**(2), 307–356 (2001)

Lecoffre, Y.: *La Cavitation* (Hermès, Paris 1994).

Peixoto, J. P., Oort, A. H.: Physics of Climate (American Institute of Physics, New York 1992).

Ross, M.: Matter under extreme conditions of temperature and pressure, *Rep. Prog. Phys.* **48**, 1–52 (1985).

Sadhal, S. S., Avyaswamy, P. S., Chang, J. N.: *Transport Phenomena with Drops and Bubbles* (Springer-Verlag, Berlin, Heidelberg, New York 1997).

Trenberth, K. E. (ed.): *Climate System Modeling* (Cambridge University Press, Cambridge 1995).

Whalley, P. B.: *Boiling, Condensation, and Gas-Liquid Flow* (Clarendon Press, Oxford 1987).

Index

Druck: Strauss Offsetdruck, Mörlenbach
Verarbeitung: Schäffer, Grünstadt